DIGITAL LOGIC AND STATE MACHINE DESIGN

DAVID J. COMER

Brigham Young University

Holt, Rinehart and Winston

New York   Chicago   San Francisco   Philadelphia
Montreal   Toronto   London   Sydney   Tokyo
Mexico   Rio de Janeiro   Madrid

Library of Congress Cataloging in Publication Data

Comer, David J.
  Digital logic and state machine design.
  Includes index.
    1. Switching circuits.   2. Digital electronics.
3. Logic design.   4. Sequential machine theory.
I. Title.   II. Title: State machine design.
TK7868.S9C66 1984 621.3815'37 83-22653
ISBN 0-03-063731-7

Printed in the United States of America

Published simultaneously in Canada

4 5 6 7   039   9 8 7 6 5 4 3 2

CBS COLLEGE PUBLISHING
Holt, Rinehart and Winston
The Dryden Press
Saunders College Publishing

DIGITAL LOGIC AND STATE MACHINE DESIGN

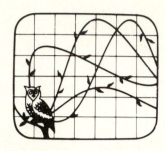

HRW
Series in
Electrical and
Computer Engineering

M. E. Van Valkenburg, Series Editor

Contents

Preface

This book is written to fill a need for a concise, practical book in the very important area of state machine and digital system design. Many system designers do not realize that state machine methods can be applied to sequential circuit design with the same validity and efficiency that Karnaugh maps apply to combinational logic circuits. These methods allow a finished design to be produced, starting with nothing more than a set of specifications.

In order to meet the primary goal of the text, the following three smaller, operational goals had to be achieved.

1. The topic of combinational circuits is prerequisite to the discussion of state machines; hence this topic is treated in the first three chapters. In order to discuss practical design problems, examples must be addressed using specific IC chips. In order to limit the textbook to an acceptable length, only TTL circuits are considered, even though other circuit families are significant.

The new IEEE/IEC symbols using dependency notation are introduced in this material on combinational logic.

2. In an effort to be both concise and practical, the theory must be filtered somewhat to apply directly to design of the state machine. Only that theory applying directly to the development of design principles is used.

Some very practical matters, often avoided by authors, are included in this text. Logic function realizations using gates, multiplexers, decoders, programmable logic arrays, and read-only memories are discussed in terms of minimizing component cost, chip count, design time, or production turnaround time. A great deal of attention is given to methods of eliminating the troublesome glitch in sequential circuits, a topic often overlooked in theoretical treatments of the subject.

3. A digital system design method is proposed using top-down methods. The term "top-down design" is first defined and then applied to various practical examples. This system design method encompasses a state machine design procedure that is also illustrated liberally by example. This design method is general enough to apply to microprocessor-based systems and, although microprocessor design is not included in this text, the method allows the designer to choose between a microprocessor controller and a state machine controller. A discussion of limitations and advantages of microprocessor systems accompanies this material.

Chapters 1 through 3 deal with combinational circuits. Chapter 4 introduces the flip-flop and considers applications of this device, such as counters and registers. Chapters 5 through 7 are devoted to clock-driven state machine design. Appendix 1 includes some simple principles of analysis and design of asynchronous systems. Appendix 2 discusses TTL, ECL, CMOS, and I²L logic components.

It is my firm belief that a thorough study of this textbook by an undergraduate student will lead to a strong capability in digital system design. This study will also allow the student to move easily into more advanced, theoretical studies of the sequential machine.

There are many people deserving acknowledgment for their assistance or influence on this project. I am truly grateful for the thorough reviews of the manuscript at all stages provided by Holt, Rinehart and Winston under the direction of the editor, Deborah Moore. The numerous comments and suggestions led to a vastly improved finished product. The use of a UNIX-based word processing computer, provided by Brigham Young University, facilitated this work. Professor Dennis Fairclough's help in making the computer system functional and available is appreciated immensely.

It is difficult to express my gratitude adequately for the support of my wife, Barbara. Her efforts go beyond the countless hours spent at the keyboard with the manuscript. Her support and encouragement are continual. She shares all effort and must therefore share any success derived from this project.

DAVID J. COMER
Provo, Utah

Introduction

It would be difficult to overstate the significance of the electronics field in world affairs. The impact on our lives of electronics-based systems such as the radio, television, and telephone is quite obvious. Beyond the obvious lie several other important applications of this imposing technology. Computers that store millions of data records in minimal space or, in seconds, make calculations that would require years to accomplish manually, greatly influence government and business operations. Satellite communication systems, space vehicles, defense systems, and industrial process control devices are other examples of electronics-based systems.

The electronics field can be divided into two general areas: analog and digital. While both areas are important, the tremendous growth of the computer industry over the last two decades has made the latter area particularly significant. Although digital electronics has been associated primarily with the digital computer, many noncomputing areas such as communications, automatic control, and instrumentation now routinely apply digital methods.

As digital applications grew in importance, digital methods of analysis and design were developed. Areas such as binary number theory, Boolean algebra, Karnaugh mapping concepts, and state machine theory have become very significant in the last few decades. Efficient design procedures based on these methods can now be applied to produce almost any type of digital system.

The purpose of this book is to introduce the necessary theory from these areas to allow a consideration of design procedures. As this theory is discussed, practical logic circuits now available on integrated circuit (IC) chips, will be introduced.

This book will depart from traditional books in certain areas. Karnaugh mapping is treated, but not to the extent included in other books. As gate arrays, multiplexers, decoders, programmable logic arrays, and read only memories become more popular in logic function generation, the importance of the Karnaugh map begins to diminish. In attempting to reflect industrial practices, the amount of coverage devoted to this and other theoretical areas has been reduced to the minimum necessary to develop design procedures.

We can broadly classify the digital design field into two subdivisions: computer design and digital circuit or system design. The latter design area is concerned with the development of component systems that perform a relatively small number of tasks. These systems may be utilized in the noncomputing applications previously mentioned, or they may serve as building blocks for a computer system. The computer design area focuses on the effective application of these component systems in the construction of a computer. This text is directed toward the circuit or system area. While computers are mentioned in connection with discussions of system applications, computer design is not covered in this text.

Within the area of digital circuits are two subclasses of subject matter. The first and simplest is the combinational circuit. This type of circuit has outputs that are functions only of the inputs, that is, the output signals at any time are uniquely determined by the input signals applied at that time. The second type of system is the sequential circuit. This important circuit contains a memory that leads to output signals that depend not only on present input signals, but also on past history of input signals. Combinational design is less complex than sequential design, and combinational circuits are used in sequential circuits. Consequently, we first consider combinational circuits in Chaps. 1 to 3 before proceeding to sequential circuits in Chap. 4.

Chapters 5 to 7 are concerned with the design of state machines. The state machine is a sequential circuit that produces control signals for the overall system. Over the years, a solid design theory for the controlling state machine has been developed. These three chapters will introduce and discuss this theory which is necessary to produce reliable, minimal digital control circuits.

CHAPTER

1

Binary Systems and Logic Circuits

Before we begin our study of digital circuits and systems, we should become aware of the reasons for using the two-level logic scheme that is so prevalent in digital systems. Thus, the first section will justify the use of binary numbers while the next section will examine several binary-related numbering systems and codes. The following sections will consider the physical relationship between different binary representations and actual logic circuits. The basic concepts needed to use available integrated circuit or IC chips are then considered.

Although several logic families are described in Appendix 2, we have chosen to use TTL (transistor-transistor logic) circuits to demonstrate logic design. This family of circuits is the most popular and versatile and, therefore, represents a reasonable choice for implementing circuit examples.

1.1 THE BLESSINGS OF BINARY

Section overview: This section indicates that the binary system can be more reliable than analog systems and that several electronic circuits or components generate binary or two-level outputs.

The binary number system forms the basis for all digital computer and most other digital system operation. That fact alone should underscore the tremendous importance of the binary number system. While it is true that many systems use hexadecimal, octal, or decimal numbering, these numbers must be coded or represented in some sort of binary-type code to the digital system. Hence, all computers use binary or a binary-related system to represent numbers and alphabetic characters.

Why is binary so universal in digital systems? There are at least two reasons for the popularity of this numbering scheme. The first is that "yes-no" or "on-off" situations are much easier to classify than are quantitative situations. It is easy to look at a light switch and determine if it is off or on. We can readily tell if a water faucet is off with no water flow or if it is on with a significant flow of water. It is more difficult to tell how much water is flowing unless we have some rather elaborate equipment. If we must determine the flow rate, we are concerned with an analog situation. The number of cubic feet per second could be a continuously changing quantity, meaning our measuring system would have to monitor and identify this changing flow rate. The reliability in classifying a digital situation as on or off is much better than measuring the value of an analog quantity such as flow rate. As we shall see, the binary system uses only two values, 0 and 1, which can correspond to an "on-off" or "yes-no" situation. Thus, a binary scheme is easy to use and has high reliability.

A second reason for using binary is that many electronic components are capable of producing only two identifiable states. A magnetic core which is a small doughnut-shaped ferrite material can be magnetized with flux flowing in one direction in the core, or the flux can be reversed to flow in the other direction. An output transistor stage can be saturated or cut off. A popular computer circuit called a flip-flop has an output that is stable only at a high voltage level or at a low voltage level. This binary nature of many electronic circuits or components makes the decision to use the binary system a very reasonable choice.

Although multilevel circuits having more than two output levels have been studied for years, binary circuits are essentially universal in computer systems.

1.2 NUMBER SYSTEMS

Section overview: This section studies the basis of the decimal, binary, octal, and hexadecimal number systems. Since digital circuits are binary in nature, all numbers represented by these circuits must use some binary-related code. Hence, binary-coded octal, binary-coded hexadecimal, and binary-coded decimal numbering methods are also reviewed. Conversion of numbers from one base to another is emphasized. The section concludes with an introduction to alphanumeric codes used to represent numbers, characters, or control information.

The decimal number system is quite useful for human transactions and is used throughout the world. This system uses the positional method to assign values to the number. That is, the position of a digit relative to the decimal point determines the power of 10 by which the digit is to be multiplied. The number 312.4 means

$$3 \times 10^2 + 1 \times 10^1 + 2 \times 10^0 + 4 \times 10^{-1}$$

In the decimal system, the number 10 is the base or radix. The digit position to the left of the radix point (decimal point if the radix is 10) is multiplied by the radix to the power of zero. The next position to the left indicates that this digit should be multiplied by the radix to the first power. In general, a number of any radix can be expressed as

$$\cdots + a_4 r^4 + a_3 r^3 + a_2 r^2 + a_1 r^1 + a_0 r^0 + a_{-1} r^{-1} + a_{-2} r^{-2} + \cdots \quad (1.1)$$

where r is the radix and a_i is the digit value. To conserve space in writing a number, we can eliminate the plus signs and allow the position of each digit to indicate the multiplier. When we do this, however, we must insert the radix point as a reference point and we must indicate the radix to be used. For example, the number

$$5 \times 8^3 + 2 \times 8^2 + 1 \times 8^1 + 3 \times 8^0 + 2 \times 8^{-1} + 6 \times 8^{-2}$$

can be expressed as

$$5213.26_8$$

using this shorthand method. The subscript indicates the radix used. We will use this method of radix identification whenever there is a possibility of confusion.

We note that this numbering system requires exactly r symbols. The decimal system uses the symbols 0 through 9. The highest symbol value is one less than the radix, since adding 1 to this value is accounted for by a 1 carried to the next most significant column.

A. The Binary System

The binary system uses the base 2; consequently only the two symbols 0 and 1 are necessary to express any binary number. The number

$$1101.101_2$$

is converted to the decimal system using Eq. (1.1):

$$1 \times 2^3 + 1 \times 2^2 + 0 \times 2^1 + 1 \times 2^0 + 1 \times 2^{-1}$$
$$+ 0 \times 2^{-2} + 1 \times 2^{-3} = 13.625_{10}$$

Table 1.1 Decimal Equivalent of 2^n

2^n	Decimal Equivalent	Binary Equivalent 11 10 9 8 7 6 5 4 3 2 1	← Col. no.
2^0	1	1	
2^1	2	1 0	
2^2	4	1 0 0	
2^3	8	1 0 0 0	
2^4	16	1 0 0 0 0	
2^5	32	1 0 0 0 0 0	
2^6	64	1 0 0 0 0 0 0	
2^7	128	1 0 0 0 0 0 0 0	
2^8	256	1 0 0 0 0 0 0 0 0	
2^9	512	1 0 0 0 0 0 0 0 0 0	
2^{10}	1024	1 0 0 0 0 0 0 0 0 0 0	

There are two popular methods of converting from decimal to binary numbers. The first method merely applies a knowledge of the powers of 2 as shown in Table 1.1. We start this procedure by identifying the highest power of 2 contained in the decimal number. We then begin to construct the binary equivalent by placing a 1 in the column that accounts for this power of 2. We then subtract from the original number the decimal equivalent of that power of 2 now included in the binary conversion. We then examine the difference and repeat this procedure. Continual repetition will lead to termination when the difference between the remaining decimal number and the power of 2 is zero.

■ *Example 1.1* Convert 789_{10} to a binary number.

□ *Solution:* The highest power of 2 contained in this number is 9 since 2^9 is 512. We start construction of our binary number by placing a 1 in the tenth column. The complete procedure is demonstrated in Table 1.2. The final number is obtained by inserting zeros in all blank columns to give 789_{10} = 1100010101_2. ■

Table 1.2 Conversion of Decimal to Binary Numbers

	789_{10}			10	9	8	7	6	5	4	3	2	1
Contains	512_{10}	=	2^9		1	1			1		1		1
Difference	277_{10}												
Contains	256_{10}	=	2^8										
Difference	21_{10}												
Contains	16_{10}	=	2^4										
Difference	5_{10}												
Contains	4_{10}	=	2^2										
Difference	1_{10}												
Contains	1_{10}	=	2^0										
Difference	0_{10}												

Another version of this procedure identifies the highest power of 2 contained in the decimal number and places a 1 in the binary number to be constructed. The difference between the original number and this power of 2 is taken and examined to see if it contains the next lower power of 2. If not, a 0 is placed to the right of the 1 in the binary number to be constructed, and the next lower power of 2 is checked. When the difference does contain a power of 2, a 1 is added to the binary number and the decimal equivalent of this power of 2 is subtracted from the remaining partially converted decimal number.

Returning to Example 1.1, we see that the 512_{10} is contained in 789_{10}. We write a 1 to start the construction of the binary equivalent. The entire procedure is demonstrated in Table 1.3.

The second method used to convert from a decimal to a binary number involves repeated division of the number by 2 and examination of the various remainders. We illustrate this method by converting 16_{10} to a binary number in Table 1.4.

The procedure is terminated when the result of zero is reached. The converted binary number is simply the remainders arranged such that the last remainder is the most significant bit, that is,

$$16_{10} = 10000_2$$

Each time a 2 is divided into the decimal number, it reflects the fact that a higher power of 2 is contained in the number. In a binary number each digit position moving from right to left represents a higher power of 2. Thus, each time we divide a 2 into the decimal number, the higher power of 2 is accounted for in the binary number by recording the remainder bit.

Table 1.3 Another Procedure for Converting to Binary

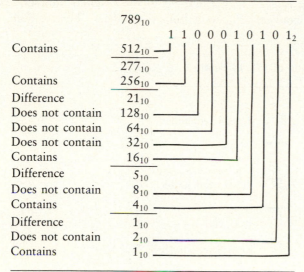

	789_{10}
	$1\ 1\ 0\ 0\ 0\ 1\ 0\ 1\ 0\ 1_2$
Contains	512_{10}
	277_{10}
Contains	256_{10}
Difference	21_{10}
Does not contain	128_{10}
Does not contain	64_{10}
Does not contain	32_{10}
Contains	16_{10}
Difference	5_{10}
Does not contain	8_{10}
Contains	4_{10}
Difference	1_{10}
Does not contain	2_{10}
Contains	1_{10}

Table 1.4 Conversion of 16_{10} to Binary

Divide by 2	$\begin{array}{r} 8 \\ 2\overline{)16} \\ \underline{16} \\ r0 \end{array}$	Record remainder	0
Divide result by	$\begin{array}{r} 4 \\ 2\overline{)8} \\ \underline{8} \\ r0 \end{array}$	Record remainder	0
Divide result by 2	$\begin{array}{r} 2 \\ 2\overline{)4} \\ \underline{4} \\ r0 \end{array}$	Record remainder	0
Divide result by	$\begin{array}{r} 1 \\ 2\overline{)2} \\ \underline{2} \\ r0 \end{array}$	Record remainder	0
Divide result by 2	$\begin{array}{r} 0 \\ 2\overline{)1} \\ \underline{0} \\ r1 \end{array}$	Record remainder	1

In the preceding example, the remainder was zero for all but the fifth division, indicating that 2^4 was contained in the original number. Now suppose we want to convert 17_{10} to a binary number. This is done in Table 1.5.

Table 1.5 Conversion of 17_{10} to Binary

Divide by 2	$\dfrac{8}{2\overline{\smash{\big)}17}}$ $\dfrac{16}{r1}$	Record remainder	1
Divide result by 2	$\dfrac{4}{2\overline{\smash{\big)}8}}$ $\dfrac{8}{r0}$	Record remainder	0
Divide result by 2	$\dfrac{2}{2\overline{\smash{\big)}4}}$ $\dfrac{4}{r0}$	Record remainder	0
Divide result by 2	$\dfrac{1}{2\overline{\smash{\big)}2}}$ $\dfrac{2}{r0}$	Record remainder	0
Divide result by 2	$\dfrac{0}{2\overline{\smash{\big)}1}}$ $\dfrac{0}{r1}$	Record remainder	1

Each division by 2 indicates 2 bits of information about the number. If the nth division results in a nonzero number, we know that the decimal number is greater than or equal to 2^n. If the remainder is 1, it means that 2^{n-1} is contained in the number. If the remainder is zero, then 2^{n-1} is not contained in the number. These points are demonstrated in Table 1.6 in converting 23_{10} to a binary number.

This method is particularly useful in converting decimal numbers greater than 2048_{10} to binary.

B. Octal and Binary-Coded Octal (BCO)

If the base or radix chosen to express a number is 8, the number system is referred to as octal. The eight symbols 0 through 7 are required to express an octal number. Conversion from octal to decimal is accomplished with Eq. (1.1); thus

$$217_8 = 2 \times 8^2 + 1 \times 8^1 + 7 \times 8^0 = 143_{10}$$

Table 1.6 Conversion of 23_{10} to Binary

1st division	$2\overline{)23}$ gives 11, 22, $r1$	Nonzero result indicates that number is $\geq 2^1$ Remainder indicates that 2^0 is contained in number	Indicated by recording a 1 in the 2^0 column
2nd division	$2\overline{)11}$ gives 5, 10, $r1$	Nonzero result indicates that number is $\geq 2^2$ Remainder indicates that 2^1 is contained in number	Indicated by recording a 1 in the 2^1 column
3rd division	$2\overline{)5}$ gives 2, 4, $r1$	Nonzero result indicates that number is $\geq 2^3$ Remainder indicates that 2^2 is contained in number	Indicated by recording a 1 in the 2^2 column
4th division	$2\overline{)2}$ gives 1, 2, $r0$	Nonzero result indicates that number is $\geq 2^4$ Zero remainder indicates that 2^3 is not contained in number	Indicated by recording a 0 in the 2^3 column
5th division	$2\overline{)1}$ gives 0, 0, $r1$	Zero result indicates that number is $\leq 2^5$ Remainder indicates that 2^4 is contained in number	Indicated by recording a 1 in the 2^4 column

$23_{10} = 10111_2$

The octal system is used in some computers and is closely related to the binary system. Of course, when the computer applies the octal system, the number must be represented by ones and zeros since the other six symbols are not available as outputs of binary circuits. Rather than applying octal code directly, the computer uses binary-coded octal (BCO). In this scheme three binary positions are used to represent one digit of the octal number. The correspondence between the two systems are shown below.

2	1	7	Octal
010	001	111	Binary-coded octal

Each 3-bit binary number can represent any number from zero to seven.

One of the interesting features of BCO is the ease of conversion to and from binary. The octal number 217_8 equals 143_{10}. Converting to binary gives

$$143_{10} = 10001111_2 = 217_8$$

Now we will compare the binary equivalent of 217_8 to the BCO representation above. Except for the leading zero in the BCO system, the two numbers are equivalent. Conversion from BCO to binary consists of concatenation of the three-digit groups and dropping any leading zeros. In order to convert from binary to octal, we must subdivide the digits to the left of the octal point into groups of three, adding any necessary zeros to the leftmost group to yield three digits in that group. The following examples demonstrate these points:

$$6524_8 = 110\ 101\ 010\ 100\ (BCO) = 110101010100_2$$

$$2.51_8 = 010.101\ 001\ (BCO) = 10.101001_2$$

$$1011101110_2 = 001\ 011\ 101\ 110\ (BCO) = 1356_8$$

We can show why the conversion from binary-coded octal to binary is so straightforward by considering the following octal and binary-coded octal number:

$$a_2 \times 8^2 + a_1 \times 8^1 + a_0 \times 8^0$$
$$= b_8 b_7 b_6 \times 8^2 + b_5 b_4 b_3 \times 8^1 + b_2 b_1 b_0 \times 8^0$$

where $b_8 b_7 b_6$ is the binary equivalent of a_2, $b_5 b_4 b_3$ is the binary equivalent of a_1, and $b_2 b_1 b_0$ is the binary equivalent of a_0. We can convert the binary-coded octal number to straight binary by converting the radix raised to various powers to binary. Noting that

$$8^0 = 1_2 \quad 8^1 = 1000_2 \quad 8^2 = 1000000_2$$

we can write the number as

$$b_8 b_7 b_6 \times 1000000 + b_5 b_4 b_3 \times 1000 + b_2 b_1 b_0 \times 1$$

Carrying out the indicated multiplications and additions gives the binary number:

b_8	b_7	b_6	0	0	0	0	0	0
+			b_5	b_4	b_3	0	0	0
+						b_2	b_1	b_0
b_8	b_7	b_6	b_5	b_4	b_3	b_2	b_1	b_0

From this result we see that the binary number is formed by concatenation of the binary-coded number.

A corollary of this result is that conversion between octal and binary can be accomplished easily by using binary-coded octal as an intermediate number. For example, binary to octal can be done as follows:

$$10110110_2 = 010\ 110\ 110\ (BCO)$$
$$= 266_8$$

C. Hexadecimal Code

A very important code in computer work is hexadecimal or hex code. This code uses 16 as the base; thus 16 symbols, 0 to 15, are required. Since positional notation is utilized, each column must have a single digit or symbol. Because of this requirement the following symbols are used:

Decimal	Hex	Decimal	Hex
0	0	8	8
1	1	9	9
2	2	10	A
3	3	11	B
4	4	12	C
5	5	13	D
6	6	14	E
7	7	15	F

The number $A29_{16}$ is converted to decimal by writing

$$A29_{16} = 10 \times 16^2 + 2 \times 16^1 + 9 \times 16^0 = 2601_{10}$$

Binary-coded hex uses four columns to represent each digit of the hex number. The binary-coded hex equivalent of $A29_{16}$ is

1010 0010 1001

Conversion between binary-coded hex and binary can be done by the method used to convert binary-coded octal. The groups of four digits are concatenated and any leading zeros dropped to form the binary number. Any base that is an integer power of 2 can be converted in this way. Again conversion between hex and binary can be facilitated by using binary-coded hex; thus

$$1E2_{16} = 0001\ 1110\ 0010\ (BCH) = 111100010_2$$

D. Alphanumeric Codes

Thus far in our discussion of binary we have considered representing only numbers with binary and binary-related codes. While numbers are important in many digital applications, alphabetic or other characters are necessary in several instances also.

There are many choices of code sets to represent alphanumeric (alphabetic and numeric) characters; however, there are two well-accepted code sets that are considered standards. The first is the ASCII code (American Standard Code for Information Interchange) which is used by most non-International Business Machines systems throughout the world. IBM systems generally use the EBCDIC code (Extended Binary Coded Decimal Interchange Code), although IBM's Personal Computer does use the ASCII code. The ASCII code is a 7-bit code that also contains several control characters useful for exchanging information between two or more digital systems. EBCDIC is an 8-bit code allowing more characters or control codes to be transmitted. Both codes are shown in Table 1.7.

Table 1.7A ASCII Code

	7-BIT ASCII BINARY	HEX		7-BIT ASCII BINARY	HEX		7-BIT ASCII BINARY	HEX		7-BIT ASCII BINARY	HEX
A	1 000 001	41	g	1 100 111	67	"	0 100 010	22	BEL	0 000 111	07
B	1 000 010	42	h	1 101 000	68	#	0 100 011	23	BS	0 001 000	08
C	1 000 011	43	i	1 101 001	69	$	0 100 100	24	CAN	0 011 000	18
D	1 000 100	44	j	1 101 010	6A	%	0 100 101	25	CR	0 001 101	0D
E	1 000 101	45	k	1 101 011	6B	&	0 100 110	26	DC1	0 010 001	11
F	1 000 110	46	l	1 101 100	6C	'	0 100 111	27	DC2	0 010 010	12
G	1 000 111	47	m	1 101 101	6D	(	0 101 000	28	DC3	0 010 011	13
H	1 001 000	48	n	1 101 110	6E	)	0 101 001	29	DC4	0 010 100	14
I	1 001 001	49	o	1 101 111	6F	*	0 101 010	2A	DEL	1 111 111	7F
J	1 001 010	4A	p	1 110 000	70	+	0 101 011	2B	DLE	0 010 000	10
K	1 001 011	4B	q	1 110 001	71	,	0 101 100	2C	EM	0 011 001	19
L	1 001 100	4C	r	1 110 010	72	-	0 101 101	2D	ENQ	0 000 101	05
M	1 001 101	4D	s	1 110 011	73	.	0 101 110	2E	EOT	0 000 100	04
N	1 001 110	4E	t	1 110 100	74	/	0 101 111	2F	ESC	0 011 011	1B
O	1 001 111	4F	u	1 110 101	45	:	0 111 010	3A	ETB	0 010 111	17
P	1 010 000	50	v	1 110 110	76	;	0 111 011	3B	ETX	0 000 011	03
Q	1 010 001	51	w	1 110 111	77	<	0 111 100	3C	FF	0 001 100	0C
R	1 010 010	52	x	1 111 000	78	=	0 111 101	3D	FS	0 011 100	1C
S	1 010 011	53	y	1 111 001	79	>	0 111 110	3E	GS	0 011 101	1D
T	1 010 100	54	z	1 111 010	7A	?	0 111 111	3F	HT	0 001 001	09
U	1 010 101	55	0	0 110 000	30	@	1 000 000	40	LF	0 001 010	0A
V	1 010 110	56	1	0 110 001	31	[	1 011 011	5B	NAK	0 010 101	15
W	1 010 111	57	2	0 110 010	32	\	1 011 100	5C	NUL	0 000 000	00
X	1 011 000	58	3	0 110 011	33	]	1 011 101	5D	RS	0 011 110	1E
Y	1 011 001	59	4	0 110 100	34	^	1 011 110	5E	SI	0 001 111	0F
Z	1 011 010	5A	5	0 110 101	35	—	1 011 111	5F	SO	0 001 110	0E
a	1 100 001	61	6	0 110 110	36	\	1 100 000	60	SOH	0 000 001	01
b	1 100 010	62	7	0 110 111	37	{	1 111 011	7B	STX	0 000 010	02
c	1 100 011	63	8	0 111 000	38	¦	1 111 100	7C	SUB	0 011 010	1A
d	1 100 100	64	9	0 111 001	39	}	1 111 101	7D	SYN	0 010 110	16
e	1 100 101	65	SP	0 100 000	20	–	1 111 110	7E	US	0 011 111	1F
f	1 100 110	66	!	0 100 001	21	ACK	0 000 110	06	VT	0 001 011	0B

Table 1.7B EBCDIC Code

	EBCDIC BINARY	HEX		EBCDIC BINARY	HEX		EBCDIC BINARY	HEX		EBCDIC BINARY	HEX
A	11 000 001	C1	l	10 010 011	93	,	01 101 011	6B	EOB	00 100 110	26
B	11 000 010	C2	m	10 010 100	94	-	01 100 000	60	EOT	00 110 111	37
C	11 000 011	C3	n	10 010 101	95	.	01 001 011	4B	ESC	00 100 111	27
D	11 000 100	C4	o	10 010 110	96	/	01 100 001	61	ETB	00 100 110	26
E	11 000 101	C5	p	10 010 111	97	:	01 111 010	7A	ETX	00 000 011	03
F	11 000 110	C6	q	10 011 000	98	;	01 011 110	5E	FF	00 001 100	0C
G	11 000 111	C7	r	10 011 001	99	<	01 001 100	4C	FS	00 100 010	22
H	11 001 000	C8	s	10 100 010	A2	=	01 111 110	7E	HT	00 000 101	05
I	11 001 001	C9	t	10 100 011	A3	>	01 101 110	6E	IFS	00 011 100	1C
J	11 010 001	D1	u	10 100 100	A4	?	01 101 111	6F	IGS	00 011 101	1D
K	11 010 010	D2	v	10 100 101	A5	@	01 111 100	7C	IL	00 010 111	17
L	11 010 011	D3	w	10 100 110	A6	\	11 100 000	E0	IRS	00 011 110	1E
M	11 010 100	D4	x	10 100 111	A7		01 101 101	6D	IUS	00 011 111	1F
N	11 010 101	D5	y	10 101 000	A8	\	01 111 001	79	LC	00 000 110	06
O	11 010 110	D6	z	10 101 001	A9	{	11 000 000	C0	LF	00 100 101	25
P	11 010 111	D7	0	11 110 000	F0	¦	01 101 010	6A	NAK	00 111 101	3D
Q	11 011 000	D8	1	11 110 001	F1	}	11 010 000	D0	NL	00 010 101	15
R	11 011 001	D9	2	11 110 010	F2	~	10 100 001	A1	NUL	00 000 000	00
S	11 100 010	E2	3	11 110 011	F3	⊂	01 001 010	4A	PF	00 000 100	04
T	11 100 011	E3	4	11 110 100	F4	¬	01 011 111	5F	PN	00 110 100	34
U	11 100 100	E4	5	11 110 101	F5			01 001 111	4F	PRE	00 100 111	27
V	11 100 101	E5	6	11 110 110	F6	ACK	00 101 110	2E	RES	00 010 100	14
W	11 100 110	E6	7	11 110 111	F7	BEL	00 101 111	2F	RLF	00 001 001	09
X	11 100 111	E7	8	11 111 000	F8	BS	00 010 110	16	RS	00 110 101	35
Y	11 101 000	E8	9	11 111 001	F9	BYP	00 100 100	24	SI	00 001 111	0F
Z	11 101 001	E9	SP	01 000 000	40	CAN	00 011 000	18	SM	00 101 010	2A
a	10 000 001	81	!	01 011 010	5A	CC	00 011 010	1A	SMM	00 001 010	0A
b	10 000 010	82	"	01 111 111	7F	CR	00 001 101	0D	SO	00 001 110	0E
c	10 000 011	83	#	01 111 011	7B	DC1	00 010 001	11	SOH	00 000 001	01
d	10 000 100	84	$	01 011 011	5B	DC2	00 010 010	12	SOS	00 100 001	21
e	10 000 101	85	%	01 101 100	6C	DC3	00 010 011	13	STX	00 000 010	02
f	10 000 110	86	&	01 010 000	50	DC4	00 111 100	3C	SUB	00 111 111	3F
g	10 000 111	87	'	01 111 101	7D	DEL	00 000 111	07	SYN	00 110 010	32
h	10 001 000	88	(	01 001 101	4D	DLE	00 010 000	10	UC	00 110 110	36
i	10 001 001	89	)	01 011 101	5D	DS	00 100 000	20	VT	00 001 011	0B
j	10 010 001	91	*	01 011 100	5C	EM	00 011 001	19			
k	10 010 010	92	+	01 001 011	4E	ENQ	00 101 101	2D			

Drill Problems: Sec. 1.2

1. Convert 683_{10} to binary, octal, and hexadecimal.

2. Convert $A2F_{16}$ to binary, octal, and decimal.

3. Convert 712_8 to binary and binary-coded decimal.

4. Convert 1000 0110 (BCD) to decimal and binary.

1.3 THE USE OF BINARY IN DIGITAL SYSTEMS

Section overview: The method of representing the binary symbols 1 and 0 by electronic circuits is considered in this section.

We now understand how to represent binary or binary-related codes by a series of ones and zeros. We can easily write 1011 on paper as the binary code for the decimal number 11. The next matter we must consider is how this same information can be represented by electronic circuits.

Electronic circuits used for digital systems are designed to generate only two possible output voltage levels. For example, the higher level may be near 5 V while the low level may be approximately 0 V. The circuits are designed to be quite tolerant to voltage level. The high level may vary from 2.7 to 5 V, while the low level may vary from 0 to 0.8 V. When a circuit output of 2.7 to 5 V is applied to a second digital circuit, this second circuit must interpret its input as the high voltage level. The circuits are also designed such that when proper inputs are applied, the output will never exist in the ambiguous region between 0.8 and 2.7 V. Of course, this region is crossed as an input or output voltage switches from one level to another. Different types of digital circuits may use different voltage levels. One digital system may use voltage levels from 0 to 2 V as the low level and 8 to 12 V as the high level. Another may use –2.1 to –1.7 V and –1.3 to –0.9 V as the two levels. Generally, system designers avoid mixing circuits with different levels, but it is sometimes unavoidable.

We have now established that digital circuits exist with two well-defined voltage levels. The binary number system requires two symbols; hence it is logical to identify a binary symbol with each voltage level. If we interpret the high level as a binary 1 and the low level as a binary 0, we are using a positive logic system. A negative logic system identifies the high level with binary 0 and the low level with binary 1. Most modern digital systems use positive logic or mixed logic systems.

Now suppose we want to represent a 4-bit binary code. Four binary circuit outputs are required. If we measure the output voltage levels of circuits A, B, C, and D to be low, high, high, and high, respectively, the code contained is 0111. This assumes that A carries the most significant bit (MSB) of the code while D carries the LSB. If we want to represent a 32-bit binary code, 32 circuits are required. Thus, many circuits are necessary to contain reasonably large numbers, an obvious disadvantage of the binary system. This disadvantage is far outweighed by the natural correspondence of binary symbols to the two levels of digital circuits and the inherent reliability of the system.

For conventional digital systems, integrated circuits are used almost exclusively, with discrete digital circuits applied only to high-power or other special designs. There are four broad classifications of integration today, their designations based on circuit or system complexity.

Small-scale integration (SSI) is applied to a chip that performs at least one basic logic function. This chip could contain gates, flip-flops, level detectors, or other useful logic circuits. Using SSI, as many as four gates can be constructed on the same chip, totaling perhaps 20 to 40 components. The next step up in complexity is medium-scale integration (MSI). This level of integration represents an order of magnitude capability increase over SSI. Logic functions requiring 20 gates and 10 flip-flops can be integrated on a single MSI chip. Large-scale integration (LSI) utilizes very complex fabrication techniques to create, perhaps, 20,000 components on a single chip. Between 100 and 1000 logic gates that can perform the logic of an entire digital system may be included on a chip constructed by LSI techniques. Very large-scale integration (VLSI) has now entered the picture as electron-beam lithography and related techniques improve. Up to 1 million devices may now be contained on a VLSI chip. The possibilities and problems associated with VLSI are truly amazing.

The first three levels of integration are widely used in the digital field. SSI is prominent in preliminary design work on systems that will later be fabricated using MSI or LSI circuits. It is also used in systems produced for a market too small to justify fabrication of MSI or LSI circuits. MSI is used for finished systems that do not require the smaller size of LSI and will be produced in relatively small quantities. LSI is directed toward finished products that require compactness or that have a large market potential. VLSI is now used to implement complex microprocessor chips but is expected to become much more prominent during the next decade. An entire computer with main memory on a single chip has now become a reality. The cost per device on a VLSI chip is in the area of a thousandth of a cent, making these chips very cost-effective in many applications.

1.4 LOGIC GATES

Section overview: The concept of a truth table to represent logic circuit behavior is introduced. The function table is also discussed. Logic gates are introduced, and their behavior is described in terms of function and truth tables.

The preceding section indicates that binary bits can be represented by the voltage level at a circuit output. We must next consider what type of circuit is actually used in digital or logic systems.

A digital system uses a building block approach. Many small operational units are interconnected to make up the overall system. The most basic unit of the system is the gate circuit. There are several different types of gate with each type behaving in a different way. OR gates, AND gates, NOR gates, and NAND gates are some of the more important gates that will be considered in this section.

We now examine two important methods of characterizing logic gates. These circuits have one output and one or more inputs. The most basic description of operation is given by the function table. This table lists all possible combinations of inputs along with resulting outputs in terms of the two levels of voltage, high and low. Figure 1.1 shows a function table for a 2-input circuit. This table indicates that if both inputs are low or both are high, the output will be low. If one input is high while the other is low, a high level results on the output line.

Inputs Output

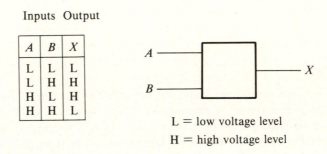

A	B	X
L	L	L
L	H	H
H	L	H
H	H	L

L = low voltage level

H = high voltage level

Figure 1.1 A function table for a logic circuit.

As we deal with logic design, it is appropriate to use 1s and 0s rather than voltage levels. Thus, we must choose either a positive logic scheme (H = 1, L = 0) or a negative logic scheme (H = 0, L = 1). Once this choice is made, we use the function table to generate a truth table. The truth table describes inputs and outputs in terms of 1s and 0s rather than voltage levels. The truth table for a given circuit will be different for a positive logic choice rather than for a negative logic choice. Figure 1.2 demonstrates this point for the function table of Fig. 1.1.

A	B	X
0	0	0
0	1	1
1	0	1
1	1	0

(a)

A	B	X
1	1	1
1	0	0
0	1	0
0	0	1

(b)

Figure 1.2 Truth tables for function table of Fig. 1.1: (a) positive logic; (b) negative logic.

We note that for a positive logic definition, the circuit produces an output of 1 when one input is 0 while the other input is 1. For the negative logic definition, an output of 1 results when both inputs are either 0 or both are 1. Function and truth tables are used by manufacturers of logic gates to specify gate operation. The manufacturer conventionally defines gates in terms of positive logic rather than negative logic.

The three basic gates required as building blocks in logic systems are the OR, AND, and NOT circuits. Two other gates that are important because they can perform the OR, AND, and NOT functions are the NAND and NOR gates. We will consider the characteristics of each of these gates in the following paragraphs.

A. OR Gates

The OR gate has one output and two or more inputs. If all inputs are caused to equal 0, the output is also equal to 0. The presence of a 1 bit at one or more inputs leads to an output of 1. The name OR gate comes from the fact that the output will equal 1 if one input or the other input or both equal 1. Figure 1.3 describes this operation in terms of a truth table. Assuming this truth table results from a positive logic definition, the function table shown is applicable. The standard symbols for a 2-input OR gate and an 8-input OR gate are also shown.

If we choose to use negative logic with the OR gate of Fig. 1.3, it no longer performs the OR function. Figure 1.4 shows the truth table resulting from a negative logic definition. This truth table corresponds to that of an AND gate as we shall see in the following paragraphs. If a single gate can correspond to an OR gate or an AND gate depending on logic definition, we must take care to avoid confusion when specifying this gate. This problem is handled simply by naming the gate according to its positive logic function. All manufacturers refer to a gate having the function table of Fig. 1.3 as an OR gate. Although this circuit can be used as an AND gate with negative logic, the manufacturers' handbooks will always list it as an OR gate.

B. AND Gates

The AND gate has one output and two or more inputs. The output will equal 0 for all combinations of input values except when all inputs equal 1. When each input is 1, the output will also equal 1. Figure 1.5 shows the AND gate, the function table, and the positive logic truth table.

We could easily show that the AND gate will function as an OR gate for negative logic, but again the gate is named for its positive logic function.

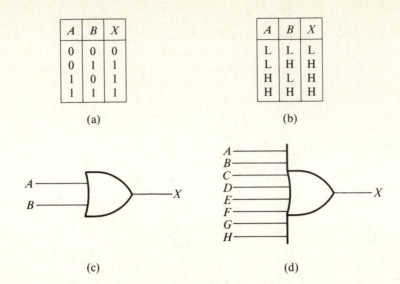

A	B	X
0	0	0
0	1	1
1	0	1
1	1	1

(a)

A	B	X
L	L	L
L	H	H
H	L	H
H	H	H

(b)

(c)

(d)

Figure 1.3 Truth tables and symbols for the OR gate: (a) truth table for positive logic; (b) function table; (c) symbol for 2-input OR gate; (d) symbol for 8-input OR gate.

A	B	X
0	0	0
0	1	0
1	0	0
1	1	1

Figure 1.4 Truth table for the OR gate for negative logic.

C. The Inverter

The inverter performs the NOT or INVERT function. This logic element has one output and one input. The output level is always opposite to the input level. Figure 1.6 shows the function table, truth table, and symbol for the inverter.

The triangle of the inverter symbol represents amplification (generally current amplification for logic circuits), while the small circle denotes an inversion of signals. We note that the inverter truth table for negative logic is the same as that for positive logic.

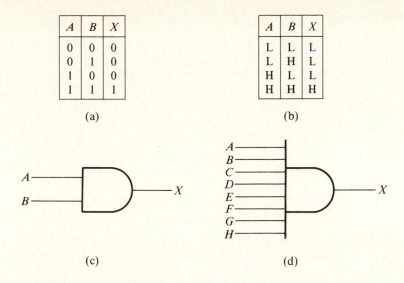

A	B	X
0	0	0
0	1	0
1	0	0
1	1	1

(a)

A	B	X
L	L	L
L	H	L
H	L	L
H	H	H

(b)

(c)

(d)

Figure 1.5 Tables and symbols for the AND gate: (a) truth table for positive logic; (b) function table; (c) symbol for 2-input AND gate; (d) symbol for 8-input AND gate.

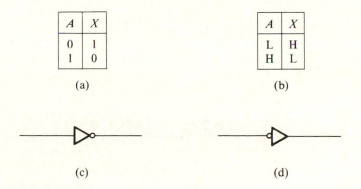

A	X
0	1
1	0

(a)

A	X
L	H
H	L

(b)

(c)

(d)

Figure 1.6 Tables and symbols for the inverter: (a) truth table; (b) function table; (c) inverter symbol; (d) alternate symbol.

At this point we are not yet aware of the kinds of logic functions that must be realized with gates. Nevertheless, we will now state a fact that will later become more meaningful, namely, that the AND gate and inverter or the OR gate and inverter each form a functionally complete set. This means that any logic function realized by logic gates can be realized with the AND and NOT functions or the OR and NOT functions only.

D. NOR Gates

From a theoretical standpoint, the OR and AND gates and the inverter are sufficient to do all necessary logic design. From a more practical standpoint, the NOR and NAND gates are very useful and are actually more universal than the other gates. Either of these gates can be shown to be functionally complete.

The NOR gate is simply an OR gate followed by an inverter. This gate is characterized by the tables and symbols of Fig. 1.7. The small circle following the OR gate again represents an inversion and changes the OR gate to the NOR gate symbol.

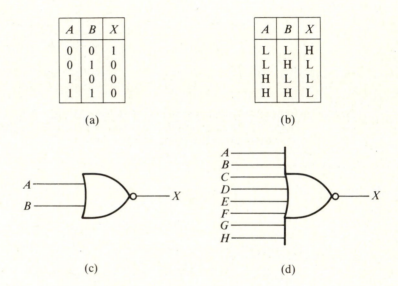

A	B	X
0	0	1
0	1	0
1	0	0
1	1	0

(a)

A	B	X
L	L	H
L	H	L
H	L	L
H	H	L

(b)

(c)

(d)

Figure 1.7 Tables and symbols for the NOR gate: (a) truth table for positive logic; (b) function table; (c) symbol for 2-input NOR gate; (d) symbol for 8-input NOR gate.

E. NAND Gates

The NAND gate is an AND gate followed by an inverter. This gate is characterized by the tables and symbols of Fig. 1.8.

F. Using NAND and NOR Gates for Inverters

A NAND or NOR gate can be used as an inverter as shown in Fig. 1.9. Although all configurations shown in Fig. 1.9 result in an output that is always at the opposite level to that of the input, the preferred configurations in some situations are those of the rightmost column. The driving circuit

A	B	X
0	0	1
0	1	1
1	0	1
1	1	0

(a)

A	B	X
L	L	H
L	H	H
H	L	H
H	H	L

(b)

(c)

(d)

Figure 1.8 Tables and symbols for the NAND gate: (a) truth table for positive logic; (b) function table; (c) symbol for 2-input NAND gate; (d) symbol for 8-input NAND gate.

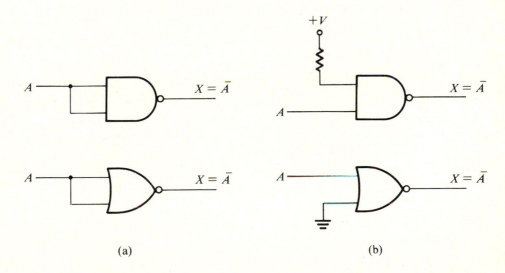

(a)

(b)

Figure 1.9 NAND and NOR gates used as inverters: (a) theoretical method; (b) practical method.

must supply current to each input it drives; thus in the left column current must be supplied to two inputs. In the right column, current is supplied or absorbed by the power supply or by ground for one input. The driving circuit now drives only a single input. In circuits where an input only drives one or two gates, either of the methods can be used.

G. Practical Value of NAND and NOR Gates

If a NAND gate is followed by an inverter, it becomes an AND gate. Since both inverters and AND gates can be constructed from NAND gates, the NAND gate is seen to be a functionally complete set in itself. So also is the NOR gate, from which inverters and OR gates can be constructed.

From a practical standpoint we note that the SSI chips containing gates may contain up to four 2-input gates per chip. If we are building a logic system using NAND or NOR chips, there will often be unused gates on some of the chips. If an invert function or an AND or OR function is required, we can often use these unused gates to produce the desired function without adding more chips to the system. If AND gates are being used, an invert function requires that an inverter chip be added regardless of the presence of unused AND gates in the system. The versatility of the NOR and NAND gates can often lead to a finished logic system requiring fewer chips than other implementations.

Drill Problems: Sec. 1.4

1. Explain the difference between a truth table and a function table.

2. Construct a truth table for two series inverters, that is, the output of inverter A connects to the input of inverter B. The circuit input is the input of inverter A, while the output of the circuit is the output of inverter B.

3. Construct a truth table for a positive logic NOR gate followed by an inverter. What logic function does this circuit perform?

1.5 LOGIC FAMILIES

Section overview: The important parameters of logic circuits that influence system performance are introduced in this section. Symbols representing required voltage, current levels, and switching times are defined. The effects of noise margin and fan-out are considered.

It is unnecessary for a logic designer to understand the details of the electronic circuits that make up the gates. The manufacturers have designed these and other logic circuits as integrated circuits on silicon chips. These circuits can be fabricated in several different configurations and from several types of device. Those chips having in common a particular device and configuration are said to belong to a logic family.

Some examples of currently useful families are transistor-transistor logic (TTL), emitter-coupled logic (ECL), complementary-symmetry MOS (CMOS), and integrated-injection logic (I^2L). All circuits of a given family must have compatible operating characteristics. The high-level voltage developed at the output of a gate must be sufficient to drive the input of any other gate to the same high level. The low-level output must pull the input of the next stage down to an acceptably low level. Certain current requirements must be met at each voltage level. In general, an entire digital system, such as a computer, will use only one or two logic-circuit families. Hundreds to thousands of SSI or MSI logic elements are connected properly to form the required subsystems of the digital system.

Sometimes it is necessary to connect logic elements that are not of the same family. When this is done, an interface between the different elements may be required. An interface consists of circuits that translate the output signals from one family to the input signals required by the other family. Certain families can be combined without interface circuits; other combinations of families are popular enough that standard interface circuits are provided within the IC families. Details of interfacing are considered in a later section of this chapter.

A. Current and Voltage Definitions

Although it is not the purpose of this textbook to duplicate a data handbook for digital ICs, some definitions are pertinent to further discussion of logic circuits.

V_{IHmin} = minimum input voltage that the logic element is guaranteed to interpret as the high logic level

V_{ILmax} = maximum input voltage that the logic element is guaranteed to interpret as the low logic level

V_{OHmin} = minimum high logic-level voltage appearing at the output terminal of the logic element

V_{OLmax} = maximum low logic-level voltage appearing at the output terminal of the logic element

I_{IHmax} = maximum current flowing into an input when a specified high logic-level voltage is applied

I_{ILmax} = maximum current flowing into an input when a specified low logic-level voltage is applied

I_{OH} = current flowing into the output when a specified high-level output voltage is present

I_{OL} = current flowing into an output when a specified low-level output voltage is presented

I_{OS} = current flowing into an output when the output is shorted and input conditions are such to establish a high logic-level output. (Current flowing out of a terminal has a negative value.)

B. Fan-Out

In a digital system a given gate may drive the inputs to several other gates. The designer must be certain that the driving gate can meet the current requirements of the driven stages at both high and low voltage levels. The number of inputs that can be driven by the gate is referred to as the fan-out of the circuit. This figure is expressed in terms of the number of standard inputs that can be driven. Most circuits of a family will require the same input current, but a few may require more. If so, the specs for such a circuit will indicate that the input is equivalent to some multiple of standard loads. For example, a circuit may present an equivalent input of two standard loads. If fan-out of a gate is specified as 10, only five of these circuits could be safely driven.

In several handbooks, the current requirements are given and fan-out can be calculated. For example, one TTL gate has the following current specs:

$$I_{IH} = 40 \ \mu A \qquad I_{IL} = -1.6 \ mA$$
$$I_{OH} = -400 \ \mu A \qquad I_{OL} = 16 \ mA$$

If this gate is to drive several other similar gates, we see that the output current capability of the stage is 10 times that required by the input. We note that the output stage can drive 400 μA into the following stages at the high level and sink 16 mA at the low level. The fan-out of this gate is 10.

C. Noise Margin

Although current requirement is the major factor in determining fan-out, input capacitance or noise margin may further influence this figure. Noise margin specifies the maximum amplitude noise pulse that will not change the state of the driven stage. This assumes that the driving stage presents a worst case logic level to the driven stage. Noise margin can be evaluated from a consideration of the voltage levels V_{IHmin}, V_{ILmax}, V_{OHmin}, and V_{OLmax}. Figure 1.10 shows two logic circuits that are cascaded.

Figure 1.10 Cascaded stages used to calculate noise margin.

If we assume that $V_{ILmax} = 0.8$ V for circuit B, this means that the input must be less than 0.8 V to guarantee that circuit B interprets this value as a low level. If circuit A has a value of $V_{OLmax} = 0.4$ V, a noise spike of less than 0.8 − 0.4 = 0.4 V cannot lead to a level misinterpretation by circuit B. The difference

$$V_{ILmax} - V_{OLmax} \tag{1.2}$$

is called the low-level noise margin.

Assuming that $V_{IHmin} = 2$ V for circuit B and $V_{OHmin} = 2.7$ V for circuit A, the high-level margin is 2.7 − 2.0 = 0.7 V. This high-level noise margin is found from

$$V_{OHmin} - V_{IHmin} \tag{1.3}$$

Since the minimum voltage developed by circuit A at the high level is 2.7 V, while circuit B requires only 2.0 V to interpret the signal as a high level, a negative noise spike of −0.7 V or less will not result in an error.

As we consider the noise margin we recognize that the values calculated in Eqs. (1.2) and (1.3) are worst case values. A particular circuit could have actual noise margins better than those calculated.

As more gate inputs are connected to a given output, the voltages generated at both high and low levels are affected as a result of increased current flow. Thus, fan-out is influenced by noise margin.

D. Switching Times

Another quantity which is used to characterize switching circuits is the speed with which the device responds to input changes. For switching circuits, the graph of Fig. 1.11 is useful in defining delay times. This figure assumes an inverting gate. There is a finite delay between the application of the input pulse and the output response. A quantitative measure of this delay is the difference in time between the point where e_{in} rises to 50 percent of its final value and the time when e_{out} falls to its 50 percent point. This quantity is called leading-edge delay t_{pHL}. The trailing-edge delay t_{pLH} is the time difference between 50 percent points of the trailing edges of the input and output signals. The propagation delay is defined as the average of t_{pHL} and t_{pLH}, or

$$t_{pd} = \frac{t_{pHL} + t_{pLH}}{2}$$

Fall and rise times are defined by 10 to 90 percent values as the output voltage swings between lower and upper voltage levels.

Propagation delay time of an integrated circuit is a function of passive delay time, rise and fall times, and the saturation storage time of the circuit's individual transistors. Since input and output capacitance will influence the integrated circuit switching times, fan-in and fan-out will also affect delay times. Switching times are sometimes specified by graphs showing the various times as functions of fan-out with specified driving conditions. Figure 1.12 shows a typical graph.

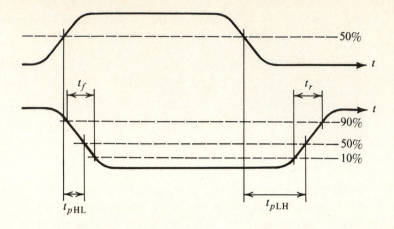

Figure 1.11 Definition of switching times.

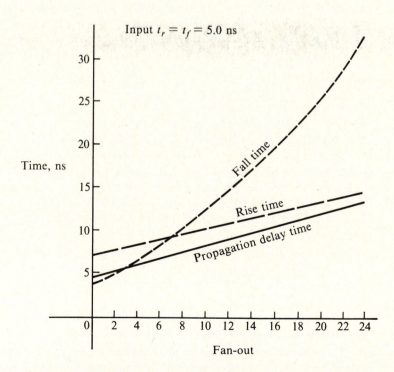

Figure 1.12 Switching times as a function of loading.

There will always be a certain amount of stray capacitance at the input and output terminals; thus, the careful designer will allow for the effect on switching time of these stray values.

An understanding of the definitions given in the preceding paragraphs allows a designer to use logic gates as building blocks in digital systems. For those that desire a greater knowledge of circuit operation of the TTL, ECL, CMOS, and I²L families, Appendix 2 can be consulted.

The TTL family has been the workhorse for many years in SSI and MSI applications. Fast and versatile, no other line offers as great a variety of circuits. The fabrication of resistors requires more chip volume than transistors do, and TTL chips use several resistors per gate. Consequently, applications in LSI circuits are somewhat limited.

CMOS has been used for all levels of integration from SSI through LSI. The high-density chips so produced are ideally applied where large memories are required, although speed of operation sometimes limits performance.

I²L and related technology is generally applied to the LSI circuits employed in large gate arrays and microprocessors. Newer methods of I²L fabrication, such as Schottky clamping, promise to increase operation speed.

MOS circuitry is heavily used for such LSI circuits as microprocessors and large-memory chips. Earlier microprocessors were fabricated exclusively with n-MOS technology, but CMOS microprocessors are now available also.

Drill Problems: Sec. 1.5

1. If the output of gate A drives an input to gate B, how should $V_{OHmin}(A)$ relate to $V_{IHmin}(B)$? How should $V_{OLmax}(A)$ relate to $V_{ILmax}(B)$?

2. If $I_{IH} = 25 \ \mu A$, $I_{IL} = -1.2$ mA, and the fan-out of the circuit is 8, what are the minimum values of I_{OH} and I_{OL}?

3. If $V_{ILmax} = 0.7$ V and the low-level noise margin is 0.5 V, what is V_{OLmax}?

4. If two inverting gates are connected in series, each having a propagation delay time of 12 ns, what is the total propagation delay between circuit input and output? Is propagation delay time additive with respect to number of series circuits? Explain.

Because of the great popularity of TTL circuits the remainder of this text will consider only this family in digital system design. This does not imply that other families are unimportant, but there is simply not enough space in the text to cover each family. The next section discusses some practical aspects of TTL gates.

1.6 TTL GATES

Section overview: This section shows some actual TTL gates that are available. The discussion then moves to two special TTL gates designed to allow several gate outputs to connect to the same line. These are the open collector gate and the three-state gate. We conclude the section with a consideration of mixing gates from different logic families. Although mixing families is generally avoided, there are rare occasions when required.

A. Available Gates

Although several thousands of gates can be integrated on a chip, SSI circuits only contain three or four gates on each chip. This limitation is imposed by the number of pins allowed on the IC package. These pins are metal contacts connected internally to the gates, extending through the package to allow external connections to be made to the gates. SSI chips generally limit the number of pins to 12, 14, or 16. The power supply requires two connections to the chip ($+5$ V and ground), and each separate gate requires an output and two (or more) inputs. Figure 1.13 shows several typical TTL chips demonstrating each type of gate discussed and an inverter chip.

The required power supply for TTL is 5 V $\pm$ 0.25 V.

B. Open Collector Gates

In digital systems it is not uncommon to drive a single input line with several different gate outputs. Figure 1.14 shows an arrangement with three NAND gates driving line A and three NAND gates driving line B. With normal TTL gates, this arrangement is inappropriate and generally results in the destruction of some of the connected gates. If the input conditions of one gate connected to point A cause the output to go low while the two other gate outputs are driven high, a conflict results. Two gates are pulling point A toward $+5$ V, and the other gate is pulling this same point toward 0 V. As explained in Appendix 2, this situation results in excess current flow through the output stages of the gate and will often destroy one or more gates.

The open collector gate can solve this problem in some applications. This gate is similar to a normal TTL gate except the pull-up transistor in the output circuit is removed (see T_4 in Fig. A2.1 of Appendix 2). The function of this transistor is to pull the output voltage up to the high level when dictated by input conditions. When this transistor is absent, the gate cannot drive the output to the high voltage level. The pull-down transistor remains in the gate. An external pull-up resistor must be added between the output and the positive power supply, to pull the output to the high voltage level when required. A single pull-up resistor is used for several gate outputs connected to a common point, as indicated in Fig. 1.15.

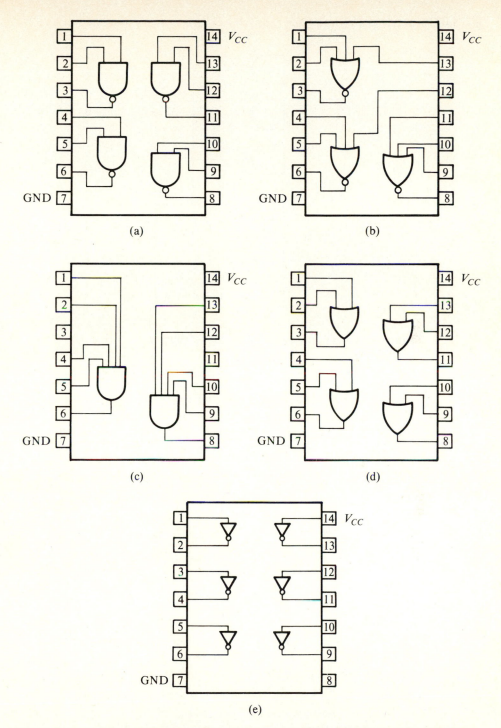

Figure 1.13 (a) A quad 2-input NAND gate chip (7400). (b) A triple 3-input NOR gate chip (7247). (c) A dual 4-input AND gate chip (7421). (d) A quad 2-input OR gate chip (7432). (e) A hex inverter chip (7404).

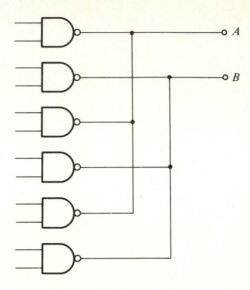

Figure 1.14 Multiple gates driving lines *A* and *B*.

The AND symbols overlaying the connection to the pull-up resistors do not represent the physical presence of gates. It is conventional to include this symbol to emphasize that only if all gates are driven to an output high condition will the circuit output go high. If a single gate is driven to an output low condition, the circuit output goes low regardless of other gate conditions. This behavior suggests an AND-type relationship and, hence, the overlaid symbol.

When a gate output is driven low, the total current absorbed by this output is supplied by the resistor rather than from the other gates. This can easily be limited to a safe value by proper selection of the resistance value.

There are upper and lower limits on the value of R. We will assume that point A in Fig. 1.15 is connected to a gate input with $I_{\text{IHmax}} = 50 \ \mu A$, $I_{\text{ILmax}} = -2.0$ mA, $V_{\text{IHmin}} = 2.7$ V, and $V_{\text{ILmax}} = 0.5$ V and that I_{OL} for each open collector gate is 20 mA and $I_{\text{OH}} = 250 \ \mu A$. When the three open collector gates are at the high level, a maximum of 250 μA leakage current flows into these gates. The resistor must supply 50 μA to the input of the driven gate plus 750 μA leakage current to the open collector outputs and the voltage must exceed 2.7 V. Since $V_{\text{CC}} = 5$ V, a 2.3 V drop across R is the maximum allowable drop. The upper limit of R is found to be

$$R_{\text{max}} = \frac{2.3 \ \text{V}}{800 \ \mu A} = 2.9 \ \text{k}\Omega$$

When one gate output is driven low, it must absorb or sink the current through the resistor plus the current from the input of the next gate, $-I_{\text{ILmax}}$,

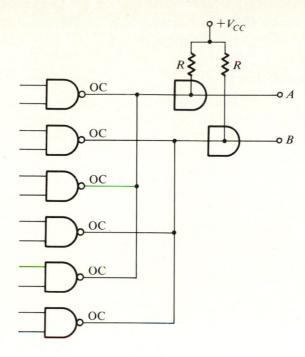

Figure 1.15 Proper method for driving lines *A* and *B* with multiple gates.

that is,

$$20 \text{ mA} \geq I_R + 2 \text{ mA}$$

This inequality limits I_R to 18 mA. Since V_{ILmax} is 0.5 V, the drop across R must be at least 4.5 V, while the current through R cannot exceed 18 mA. The minimum value of R is then

$$R_{\text{min}} = \frac{4.5 \text{ V}}{18 \text{ mA}} = 250 \ \Omega$$

A typical resistance value might be 1 kΩ.

The general equations for R_{max} and R_{min} are given by

$$R_{\text{max}} = \frac{V_{CC} - V_{\text{IHmin}}}{N_1 I_{\text{OH}} + N_2 I_{\text{IH}}}$$

and

$$R_{\text{min}} = \frac{V_{CC} - V_{\text{ILmax}}}{I_{\text{OL}} + N_2 I_{\text{IL}}}$$

where N_1 is the number of open collector gates and N_2 is the number of gate inputs driven.

The 7403 is an example of a NAND gate with open collectors. This chip includes four 2-input gates.

C. Three-State Outputs

Another circuit designed to allow several gate outputs to drive a single input is the three-state device. Three-state output gates are often used with microprocessor circuits and have become very popular as this device has grown in importance. In three-state circuits, an additional input, called the enable, is provided on the chip. The symbol for a three-state AND gate is shown in Fig. 1.16(a).

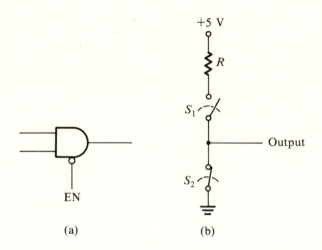

(a) (b)

Figure 1.16 (a) A 2-input AND (three-state) symbol. (b) An equivalent circuit for the three-state output.

When the enable input is low, the gate functions as a normal gate. When the enable input goes high, the gate becomes inactive with the inputs having no effect on the output. In addition, the impedance from output to ground becomes very high, falling in the $M\Omega$ range. Taking the enable line high effectively disconnects the gate output from the line it is physically connected to.

Since a normal gate can take on two voltage levels or states, this high-impedance state is called a third state. The three-state gate can be visualized in terms of the circuit shown in Fig. 1.16(b). When the gate is enabled, one switch is open and the other is closed. For the enabled gate, the condition of output high is equivalent to switch S_1 closed and S_2 open. Output low is equivalent to S_1 open and S_2 closed. When the enable input is driven high to disable the circuit, it is equivalent to opening both switches. The output is no longer driven by the gate and assumes a voltage determined by other circuits connected to this point. The switch S_1 corresponds to the pull-up transistor T_4 of the TTL gate (see Fig. A2.1 of Appendix 2) and S_2 corresponds to the pull-down transistor T_3. In a normal gate, only one of these transistors is off at any given time, but the three-state gate has additional circuitry that drives both transistors off simultaneously in the disabled state.

If the output terminal is connected to some point common to other gate outputs, another gate can pull the common point low without being required to sink current from the disabled gates. This gate should be used only in applications that allow a single gate to be enabled at any given time. If more than one is enabled simultaneously, destruction of a gate can occur.

Three-state outputs are also used for certain latches, memory circuits, counters, registers, and other circuits that require a third high-impedance state at the outputs. This feature is available in several MOS logic systems and is often used in connection with microprocessor input/output (I/O) buses.

Figure 1.17 shows a microprocessor with several I/O devices connected to a bidirectional data bus. When data presented by a given device are to be read into the microprocessor, the enable line is taken low so that the device can drive the microprocessor bus. At this time all other devices must be disabled, presenting high impedances to each line of the bus. With this configuration a single set of bus lines can serve a great number of devices that could include memories, input and output registers, and other digital devices.

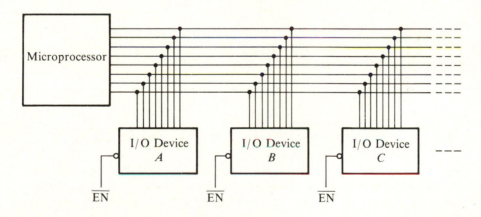

Figure 1.17 Microprocessor data bus organization.

D. Logic Interfacing

The elements of each logic family are compatible with all other devices of that family. There are occasions, however, when it is useful or necessary to use two different logic families in the same system. Because voltage level and current requirements are usually different for each family, some considerations must be given to the connections between families. The circuits required to allow one family to drive another are called interface circuits.

In order to interface circuits from different logic families, several conditions must be satisfied. The value of V_{OHmin} for the driving family must exceed V_{IHmin} of the loading family. The low-level output voltage of the driving family, V_{OLmax}, must be less than V_{ILmax} of the loading family. The driving family must not produce a higher output voltage than the loading family can accept. Output current values of the driving family at both voltage levels must satisfy input current requirements of the loading family. These conditions are summarized below.

Driving Family		Loading Family
V_{OHmin}	$>$	V_{IHmin}
V_{OLmax}	$<$	V_{ILmax}
V_{OHmax}	$<$	Maximum input voltage
$-I_{OHmax}$	$>$	I_{IHmax}
I_{OLmax}	$>$	$-I_{ILmax}$

I_{OHmax} and I_{ILmax} will have negative values, since they flow out of respective terminals. If voltage conditions are satisfied, but current conditions are not, current amplifier or buffer stages must be used. If current requirements are met, but voltage requirements are not, a level shifter with a required current gain near unity must be used. When all conditions are met, no additional circuitry is required. Of course, if the driving circuitry is loaded by several

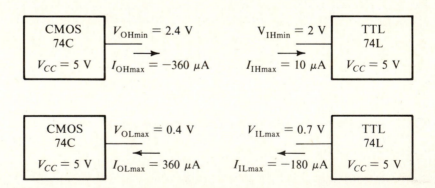

Figure 1.18 Demonstration of CMOS 74C compatibility with TTL 74L.

circuits, the source and sink current capability of the driving gate must exceed the sum of the loading current requirements.

Figure 1.18 shows the compatibility of CMOS series 74C with TTL series 74L. The noise margin for the situation of a 74C gate driving a 74L gate is found to be $0.7 - 0.4 = 0.3$ V for the low level and $2.4 - 2 = 0.4$ V for the high level.

When TTL series 74L becomes the driving circuit and CMOS 74C becomes the loading circuit, voltage compatibility does not quite exist: $V_{OHmin} = 2.4$ V, but $V_{IHmin} = 3.25$ V; and $V_{OLmax} = 0.4$ V, while $V_{ILmax} = 0.8$ V. The high-level output voltage of the 74L will not meet the required input value of the 74C. Level shift circuits are required to remedy this situation.

Drill Problems: Sec. 1.6

1. If $I_{IHmax} = 40$ μA, $I_{ILmax} = -1.6$ mA, $I_{OL} = 16$ mA, $I_{OH} = 250$ μA, $V_{IHmin} = 2.8$ V, and $V_{ILmax} = 0.6$ V, calculate the minimum and maximum values for the pull-up resistor to allow 3 open-collector gates to drive an equivalent gate input.

2. When a three-state gate output connects to some point A, what voltage level exists at this point when the gate is disabled? Explain.

SUMMARY OF SIGNIFICANT POINTS IN CHAP. 1

1. When using electronic circuits, the binary system can be more accurate and reliable than analog circuits in representing numbers.
2. Digital systems use binary or binary-related codes. In order to design these systems, we must become familiar with codes such as binary, binary-coded octal, binary-coded hex, binary-coded decimal, and alphanumeric codes.
3. Most electronic circuits represent the binary symbols 0 and 1 by voltage levels. For example, all voltages less than 0.8 V may correspond to binary 0, while all voltages greater than 2.4 V may correspond to binary 1.
4. Gates are important circuits in digital systems as will be seen in succeeding chapters. OR, AND, NOR, and NAND gates can be characterized by a corresponding truth table.
5. Manufacturers name gates according to the functions performed when using a positive logic definition. If negative logic is used, the function performed by the gate changes. The function table is not a function of logic level definition and can be used to construct a truth table after the logic level definition is made.

6. Certain parameters that characterize logic gates must be known to do effective logic design. Among these are voltage and current level requirements, noise margin, switching times, and fan-out figures.

7. The TTL logic family is very versatile and popular. Several standard gates are available such as OR, AND, NOR, and NAND gates with 2, 3, 4, or 8 inputs. Open-collector and three-state gates are also available and are used in applications requiring several gate outputs to drive a common line. Generally, only one gate output will be active at any given time in this situation.

8. For those occasions when more than one logic family must be used in a system, proper interfacing must take place.

REFERENCES AND SUGGESTED READING

1. D.J. Comer, *Electronic Design with Integrated Circuits*. Reading, Mass.: Addison-Wesley, 1981, chap. 3.
2. H.O. Daley, *Fundamentals of Microprocessors*. New York: Holt, Rinehart and Winston, 1983, chaps. 1 and 2.
3. Engineering Staff, *The TTL Data Book for Design Engineers*. Dallas: Texas Instruments, 1983.
4. R.S. Sandige, *Fundamentals of Digital Analysis*. New York: McGraw-Hill, 1978, chap. 7.

CHAP. 1 PROBLEMS

Sec. 1.1

1.1 Explain why the binary number system is used for most digital systems.

*__1.2__ Assume a circuit output can vary from 0 to 5 V. A detection circuit is capable of distinguishing levels that vary by 20 mV. How many different numbers can be represented by this system?

Sec. 1.2

1.3 Convert the following decimal numbers to binary. Results should be accurate to within 0.01_{10}.

    a. 682   **c.** 25.35
    b. 121.78 **d.** 1.684

*__1.4__ Convert the following decimal numbers to BCD.

    a. 841 **b.** 6875

---

* Answers to starred problems are given in Answers to Selected Problems at the back of the book.

1.5 Convert the decimal numbers of Prob. 1.3 to octal and binary-coded octal.

1.6 Convert the decimal numbers of Prob. 1.3 to hex and binary-coded hex.

*__1.7__ Convert the following binary numbers to decimal and to BCD.
   a. 11111111   **c.** 1.011
   b. 10010110   **d.** 11.1101

1.8 Convert the binary numbers of Prob. 1.7 to octal and binary-coded octal.

1.9 Convert the binary numbers of Prob. 1.7 to hex and binary-coded hex.

1.10 Write the ASCII code for the message

   THIS CODE IS NEAT.

Sec. 1.4

*__1.11__ Two inputs A and B are inverted and applied to an OR gate. Construct the corresponding truth table for positive logic.

1.12 Two inputs A and B are inverted and applied to an OR gate. The gate output is connected to an inverter. Construct a truth table for the inverter output for positive logic.

1.13 An input A is inverted and applied to an OR gate. The other input B is applied directly to the gate. Construct the corresponding truth table for positive logic.

1.14 Two inputs A and B are inverted and applied to an AND gate. Construct the corresponding truth table for positive logic.

*__1.15__ An input A is inverted and applied to a NAND gate. The other input B is applied directly to the gate. Construct the corresponding truth table for positive logic.

1.16 Show how to connect the 7402 NOR gate chip to realize the circuit shown in Fig. P1.16 with a single chip.

1.17 Show how to connect the 7400 AND gate chip to realize the circuit shown in Fig. P1.17 with a single chip.

*__1.18__ Show how to connect inverters to a positive logic OR gate to convert to a positive logic AND gate.

1.19 Show how to connect 3 positive logic, 2-input AND gates to create a 3-input AND gate. Show how to connect any unused inputs.

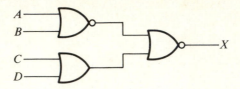

Figure P1.16

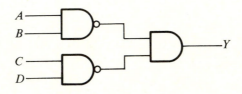

Figure P1.17

Sec. 1.5

1.20 The current specs on the 7408-quad 2-input AND gate are I_{IHmax} = 40 μA, I_{ILmax} = −1.6 mA, I_{OH} = −800 μA, and I_{OL} = 16 mA. How many similar inputs can one 7408 gate drive?

*__1.21__ Repeat Prob. 1.20 for the 7428-quad 2-input NOR buffer. The current specs are I_{IHmax} = 40 μA, I_{ILmax} = −1.6 mA, I_{OH} = −2400 μA, and I_{OL} = 48 mA.

Sec. 1.6

1.22 Assume that two TTL gates must drive the same bus line, but never simultaneously. Compare the use of open collector gates to the use of three-state output gates.

1.23 Calculate the pull-up resistor of an open-collector TTL gate that drives three similar inputs. The current specs for the gate are I_{IHmax} = 50 μA, I_{ILmax} = −2.0 mA, I_{OH} = −250 μA, and I_{OL} = 20 mA. Assume V_{IHmin} = 2.7 V and V_{ILmax} = 0.5 V.

1.24 The output of gate A drives an input of gate B. The specs shown pertain to the gates.

Gate	V_{IHmin}	V_{ILmax}	V_{OHmin}	V_{OLmax}
A	2.7 V	0.5 V	3.2 V	0.3 V
B	2.8 V	0.7 V	3.4 V	0.4 V

Gate	I_{IHmax}	I_{ILmax}	I_{OHmax}	I_{OLmax}
A	50 μA	−2.0 mA	−250 μA	20 mA
B	300 μA	−4.0 mA	−400 μA	24 mA

Do you anticipate any problems with this arrangement? Why? If gates A and B are interchanged in the circuit, would there be any problems? Explain.

1.25 Explain what problems result from selecting an open-collector pull-up resistor too large? Too small?

CHAPTER 2

Boolean Algebra and Mapping Methods

The introduction to this text mentions the subdivision of logic circuits into two types: combinational and sequential. Combinational circuits have outputs that equal 1 only for certain combinations of input variables. Other input combinations cause the output to equal 0. The output is a direct function of the input values. There is no time dependence nor is there a dependence on some previously applied input combination that is no longer present. Sequential circuits have outputs that depend on present input, time, and past input history. These circuits are more complex than combinational circuits and use combinational circuits as building blocks. Before we can proceed to sequential design, we must understand combinational design. Thus, this chapter and Chap. 3 will be devoted to a study of combinational logic theory and application.

In designing any digital system there are three objectives that an effective design procedure should achieve. These objectives are to (1) build a system that operates within the given specifications, (2) build a reliable system, and (3) minimize resources.

The principles to be discussed in this chapter allow objectives 1 and 2 to be met in a straightforward manner. Minimization of resources is a more complex issue and one that depends on several variables. For example, one might want to minimize the number of gates used in a circuit or the number of inputs. Often the number of IC chips used is a more important quantity to minimize than is the number of gates. There are other occasions when chip cost is the most significant quantity, and a system implemented with less chips may be more expensive due to higher individual chip cost. Other situations may call for a minimization of the designer's time rather than minimum component cost. Still other cases may call for the entire circuit to be fabricated on a single chip since many firms own IC fabrication facilities or use custom fabrication companies. An important quantity to minimize in some instances is the time between start of design and availability of finished product. Minimization of this time often earns an edge over slower competitors in the same field.

Obviously, no design procedure can minimize all resources simultaneously. We will direct the following discussions toward the minimization of gates and inputs which is a strength of Boolean algebra and Karnaugh map techniques. In later chapters we will indicate alternate methods of logic design that can minimize other quantities listed above.

2.1 BOOLEAN ALGEBRA

Section overview: Boolean algebra is introduced in this section followed by a discussion of several important algebraic relations. The basis for reduction of logic functions to simpler forms is considered. The final topic of the section is that of assertion level and dependency notation.

This type of algebra for logic circuits was named after its inventor George Boole (1815–1864). Although available for several years, Boolean algebra was used little until the invention of the digital computer. It now serves as the basis of almost all design and analysis methods in digital systems.

We introduce this subject by considering a practical problem: that of identifying the presence of certain digital codes on a set of lines. A system may be transmitting 4 bits of binary information over 4 lines to another system. With 4 lines, 16 unique codes could be represented. Each code transmitted will exist for a certain length of time after which a new code can occur. The receiving system may need to identify the presence of certain transmitted codes. As an example, suppose the system is to identify the occurrence of the codes representing the decimal numbers 1, 4, 5, 9, 11, and

12. Each time one of these codes appears on the 4 lines, a circuit is to generate an output of 1. When any other code is present, the output should be 0. One method of implementing this system is shown in Fig. 2.1.

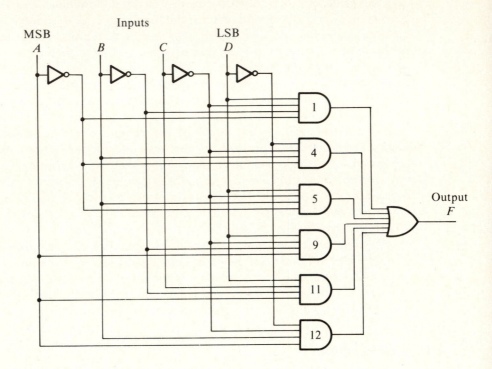

Figure 2.1 A logic system.

The input code for decimal 1 is 0001. The only time we will have this combination of variables present simultaneously is when the code for 1 is on the lines. In order to check this simultaneous condition, we use a 4-input AND gate, inverting all lines applied to the gate except the least significant bit. Each of the gate inputs will equal 1 only when 0001 is present on the lines. The remaining five AND gates perform the same function for the codes 4, 5, 9, 11, and 12. The OR gate causes the circuit output to equal 1 when any of these codes appear on the lines.

The techniques to be considered in this chapter, based on Boolean algebra, will allow us to accomplish two important objectives in solving this problem. First, we will learn how to express the problem succinctly in terms of an output equation rather than several sentences. Second, we can apply minimization methods to reduce the number of gates and inputs required by the system. Although other advantages could be cited, these two are significant enough to justify further study of Boolean algebra.

We note that there are only two constants in a 1-bit binary system, 0 or 1. A given input or output line is designated by some variable name, for example, A. The four input lines in Fig. 2.1 are labeled A, B, C, and D proceeding from MSB to LSB. The output is designated F. At any given time, each of these lines will equal either 0 or 1. When we assign a variable name to a line, we imply a value for the complemented variable. If A takes on a value of 0, $\overline{A}$ must equal 1. If A takes on a value of 1, $\overline{A}$ must equal 0. This allows us to write binary codes in terms of symbols. The code for decimal 9 is 1001. In terms of symbols we can write $A\overline{B}\overline{C}D$. When the code 1001 is present, an AND gate with inputs A, $\overline{B}$, $\overline{C}$, and D would produce an output of 1. We will expand on the use of this result later.

A. OR Relations

We discussed the operation of the 2-input OR gate in terms of a truth table in Chap. 1. The same information given by the truth table is represented by the Boolean expression

$$A + B = X$$

where the variables A, B, and X are defined in Fig. 2.2. The plus sign stands for the OR symbol in Boolean algebra. The expression is read "A or B equals X" and implies that if A or B (or both) equals 1, X will also equal 1.

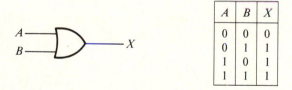

A	B	X
0	0	0
0	1	1
1	0	1
1	1	1

Figure 2.2 An OR gate with truth table.

There are several Boolean identities associated with the OR gate that can be quite useful. These equations, which can be easily verified by a truth table, are listed as Eqs. (2.1) through (2.5).

$$A + 0 = A \tag{2.1}$$

$$A + 1 = 1 \tag{2.2}$$

$$A + A = A \tag{2.3}$$

$$A + B = B + A \tag{2.4}$$

$$A + B + C = (A + B) + C = A + (B + C) \tag{2.5}$$

Equation (2.4) is the commutative law for the OR relation. Equation (2.5) is the associative law for the OR relation. The first equation can be proven from the truth table by setting $B = 0$. Rows 1 and 3 show that $X = A$ for this case. If $B = 1$, rows 2 and 4 prove that $A + 1 = 1$ as indicated by Eq. (2.2). Equation (2.3) is proven by considering rows 1 and 4 in which A and B are equal. If A and B values are exchanged in the truth table, we see that the output is unchanged, and thus Eq. (2.4) is demonstrated. The associative law can also be proven with a truth table.

We should recognize the physical meaning of these OR relations. Figure 2.3 demonstrates the implementation of each equation using positive logic.

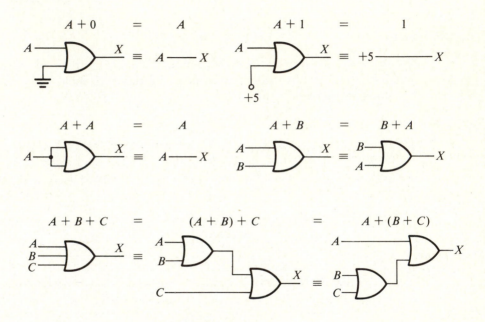

Figure 2.3 Implementation of OR relations.

B. AND Relations

A 2-input AND gate is shown in Fig. 2.4 along with its truth table. The Boolean expression meaning the AND operation of A and B is

$$A \cdot B = X$$

or

$$AB = X$$

This expression states that if both inputs A and B are equal to 1, output X will equal 1. All other combinations of A and B result in $X = 0$. Some

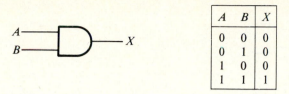

A	B	X
0	0	0
0	1	0
1	0	0
1	1	1

Figure 2.4 Two-input AND gate and truth table.

important Boolean AND identities are listed below and demonstrated in Fig. 2.5.

$$A0 = 0 \qquad (2.6)$$

$$A1 = A \qquad (2.7)$$

$$AA = A \qquad (2.8)$$

$$AB = BA \qquad (2.9)$$

$$ABC = (AB)C = A(BC) \qquad (2.10)$$

Again these identities can be proven by the AND gate truth table. Equation (2.9) is the commutative law for the AND relation. Equation (2.10) is the associative law for the AND relation. The physical implementations of these relations, shown in Fig. 2.5, demonstrate the savings that can result from using the equations.

C. Other Relations

There are several other relations that are useful in logic design. If we invert or complement a variable twice, the result equals the input variable. That is,

$$\overline{\overline{A}} = A \qquad (2.11)$$

A variable ORed with its complement equals 1, since one of these inputs will equal 1. We can write

$$A + \overline{A} = 1 \qquad (2.12)$$

If a variable is ANDed with its complement, the result is always 0 since one of the inputs will equal 0. This gives the equation

$$A\overline{A} = 0 \qquad (2.13)$$

Another form that is encountered in logic design is given by $A + \overline{A}B$ or $\overline{A} + AB$. The truth table of Fig. 2.6 can be used to show that

$$A + \overline{A}B = A + B \qquad (2.14a)$$

and

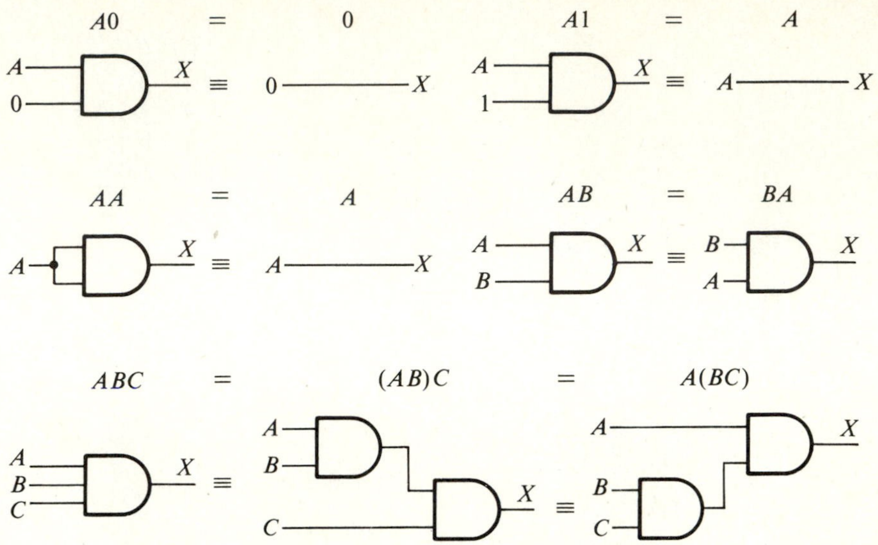

Figure 2.5 Implementation of AND relations.

$$\overline{A} + AB = \overline{A} + B \qquad\qquad (2.14b)$$

The last relation we will consider is called the distributive law. It can be expressed as

$$A(B + C) = AB + AC \qquad\qquad (2.15)$$

Figure 2.7 shows the implementation of this relation.

A	B	$\overline{A}$	$A + \overline{A}B$	$A + B$	$\overline{A} + AB$	$\overline{A} + B$
0	0	1	0	0	1	1
0	1	1	1	1	1	1
1	0	0	1	1	0	0
1	1	0	1	1	1	1

Figure 2.6 Truth table to prove Eq. (2.14).

We note again that one realization of the expression uses only two gates while the other form uses three. Figure 2.8 proves this relation with a truth table.

A very significant relation can be considered at this point. If we encounter the equation

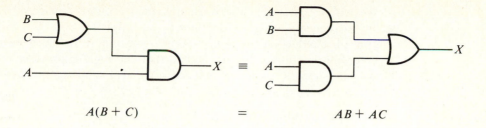

$$A(B + C) \qquad\qquad = \qquad\qquad AB + AC$$

Figure 2.7 Implementation of distributive law.

A	B	C	$B + C$	$A(B + C)$	AB	AC	$AB + AC$
0	0	0	0	0	0	0	0
0	0	1	1	0	0	0	0
0	1	0	1	0	0	0	0
0	1	1	1	0	0	0	0
1	0	0	0	0	0	0	0
1	0	1	1	1	0	1	1
1	1	0	1	1	1	0	1
1	1	1	1	1	1	1	1

Figure 2.8 Truth table to prove distributive law.

$$F = A\bar{B} + \bar{A}\bar{B}$$

we can apply the distributive law to get

$$F = \bar{B}(A + \bar{A})$$

Equation (2.12) is then applied to result in

$$F = \bar{B}$$

The starting equation contains two terms that are logically adjacent. The difference between $A\bar{B}$ and $\bar{A}\bar{B}$ lies in the variable A. In one term A appears, while $\bar{A}$ appears in the other term. The remaining variable in both terms is $\bar{B}$. If only a single variable changes between two terms, the terms are said to be logically adjacent. The expression

$$X = AB\bar{C}\bar{D} + AB\bar{C}D$$

demonstrates two 4-variable terms that are logically adjacent. The significance of logical adjacency will become apparent when we discuss the Karnaugh map later in the chapter.

We will see later that the Karnaugh map is more efficient than Boolean algebra in reducing a logic expression to its simplest form. We will show an example of a reduction problem at this point simply to illustrate the use of the preceding relations. Let us assume that the logic expression $F = A\bar{B} + \bar{A}\bar{B} + AB\bar{C} + \bar{A}B\bar{C}$ is to be reduced. We will tabulate the steps that can be used to reduce this expression.

$$F = A\bar{B} + \bar{A}\bar{B} + AB\bar{C} + \bar{A}B\bar{C}$$

$$= \bar{B}(A + \bar{A}) + B\bar{C}(A + \bar{A}) \quad \text{by Eq.(2.15) (distributive law)}$$

$$= \bar{B} + B\bar{C} \quad \text{by Eq.(2.12)}$$

$$= \bar{B} + \bar{C} \quad \text{by Eq.(2.14b)}$$

This final result can be implemented by a single 2-input OR gate (ignoring inverters), whereas the original expression requires two 2-input AND gates, two 3-input AND gates, and one 4-input OR gate.

There are several manipulations that can be done in Boolean algebra that require either creativity or guess work. For example, suppose we are asked to prove the equality

$$AB + BC + \bar{A}C = AB + \bar{A}C$$

One approach to this is to AND the quantity $B + 1$ with $\bar{A}C$. Since $B + 1 = 1$ and $\bar{A}C1 = \bar{A}C$, the value of the equation is unchanged. We continue with

$$AB + \bar{A}C = AB + \bar{A}C(B + 1)$$

$$= AB + \bar{A}CB + \bar{A}C \quad \text{by Eq.(2.15)}$$

$$= B(A + \bar{A}C) + \bar{A}C \quad \text{by Eq.(2.15)}$$

$$= B(A + C) + \bar{A}C \quad \text{by Eq.(2.14a)}$$

$$= AB + BC + \bar{A}C$$

The proper approach to proving a Boolean relationship is not always obvious as demonstrated by the previous proof. This type of ambiguity does not exist in Karnaugh mapping methods as we shall later see.

Drill Problems: Secs. 2.1A to 2.1C

1. Implement the expression $X = (A + B)(\bar{A} + \bar{B})$ with two OR gates, an AND gate, and three inverters.

2. Implement the expression X in problem 1 using two AND gates, an OR gate, and two inverters.

3. Reduce the expression $X = \bar{A} + A\bar{B} + BC$ to simplest terms.

4. Reduce the expression $F = (A + \bar{B} + \bar{C})(\bar{A} + B + C)\,CB$ to simplest terms.

5. Prove that $C = AC + \bar{A}C$.

6. Prove that $AB + A\bar{B} + \bar{A}C = A + C$.

D. DeMorgan's Laws

These two laws allow us to convert between forms of logic equation and also convert between types of gate used. DeMorgan's laws are given by

$$\overline{ABC \cdots N} = \bar{A} + \bar{B} + \bar{C} + \cdots + \bar{N} \tag{2.16}$$

and

$$\overline{A + B + C + \cdots + N} = \bar{A}\bar{B}\bar{C} \cdots \bar{N} \tag{2.17}$$

We can state Eq. (2.16) in words as follows:

When N variables are ANDed and then inverted, the only combination leading to a 0 output occurs when all N variables equal 1. If we invert the N variables and OR the result, the only combination leading to a 0 output again occurs when all N variables equal 1.

The second law can be stated as:

When N variables are ORed and then inverted, the only combination leading to a 1 output occurs when all N variables equal 0. If we invert the N variables and AND the result, the only combination leading to a 1 output again occurs when all N variables equal 0.

These word statements essentially prove the results without the necessity of using truth tables. A physical implementation of the two laws again leads to significant practical results.

Figure 2.9 shows these equivalencies for 3-input variables.

In terms of gates we see that a NAND gate is equivalent to an OR gate with inverted inputs and a NOR gate is equivalent to an AND gate with inverted inputs. As a memory aid we note that any gate, AND or OR, followed by an inverter is equivalent to the opposite type gate, OR or AND, with inverted inputs.

Application of these two laws along with the Boolean identities allows a designer to implement a given expression in terms of the gates available. In constructing a system, several unused gates may be available on the IC chips. If these are NANDs, for example, we can use them for other types of gates. Obviously, if an AND gate is required, we can follow a NAND by an inverter and produce this function. In more complex cases it is useful to use DeMorgan's laws to synthesize the desired expression.

Let us demonstrate the key points in this method. Suppose we have a function $X = A(B + CD)$ that we must implement with NOR gates and

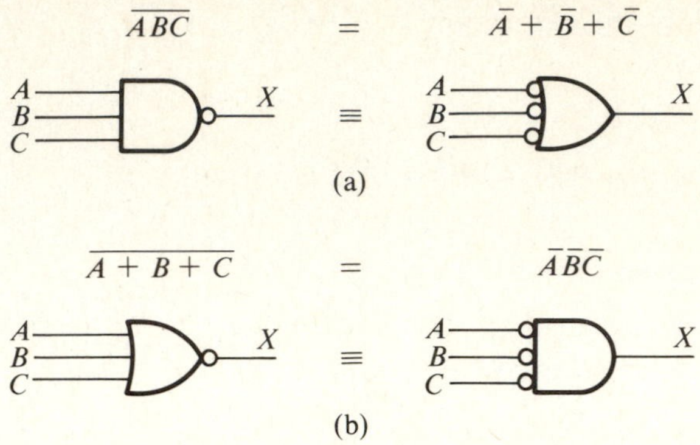

$$\overline{ABC} \qquad = \qquad \bar{A} + \bar{B} + \bar{C}$$

(a)

$$\overline{A + B + C} \qquad = \qquad \bar{A}\bar{B}\bar{C}$$

(b)

Figure 2.9 Implementation of DeMorgan's laws: (a) first law; (b) second law.

inverters. The form of the function we should obtain is one that can be realized directly with NOR gates. An expression such as

$$Y = \overline{\overline{(R + S)} + \overline{(P + Q)}}$$

results from two stages of NOR gates as shown in Fig. 2.10. We can approach this form for the function X by applying DeMorgan's laws to obtain

$$X = \overline{\overline{\bar{A} + \bar{B} + CD}} = \overline{\bar{A} + \bar{B} \cdot \overline{CD}} = \overline{\bar{A} + \bar{B}(\bar{C} + \bar{D})}$$

$$= \overline{\bar{A} + \overline{BC} + \overline{BD}} = \overline{\overline{\bar{A} + \bar{B} + C} + \overline{B + D}}$$

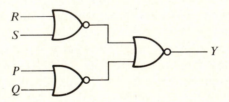

Figure 2.10 Two-stage gating.

The final expression can be realized as shown in Fig. 2.11.

Rather than striving to put an expression into NOR or NAND form, we can convert directly to the gates required. The expression $X = A(B + CD)$ can be realized as shown in Fig. 2.12. If we want to convert to NOR gates and inverters, we first note that the OR gate is similar to a NOR. By adding two inversions at the output of this OR, we can convert to a NOR gate without changing the function of the circuit as shown in Fig. 2.13(a). To change the AND gates to NOR gates, we must introduce inverters at the gate

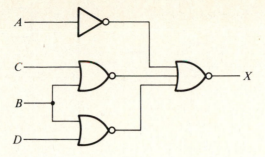

Figure 2.11 Implementation of $X = A(B + CD)$ using NOR gates.

inputs. To preserve the logic function, we introduce two inverters at each desired point as shown in Fig. 2.13(b). Next the input inverters are moved to the outputs of the two gates while the AND symbol is changed to the OR. The result is shown in Fig. 2.13(c). We note that this realization is not unique when compared with the earlier realization of the same function as shown in Fig. 2.11.

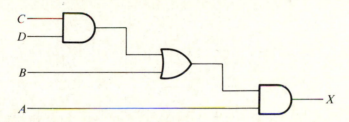

Figure 2.12 Realization of $A(B + CD)$.

Figure 2.14 shows the development of the NAND gate realization of the function $X = A(B + CD)$.

DeMorgan's laws can also be used to reduce logic expressions. For example, given the expression $X = ABC + \bar{A}CD + \bar{B}CD$, we could reduce as follows:

$$X = ABC + \bar{A}CD + \bar{B}CD$$
$$= C(AB + \bar{A}D + \bar{B}D) \qquad \text{by Eq.} (2.15)$$
$$= C(AB + [\bar{A} + \bar{B}]D) \qquad \text{by Eq.} (2.15)$$
$$= C(AB + \bar{A}\bar{B}D) \qquad \text{by Eq.} (2.16)$$
$$= C(AB + D) \qquad \text{by Eq.} (2.14a)$$
$$= ABC + DC \qquad \text{by Eq.} (2.15)$$

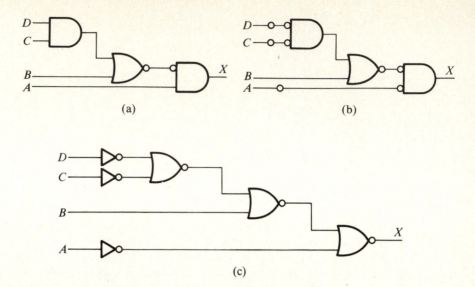

Figure 2.13 Steps in producing a NOR gate realization of X.

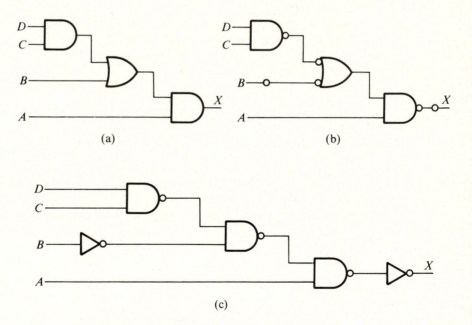

Figure 2.14 Steps in producing a NAND gate realization of X.

Drill Problems: Sec. 2.1D

1. Express F in terms of variables and single-variable complements. ($\overline{A}$ is a single-variable complement; $\overline{A + B}$ is not.)

a. $F = \overline{AB + CD}$

b. $F = \overline{\overline{AB} + CD}$

c. $F = \overline{\overline{AB} + \overline{CD}}$

d. $F = \overline{\overline{\overline{AB}} + \overline{\overline{CD}}}$

2. Reduce to simplest form $X = \overline{\overline{\overline{ABC}} + \overline{\overline{AB}}}$.

E. Assertion Levels and Dependency Notation

We introduced the small circle as the inversion symbol in Chap. 1. This symbol is in the process of being displaced by a small triangle. In fact, there is a new IEEE standard, the revised IEEE Std 91/ANSI Y32-14, that proposes a new logic symbology for all standard logic circuits. This new standard was developed by the International Electrotechnical Commission (IEC). The newer data books have included the symbols from this standard along with the older symbols. These newer symbols are called the IEEE/IEC symbols in the data books. Since standards are only accepted over a period of years, the older inverter symbol will be used by some manufacturers during the next few years. Others will use the new symbol to be discussed in this section. We will use this new symbol throughout the remainder of the text.

One of the fundamental ideas making the new standard more powerful is that of dependency notation. This notation provides the means of noting relationships between inputs, outputs, or inputs and outputs without showing all elements and interconnections involved. For a full understanding of the concept of dependency notation, the above cited standard should be consulted.

We will present here only a few ideas and symbols related to this symbology. Logic polarity indicators are used to eliminate the need for specifying whether positive logic or negative logic definitions apply. This leads to an advantage in handling mixed logic systems. When used for an input, the triangular polarity indicator of Fig. 2.15 tells us that a low applied level will lead to an internal state of 1 for the device. When used for an output, the triangle indicates that the internal 1 state will produce a low external level.

The IEEE/IEC symbol for a quad 2-input NAND gate is shown in Fig. 2.16(a). We will not use this type of symbol for gates. Instead, the symbol of Fig. 2.16(b) will be used since we will often need to consider individual gates. The character & is used to indicate the AND operation while ≥ 1 would indicate the OR operation for the standard symbols.

Figures 2.16(c) and (d) show two other symbols of importance to gate circuits. These are the open-collector output and the three-state output symbols which appear inside the gate outline near the output terminal when used.

Figure 2.15 Logic polarity indicator.

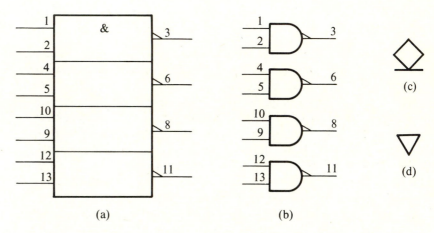

Figure 2.16 (a) IEEE/IEC symbol for quad 2-input NAND chip. (b) Symbol to be used in this text. (c) Open-collector output symbol. (d) Three-state output symbol.

We often speak of the 1 level as being the active or asserted level. The meaning of "to assert" is to put in force or to cause a positive action. In logic circuits we refer to high asserted or low asserted signals. If a terminal is a low assertion input, the circuit requires this input to be low to activate the circuit. High assertion inputs can activate or assert the circuit only for high-level input voltages. A low asserted output indicates that when the circuit is active, the output exists at the low voltage level.

The indicator light of Fig. 2.17 demonstrates a circuit with a low assertion input. The resistor-lamp combination will be active only when current flows through this circuit. Since the upper terminal is connected to $+5$ V, no current will flow if the input to this circuit is also at 5 V. Thus, when the logic circuit output is near the high level of 5 V, the indicator lamp is not asserted. In order to cause a voltage drop across the resistor and lamp, the logic circuit must drive the point X to the 0 V level. The input to the lamp is then a low asserted input, causing an action when driven low.

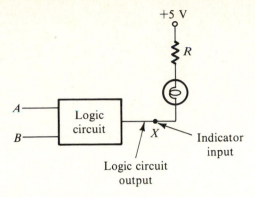

Figure 2.17 A low assertion input.

Several modern logic systems involve both high and low asserted signals and are then mixed logic systems rather than strictly positive or strictly negative.

Now let us suppose that the logic circuit driving the lamp is a gate that is to assert the lamp input only when inputs A and B are both high. This requires a gate having an output that goes low when both inputs are high. Obviously, a NAND gate will work in this circuit. The two NAND gate inputs are signals that are asserted high while the output is asserted low. The logic function that is being performed is an AND function, looking for the condition of $A = B = 1$. When this condition occurs, the gate causes a positive action, asserting the output. If we indicate low-assertion levels by the logic polarity indicator, we could represent the circuit of Fig. 2.17 as shown in Fig. 2.18(a).

The indicator at the lamp input does not represent an inverter. It simply indicates that this input must be driven low to assert the lamp. The triangle on the AND gate output can represent an inversion, converting the AND to a NAND. We should consider the other view of this circuit, that is, the gate is performing the AND function on high assertion inputs to produce a low asserted output.

From DeMorgan's laws we recognize that we could also represent the circuit as shown in Fig. 2.18(b). Although technically correct, the circuit of part (b) does not directly indicate what function the gate is actually performing. It also does not match the assertion levels of the variables A and B to the assertion levels of the gate inputs. Nor does it match assertion levels of the gate output to the indicator lamp input.

More information is conveyed by a schematic using a symbol choice that reflects the actual function being performed by each circuit. The key to choosing the correct gate symbols is to match assertion levels as far as possible. Figure 2.18(a) uses a gate with high assertion level inputs, matching the

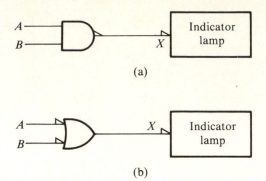

(a)

(b)

Figure 2.18 (a) A low assertion lamp circuit. (b) An equivalent of the circuit in (a).

variables A and B. A low gate output assertion level matches the input level required to assert the lamp. When all assertion levels are matched, we can read from the logic symbol the function being performed. From Fig. 2.18(a) we see that A ANDed with B will assert the light. Figure 2.18(b) does not immediately express this information.

As another example of choosing symbols, consider the logic circuits of Fig. 2.19. The two circuits are equivalent, but the first does not immediately indicate the function being performed while the second does. The two input gates perform the AND function with outputs asserted low. Since these outputs are asserted low, they should drive low assertion inputs to the last gate. Figure 2.19(b) chooses the correct symbol for this gate and we can directly see that the OR function is performed with output asserted high. We can now write by inspection that

$$X = AB + CD$$

We could arrive at this same conclusion from the first circuit only after some algebraic manipulation. We would write

$$X = \overline{\overline{AB} \cdot \overline{CD}}$$

and then apply DeMorgan's first law to obtain the earlier result. Obviously, the method of matching assertion levels is preferable when dealing with logic diagrams since the resulting functions are directly observable.

Figure 2.20 offers a more complex example of this method. We first note that gates 1 and 2 have low asserted outputs; thus gate 4 is changed to a low asserted input gate. This results in an AND gate with a low asserted output. Since both gates 3 and 4 drive gate 5, we also want gate 3 to match the low asserted output of gate 4. The inputs to gate 3 are high assertion signals; thus the NAND gate form is appropriate. After changing this gate

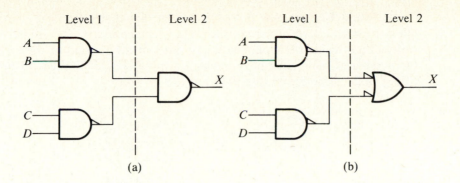

Level 1   Level 2   Level 1   Level 2

(a)   (b)

Figure 2.19 (a) Improper symbol choice; (b) proper choice.

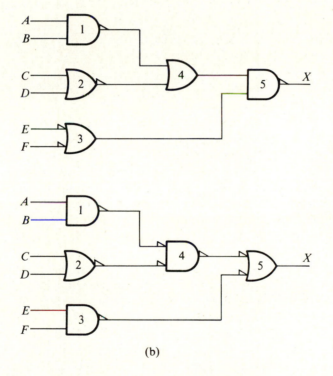

(b)

Figure 2.20 (a) Given logic circuit; (b) revised symbols.

to its NAND equivalent, we then match gate 5 to its driving signals to arrive at the circuit of Fig. 2.20(b). Since all assertion levels are now matched, we can write by inspection that

$$X = AB \cdot (C + D) + EF$$

It would be very nice indeed if we encountered only logic circuits that allowed matching of all assertion levels. Unfortunately, the situation some-times occurs when the inputs to a gate are not all asserted at the same level. In these cases, it is not obvious which symbol should be used. The circuits of Fig. 2.21 demonstrate this situation.

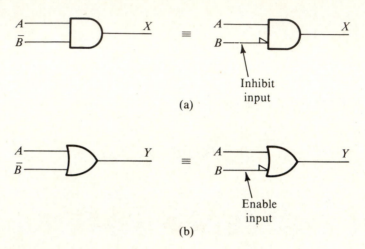

(a)

(b)

Figure 2.21 (a) An inhibit circuit. (b) An enable circuit.

Figure 2.21(a) demonstrates an inhibit circuit. If $B = 1$, the gate output will be unaffected by changes in A. For this case the effect of A on the out-put is inhibited. If B goes to 0, then $X = A$. The circuits represented in Fig. 2.21(b) are enable circuits. If B is asserted high or equal to 1, the out-put Y is given by $Y = A$. If B goes to 0, the output remains at 1 regard-less of the value of A.

A final example will demonstrate the usefulness of mixing positive and negative logic to obtain matched assertion levels. Suppose we have a circuit with an input called CLEAR that must be asserted. If both A and B are asserted high, the circuit of Fig. 2.22(a) results. Since A and B must be ANDed, the AND symbol is used. The CLEAR input must be asserted low; so the triangle is drawn at the gate output, creating a NAND gate. The tri-angle on the CLEAR input simply indicates that this variable must be asserted low. Using the correct symbol allows the expression for CLEAR to be written CLEAR $= AB$.

In Fig. 2.22(b) low inputs are required to assert CLEAR. The inputs to the gate must now be driven with $\bar{A}$ and $\bar{B}$. When A and B are high, both $\bar{A}$ and $\bar{B}$ will be low, driving the gate output low to assert CLEAR. We note that the gate reduces to an OR gate driving the CLEAR input. Again the expression for CLEAR is CLEAR $= AB$, but $\bar{A}$ and $\bar{B}$ are applied to the circuit inputs.

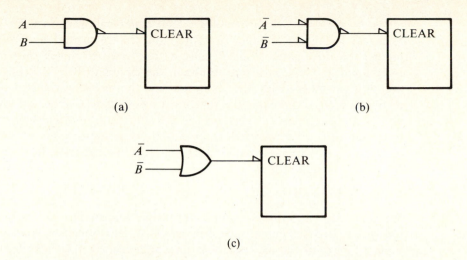

Figure 2.22 (a) *A* and *B* asserted high. (b) *A* and *B* asserted low. (c) An equivalent of the circuit in (b).

The AND gate and inverters of Fig. 2.22(b) could be converted to the OR of Fig. 2.22(c). This circuit does not immediately reflect the single condition required to assert the CLEAR. Since inputs imply low assertion of *A* and *B* and CLEAR indicates a required low assertion signal, we would choose to use the circuit that includes the AND gate plus inverters of Fig. 2.22(b). As mentioned previously, this equivalent circuit immediately tells us that CLEAR = *AB*.

We will emphasize the fact that manufacturers name their gates based on a positive logic definition. A NAND gate assumes that high assertion inputs are required to produce the low-level output. We recognize from the preceding discussion that a redefinition of levels results in a new function performed by the gate. The general practice of basing the gate name on a positive logic definition removes any ambiguity that could result from other choices of assertion level. Of course, in dealing with circuits we will mix positive and negative logic to match assertion levels.

The first step in applying this method consists of identifying any required assertion levels in the circuit. The lamp indicator of Fig. 2.17 demonstrates the requirement for a low assertion input. These required signals can be called defined assertion levels. All remaining undefined input and output lines are then manipulated using DeMorgan's laws to match the defined assertion-level lines. If all output levels match corresponding input levels, the function of the circuit can be immediately identified.

It is not necessary to match low assertion levels with complemented input variables. Variables that are not matched appear in the result in complemented form. Figure 2.23 is used to demonstrate the application of this method. Since the system requires a high assertion input, gate 3 is converted to match its output to input X. After this change, gate 3 has low assertion inputs; therefore gates 1 and 2 must be converted to match output levels to these inputs. This conversion results in the circuit of Fig. 2.23(b). All levels are now matched with the exception of the input variables. Once this matching has taken place, the function of each gate is indicated by its symbol. Gate 1 and gate 3 perform the OR function, and gate 2 performs the AND function. The expression for X is now written by inspection, complementing those variables that do not match input assertion level, namely, B, C, and D. The result is

$$X = (A + \bar{B}) + \overline{CD}$$

We note that A appears in uncomplemented form in this expression even though the actual input is complemented. As mentioned previously, matched inputs appear in the expression as uncomplemented values while unmatched inputs appear in complemented form.

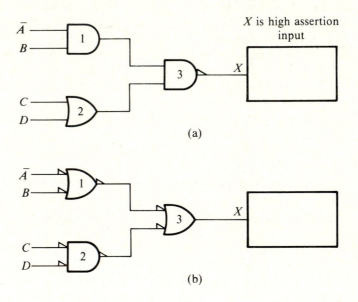

Figure 2.23 (a) A circuit with unmatched levels. (b) The matched equivalent.

Drill Problems: Sec. 2.1E

1. Match assertion levels so that X can be written by inspection.

2. Repeat problem 1 if X is a low assertion signal.

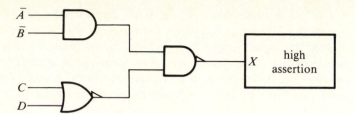

Figure DP2.1E-1

2.2 KARNAUGH MAPS

Section overview: The Karnaugh map can be used to minimize logic functions allowing the function to be implemented with the minimum number of gates. This method is considered in this section. The section is concluded with a discussion of static and dynamic hazards.

Boolean algebra allows us to express logic functions in a concise form and allows us to reduce expressions to simpler equations. With more complex expressions, the result of algebraic manipulation may not lead to a unique reduced expression. In order to demonstrate this point, consider the equation

$$X = \bar{A}\bar{B}\bar{C} + \bar{A}B\bar{C} + AB\bar{C}$$

If the distributive law is applied to the first two terms, the expression reduces to

$$X = \bar{A}\bar{C}(\bar{B} + B) + AB\bar{C} = \bar{A}\bar{C} + AB\bar{C}$$

Another form of reduction is found by applying the distributive law to the second two terms, giving

$$X = \bar{A}\bar{B}\bar{C} + B\bar{C}(\bar{A} + A) = \bar{A}\bar{B}\bar{C} + B\bar{C}$$

The expression for X could be reduced slightly further by noting that $\bar{A}\bar{B}\bar{C} = \bar{A}B\bar{C} + \bar{A}\bar{B}\bar{C}$ from Eq. (2.3). We then expand X to be

$$X = \bar{A}\bar{B}\bar{C} + \bar{A}B\bar{C} + \bar{A}B\bar{C} + AB\bar{C}$$

before reducing to

$$X = \bar{A}\bar{C} + B\bar{C}$$

This expression reduces the number of variables involved by one. Actually we could have manipulated the earlier reductions to reach this same result, but an important point should now be obvious. Algebraic manipulation, if not performed in an orderly way, may lead to nonminimal results.

The Karnaugh map, or K-map, applies an orderly procedure to the reduction of logic expressions leading to a truly minimized result. K-maps provide the most popular technique for reduction of moderately complex logic systems. Unfortunately, systems involving several variables (over five) are difficult to handle with Karnaugh methods. In these situations the computer becomes necessary to minimize the logic function [2].

A. Standard Sum-of-Products Form

In order to apply the K-map method, a function should be expressed in standard sum-of-product (SSOP) form. An expression such as

$$AB + BC + \bar{B}D$$

is in sum-of-product (SOP) form since the products (ANDs) of individual terms are first formed, followed by the summation of these products. Another form that can occur in logic expressions is the product-of-sums (POS). This form is exemplified by the expression

$$(A + B + C)(\bar{B} + \bar{C})$$

A POS expression can be converted to the SOP form by using the distributive law. The preceding expression would then become

$$A\bar{B} + B\bar{B} + C\bar{B} + A\bar{C} + B\bar{C} + C\bar{C} = A\bar{B} + C\bar{B} + A\bar{C} + B\bar{C}$$

Although the last equation is in SOP form, it is not in SSOP form. This form requires that each product term contain all variables involved. In the previous expression, the variables A, B, and C are involved; thus each product term should include these variables or complemented forms of these variables. To expand the SOP form into SSOP, we note that we can AND (or multiply) any term by 1 without changing the value of the term. Furthermore, since $X + \bar{X} = 1$, we could multiply any term with $C + \bar{C}$, $B + \bar{B}$, or $A + \bar{A}$ without affecting the value of the function. To put the earlier expression in SSOP form, we proceed as shown in the following equation:

$$A\bar{B}(C + \bar{C}) + C\bar{B}(A + \bar{A}) + A\bar{C}(B + \bar{B}) + B\bar{C}(A + \bar{A})$$

$$= A\bar{B}C + A\bar{B}\bar{C} + A\bar{B}C + \bar{A}\bar{B}C + AB\bar{C} + A\bar{B}\bar{C} + AB\bar{C} + \bar{A}B\bar{C}$$

$$= A\bar{B}C + A\bar{B}\bar{C} + \bar{A}\bar{B}C + AB\bar{C} + \bar{A}B\bar{C} + AB\bar{C}$$

Each product term with a missing variable is ANDed with the sum of the variable and its complement. All duplicate product terms are eliminated since $X + X = X$ or

$$A\bar{B}C + A\bar{B}C = A\bar{B}C$$

If both B and C were missing, leaving only A as a term in an expression, $(C + \bar{C})$ and $(B + \bar{B})$ must multiply A. For example,

$$A + \overline{A}\overline{B}\overline{C} = A(B + \overline{B})(C + \overline{C}) + \overline{A}\overline{B}\overline{C}$$

$$= ABC + AB\overline{C} + A\overline{B}C + A\overline{B}\overline{C} + \overline{A}\overline{B}\overline{C}$$

The SSOP terms in an expression are called minterms. Minterms are numbered according to the decimal code they represent. The above expression contains minterms m_0, m_4, m_5, m_6, and m_7. The equation

$$F = \overline{A}\overline{B}C + \overline{A}BC + A\overline{B}\overline{C}$$

tells us that $F = 1$ when minterm m_1, m_3, or m_4 is present and implies that $F = 0$ when minterm m_0, m_2, m_5, m_6, or m_7 is present. We will use minterm notation as we discuss mapping of logic functions.

Drill Problems: Sec. 2.2A

1. Put into SOP form $X = AB + (C + \overline{B})(AB + \overline{C})$.

2. Put X of problem 1 into SSOP form.

3. Write the expression for F if F is to equal 1 when any of the minterms m_0, m_3, m_4, or m_6 is present.

B. Constructing a *K*-map

Now we turn our attention to the construction of K-maps. All possible terms of an expression can be represented by a location on a map. The two variables A and B have four possible combinations which can be represented by the maps of Fig. 2.24. The second map is more conventional than the first with each square corresponding to a unique set of values for A and B. The top left square represents the values $A = 0$ and $B = 0$, the bottom left represents $A = 0$ and $B = 1$, the top right represents $A = 1$ and $B = 0$, and the bottom right represents $A = 1$ and $B = 1$. Each location represents a minterm and can be numbered as shown.

00	10
01	11

$\diagdown A$	0	1
B		
0	0	2
1	1	3

Figure 2.24 Maps showing all possible combinations of a two-variable expression.

When plotting a function on a K-map we recognize that a logical function is specified in terms of those combinations of variables that lead to a value of 1 for the function. The meaning of the equation

$$F = AB + \overline{A}B$$

is that F will be 1 when $B = 1$ and $A = 1$ ($BA = 1$) or when $B = 1$ and $A = 0$ ($B\overline{A} = 1$). The equation also implies that $F = 0$ for the values $B = 0$ and $A = 0$ ($\overline{B}\overline{A} = 1$) or $B = 0$ and $A = 1$ ($\overline{B}A = 1$). Figure 2.25 plots this information.

Figure 2.25 A Karnaugh map.

Drill Problems: Sec. 2.2B

1. Plot the K-map for $X = AB + A\overline{B}\overline{C}$.

2. Plot the K-map for $Y = AB + A\overline{B}\overline{C}D + \overline{A}\overline{B}\overline{C}\overline{D}$.

3. Plot the K-map for $Z = ABC + \overline{A} + A\overline{C}\overline{B}$.

C. Reducing an Expression

In reducing the expression F to the simplest terms, we look for adjacent locations of the map containing values of 1. These locations form a couple. A couple consists of two terms that are logically adjacent. As indicated earlier, this means the two terms would be identical except that a single variable appears in its complemented form in one term and in its uncomplemented form in the other term. The expression

$$F = AB + \overline{A}B$$

comprises a couple. From our experience with Boolean algebra we recognize that a couple can be reduced; thus

$$F = BA + B\overline{A} = B(A + \overline{A}) = B$$

A three-variable term has eight possible combinations leading to a K-map as shown in Fig. 2.26. Note the arrangement of the variables AB. Rather than counting from 00 to 11 in numerical sequence, the order is such that adjacent squares are always logically adjacent. Only one variable is complemented with the other two unchanged as we move from one square to an adjacent one. Furthermore, logical adjacency also exists between the extreme squares of a row or column. The location $A\overline{B}\overline{C}$ (top right) is adjacent to the location $\overline{A}\overline{B}\overline{C}$ (top left).

AB
C
00 01 11 10
0 0 0 1 1
1 1 0 1 1

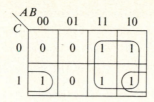

$$F = \bar{A}\bar{B}C + A\bar{B}\bar{C} + A\bar{B}C + AB\bar{C} + ABC$$

Figure 2.26 A three-variable *K*-map.

In the map of Fig. 2.26 there is a quad which contains four 1s, each logically adjacent to two others, and there is a couple. This quad of four terms can be reduced to A. We can note this by factoring the four terms to give

$$\begin{aligned} A\bar{B}\bar{C} + A\bar{B}C + AB\bar{C} + ABC &= A(\bar{B}\bar{C} + \bar{B}C + B\bar{C} + BC) \\ &= A(\bar{B}[\bar{C} + C] + B[\bar{C} + C]) \\ &= A(\bar{B} + B) = A \end{aligned}$$

The four terms in parentheses encompass all possible combinations of the variables B and C. This tells us that, regardless of the values of B and C, the function equals 1 when A is 1. Thus, the term reduces to A.

A simple way to arrive at the reduced expression without using algebra is to note which variable or variables of a group remain unchanged. From the K-map of Fig. 2.26, the quad shows both B and C changing from 0 to 1 as we move to different locations of the quad while A remains constant at 1. The couple shows A changing from 0 to 1 while B remains constant at 0 and C remains constant at 1. The reduced expression is then $A + \bar{B}C$.

Because of the arrangement of the map to achieve logical adjacency for physically adjacent squares, the minterm numbers do not proceed sequentially. Figure 2.27 shows the three-variable map with numbered locations corresponding to minterms.

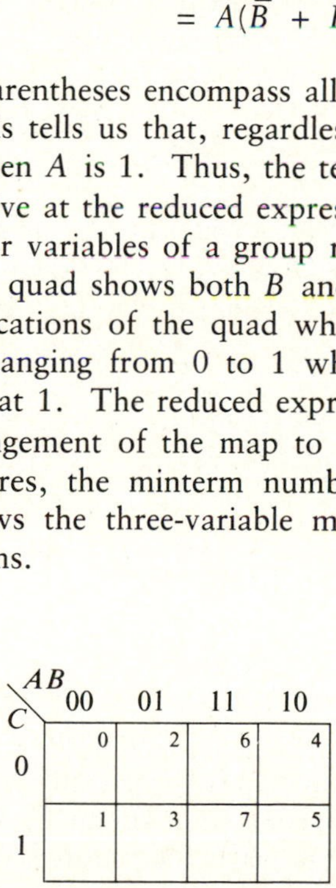

AB
C
00 01 11 10
0 0 2 6 4
1 1 3 7 5

Figure 2.27 Minterm numbering for a three-variable map.

For expressions involving more terms, the K-map provides an orderly means of locating all terms that can be reduced in the manner described. Consider the four-variable function:

$$F = ABCD + A\bar{B}CD + A\bar{B}\bar{C}D + \bar{A}BCD + \bar{A}BC\bar{D} + \bar{A}\bar{B}CD + \bar{A}BC\bar{D}$$

With four variables there are 16 possible combinations; thus, the K-map must contain 16 locations. Figure 2.28 shows the standard arrangement of a four-variable K-map. After plotting F, we note several adjacent locations containing a 1. The row containing four ones tells us that a 1 occurs whenever $C = D = 1$, regardless of the values of A or B. The second column indicates that a 1 should also occur in the expression whenever $\bar{A}B\bar{D} = 1$. The fourth column indicates the occurrence of a 1 in the expression when $A\bar{B}\bar{D} = 1$. Thus, the entire expression can be reduced to

$$F = CD + \bar{A}B\bar{D} + A\bar{B}\bar{D}$$

If F is implemented by hardware, seven 4-input AND gates plus a 7-input OR would be required for the original expression. Alternatively the reduced expression could be implemented using two 3-input AND gates plus a 2-input AND gate, along with a 3-input OR gate. Obviously the reduced expression is much more economical.

CD \ AB	00	01	11	10
00	0	1	0	0
01	0	0	0	1
11	1	1	1	1
10	0	1	0	0

$$F = ABCD + A\bar{B}CD + A\bar{B}\bar{C}D + \bar{A}BCD + \bar{A}BC\bar{D} + \bar{A}\bar{B}CD + \bar{A}BC\bar{D}$$

Figure 2.28 A four-variable K-map.

The basic idea in reducing an expression consists of first plotting the K-map, then forming groups containing 2, 4, 8, or 16 adjacent 1s. A couple allows two original terms to be reduced to one smaller term, a quad allows four original terms to be reduced to one smaller term, and an octuplet reduces eight terms to a single smaller term. Given a choice we always form the largest group possible to achieve maximum reduction. A square containing a 1, surrounded by adjacent zero squares, indicates a term in the original expression that must be included in the final expression with no further reduction. We note that the two squares at opposite ends of a row or column are adjacent locations. Some examples of these ideas are demonstrated in Fig. 2.29. In the first map, the couple is formed from locations at opposite ends of a row. The second map shows a quad that can be represented by a single two-variable term. The third map also contains a quad, although it is not so obvious as in the preceding map. In addition, this map contains a four-variable term that cannot be reduced. The last map demonstrates that one square containing a 1 can be used to form more than one couple. With these basic ideas, a great deal of reduction can be performed.

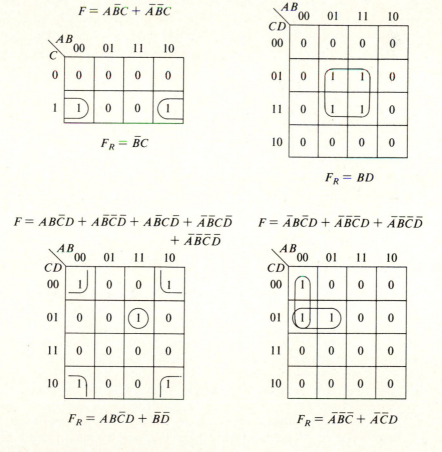

Figure 2.29 Some examples of reduction.

Drill Problems: Sec. 2.2C

1. Reduce to simplest form $X = AB + A\overline{B}\overline{C} + \overline{A}\overline{B}\overline{C}$.

2. Reduce to simplest form $Y = ABC + A\overline{B}\overline{C}D + \overline{A}\overline{B}\overline{C}D$.

3. Reduce to simplest form $Z = ABC\overline{D} + AB\overline{C}D + A\overline{B}CD + \overline{A}BC\overline{D} + \overline{A}\overline{B}CD$.

D. A Closer Look at Minimization

It is possible that the application of the steps previously outlined may lead to redundant or unnecessary terms in the final expression. In order to obtain a true minimum expression, all redundant terms must be eliminated. The source of these redundant terms can be understood after a consideration of prime implicants.

When a K-map is plotted, all possible largest groups are circled to include all 1s on the map. These groups will contain 1, 2, 4, 8, or 16 mutually adjacent 1s for a four-variable map as shown in Fig. 2.30.

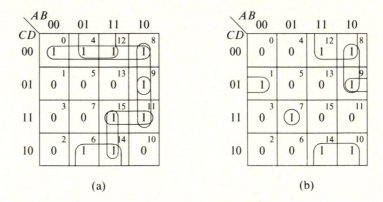

(a)   (b)

Figure 2.30 K-maps with groups that represent prime implicants.

The term that represents each group is called a prime implicant. The prime implicants for the map of Fig. 2.30(a) are $\overline{C}\overline{D}$, $B\overline{D}$, $A\overline{B}\overline{C}$, $A\overline{B}D$, ACD, and ABC. For Fig. 2.30(b) the prime implicants are $\overline{A}BCD$, $A\overline{B}\overline{C}$, $A\overline{D}$, and $\overline{B}CD$. Not all prime implicants are necessary in the final expression since certain prime implicants include 1s that have all been included as parts of other prime implicants. The prime implicant covering locations 8 and 9 in either map is an example of a redundant implicant. For absolute minimization of a function, all redundant implicants must be eliminated. The remaining prime implicants are called essential implicants. The final expression of a logic function will contain only essential implicants.

The map of Fig. 2.30(a) will result in a minimized expression of

$$F = \overline{C}\overline{D} + B\overline{D} + A\overline{B}D + ABC$$

Two prime implicants were redundant in this map. This result, although minimum, may not be unique. Two other possible expressions are

$$F = \overline{CD} + B\overline{D} + A\overline{BC} + ACD$$

and

$$F = \overline{CD} + B\overline{D} + A\overline{B}D + ACD$$

We will leave it to the reader to verify that the groups chosen for the last two expressions include only essential implicants.

The map of Fig. 2.30(b) leads to a minimized expression of

$$F = A\overline{D} + \overline{B}CD + \overline{A}BCD$$

There are no other options for this minimized expression.

We will now return to the example that opened the discussion of Boolean algebra in Sec. 2.1. This problem will be restated in the following example.

■ **Example 2.1** Four variables are used to represent a 4-bit binary number. When certain numbers occur, a logical 1 should be generated. The numbers for which the output should indicate 1 are 1, 4, 5, 9, 11, and 12. Assume that each variable is also present in complemented form.

☐ **Solution:** Logical 1 is to be generated for any of the following combinations.

Decimal	Binary	Four-Variable Term
1	0001	$\overline{A}\overline{B}\overline{C}D$
4	0100	$\overline{A}B\overline{C}\overline{D}$
5	0101	$\overline{A}B\overline{C}D$
9	1001	$A\overline{B}\overline{C}D$
11	1011	$A\overline{B}CD$
12	1100	$AB\overline{C}\overline{D}$

Therefore, the expression is

$$F = \overline{A}\overline{B}\overline{C}D + \overline{A}B\overline{C}\overline{D} + \overline{A}B\overline{C}D + A\overline{B}\overline{C}D + A\overline{B}CD + AB\overline{C}\overline{D}$$

This expression was realized directly with six 4-input AND gates plus one six-input OR in Fig. 2.1.

In order to minimize this expression, the K-map of Fig. 2.31(a) is plotted. The three couples circled in the figure are the essential implicants necessary to represent the expression giving

$$F_R = B\overline{C}\overline{D} + \overline{A}\overline{C}D + A\overline{B}D$$

Note that the numbered minterm locations can simplify the plotting process.

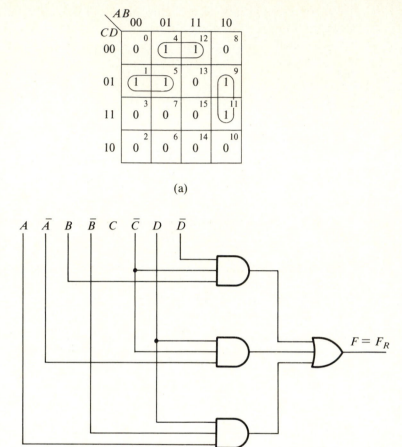

(a)

(b)

Figure 2.31 (a) *K*-map for Example 2.1. (b) Implementation of the reduced expression.

This reduced expression is implemented in Fig. 2.31(b) and requires only 4 total gates and 12 total inputs, compared with 7 total gates and 30 total inputs to implement the original expression. It is left as an exercise for the student to realize this expression with NAND-NOR gate logic. ■

In several applications we can take advantage of "don't-care" conditions to minimize an expression. Suppose we are given a problem of implementing a circuit to generate a logical 1 when a 2, 7, or 15 appears on a four-variable input. A logical 0 should be generated when 0, 1, 4, 5, 6, 9, 10, 13, or 14 appears. The input conditions for the numbers 3, 8, 11, and 12 never occur in this system. When we know that certain combinations never occur, we don't care whether these inputs generate a one or a zero on the K-map. We can interpret these "don't-care" conditions as either ones or zeros, whichever allows us to further minimize the reduced function. A "Θ" can be used to represent the "don't-care" combinations as shown in the map of Fig. 2.32. The two "don't-care" conditions in the third row are taken as 1s while the other two are taken as 0s. This allows the reduced expression to be written

$$F_R = CD + \overline{A}\overline{B}C$$

Had all "don't-care" conditions been interpreted as 0s, the expression would have been

$$F = BCD + \overline{A}\overline{B}C\overline{D}$$

Obviously, we must consider these conditions in order to truly minimize the system.

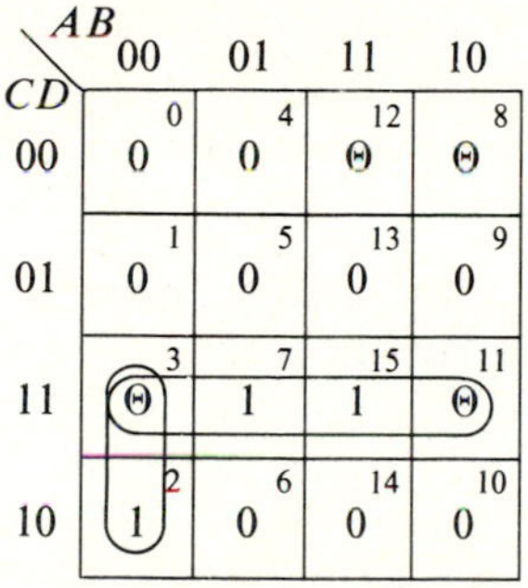

Figure 2.32 K-map with "don't care" conditions.

After the reduced expression is obtained, it may be possible to further minimize the number of gates used. Consider an expression such as

$$X = A\overline{B} + C\overline{B}$$

This can be factored to give

$$X = \overline{B}(A + C)$$

Implementing the first form of X requires two 2-input AND gates and a 2-input OR. The second expression can be realized with one 2-input OR gate followed by a 2-input AND. Thus, we should consider factoring the reduced expression prior to implementation of the circuit. (See Prob. 2.28.)

We will emphasize that we should always include any logical 1 within the largest group of 1s possible. For example, consider the maps of Fig. 2.33.

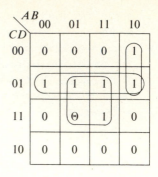

Figure 2.33 Map reductions using different groupings.

The expression for the first map is

$$F = \overline{C}D + BCD + A\overline{B}\overline{C}\overline{D}$$

By including each 1 in the largest group possible, as in the second map, the expression becomes even simpler. It is

$$F_R = \overline{C}D + BD + A\overline{B}\overline{C}$$

■ **Example 2.2** Write the simplest expression possible for the function represented by the map of Fig. 2.34(a).

☐ **Solution:** All prime implicant groups are shown in Fig. 2.34(a). In selecting the essential implicants, we should strive to achieve the highest number of redundant groups possible. Figure 2.34(b) indicates an incorrect choice of groups while Fig. 2.34(c) shows the proper choice. In this map, one quad becomes redundant since all 1s of this group are included in other necessary groups. The resulting expression is

$$F = \overline{C}\overline{D} + \overline{A}\overline{B}D + \overline{A}BC + ABD + A\overline{B}\overline{C} \quad ■$$

Another check that should be made before selecting a final expression is the possibility of realizing the complemented function rather than the function. Each location of the map that contains a 0 indicates that $F = 0$ for this term. This is equivalent to saying that F is not 1 or $F = \overline{1}$. Complementing both sides of this equation results in $\overline{F} = 1$. Thus, $\overline{F} = 1$ results from the terms corresponding to $F = 0$. We can write an expression for $\overline{F}$ in terms of the 0 locations using the same ideas to group the 0s as we did previously to group the 1s.

Returning to Fig. 2.34, we can write

$$\overline{F} = \overline{A}B\overline{C}D + A\overline{B}\overline{C}D + \overline{A}BC\overline{D} + ABC\overline{D}$$

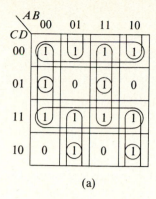

(a)

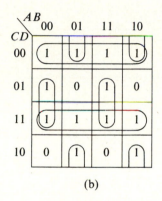

(b)

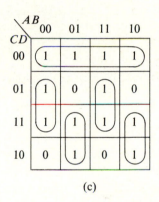

(c)

Figure 2.34 Two realizations of an expression: (a) prime implicants; (b) incorrect implicant choice; (c) correct choice.

Since we ultimately want to realize F rather than $\bar{F}$, both sides of this equation are complemented, giving

$$\bar{\bar{F}} = F = \overline{\overline{A}B\overline{C}D \ + \ A\overline{B}\overline{C}D \ + \ \overline{A}\overline{B}C\overline{D} \ + \ ABC\overline{D}}$$

In some instances, the expression resulting from this method may be simpler than that resulting from realizing F directly. Groups of zeros can be used to minimize the $\bar{F}$ expression just as groups of ones are used to reduce F.

■ **Example 2.3** Minimize the expression

$$F = \overline{A}B\overline{C}\overline{D} \ + \ \overline{A}B\overline{C}D \ + \ \overline{A}BCD \ + \ \overline{A}BC\overline{D} \ + \ A\overline{B}\overline{C}\overline{D} \ + \ A\overline{B}\overline{C}D$$

$$+ \ A\overline{B}C\overline{D} \ + \ A\overline{B}CD \ + \ AB\overline{C}\overline{D} \ + \ ABC\overline{D}$$

□ *Solution:* The Karnaugh map is shown in Fig. 2.35. Grouping the ones as shown results in

$$F_1 = \overline{A}B + A\overline{B} + A\overline{D}$$

If we use the two groups of zeros, we obtain

$$\overline{F}_2 = \overline{A}\overline{B} + ABD$$

and

$$F_2 = \overline{\overline{A}\overline{B} + ABD}$$

which is the simplest possible realization for F. ∎

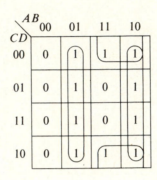

Figure 2.35 *K*-map for Example 2.3.

A five-variable map is shown in Fig. 2.36. In this situation, couples or quads can be formed in corresponding locations of the two parts of the map.

Generally, expressions involving seven or more variables are reduced by means of computer programs.

Drill Problems: Sec. 2.2D

1. Reduce to simplest form

$$X = \overline{A}B\overline{C}\overline{D} + \overline{A}B\overline{C}D + AB\overline{C}\overline{D} + AB\overline{C}D$$
$$+ A\overline{B}\overline{C}D + ABC\overline{D} + \overline{A}BCD + \overline{A}\overline{B}\overline{C}\overline{D}$$

2. A system sends *BCD* code over four lines *A* (MSB), *B*, *C*, and *D*. Design a minimal circuit to generate a 1 when the codes for 1, 5, 6, 7, or 9 are present. *Hint:* Remember to use "don't cares."

3. Reduce X from problem 1 by reducing $\overline{F}$ and then complementing.

$$F = \bar{A}B\bar{C}\bar{D}\bar{E} + AB\bar{C}\bar{D}\bar{E} + \bar{A}\bar{B}\bar{C}D\bar{E} + \bar{A}BC\bar{D} + \bar{A}\bar{B}C\bar{D}E +$$
$$ABC\bar{D}\bar{E} + ABC\bar{D}E + \bar{A}B\bar{C}DE$$

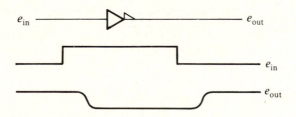

E = 0 (AB over CD map)

CD\AB	00	01	11	10
00	0	1	1	0
01	0	1	0	0
11	0	0	0	0
10	0	1	1	0

E = 1 (AB over CD map)

CD\AB	00	01	11	10
00	0	0	0	0
01	0	1	0	0
11	0	0	0	0
10	1	1	1	0

Couple

Quad

$$F_R = B\bar{D}\bar{E} + BC\bar{D} + \bar{A}B\bar{C}D + \bar{A}C\bar{D}E$$

Figure 2.36 A five-variable map.

E. Hazard Covers

In our previous discussions we have assumed ideal switching characteristics for all gates and inverters. In practice, there is a propagation delay associated with each logic element as explained in Sec. 1.5. Each waveform also has finite rise and fall times as indicated by the output waveform of Fig. 2.37. In certain situations this delay can lead to undesirable results as outlined in the following paragraphs.

e_{in} —▷— e_{out}

e_{in}

e_{out}

Figure 2.37 Practical output waveform.

In order to demonstrate the problem of propagation delay time in combinational circuits, we will consider the K-map of Fig. 2.38. For the input conditions of $A = B = 1$ and C changing from 1 to 0, we can plot the timing chart of Fig. 2.39. Although the equation for F indicates that F

should remain equal to 1 as C switches, we see that in reality F may contain a short excursion toward zero. This unwanted "glitch" is called a static hazard because the signal was supposed to remain static at the 1 level. Since C drops from 1 to 0 prior to $\bar{C}$ rising from 0 to 1, there is a short period during which neither AND condition is satisfied.

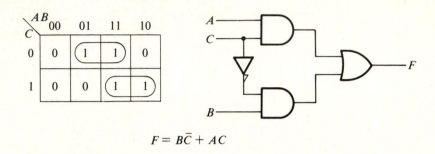

$$F = B\bar{C} + AC$$

Figure 2.38 *K*-map and logic circuit.

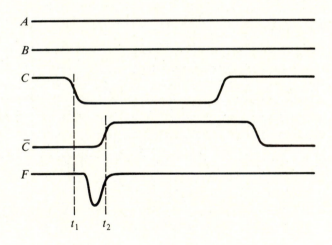

Figure 2.39 Timing chart for the circuit of Fig. 2.38.

When C switches from 0 to 1, the anomaly does not occur for this particular function. The circuit of Fig. 2.40 shows a system that generates a static hazard of the opposite polarity assuming the propagation delay of the inverter is greater than that of the AND gate. The circuit of Fig. 2.38 generates a static 0 hazard while the hazard shown in Fig. 2.40 is a static 1 hazard.

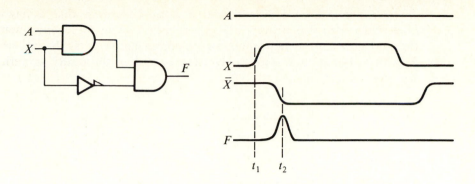

Figure 2.40 A logic circuit with a static hazard.

We might suppose that static hazards are inherently dangerous to a given design since an unwanted signal appears on the output line. In certain popular types of logic design, the static hazard produces no undesirable effects. In particular, synchronous sequential circuits (Chap. 5) are designed to ignore the static hazards that may be present. Other types of systems can produce erroneous outputs when these hazards occur, however, and in these instances the hazard must be eliminated.

In order to eliminate the hazard in the circuit of Fig. 2.38, we note that the glitch occurs as C changes from 1 to 0 when $A = B = 1$. We can make the output independent of the value of C, for these values of A and B, by including the redundant implicant or couple as shown in Fig. 2.41. This couple is referred to as a hazard cover and results in one more required gate in the actual circuit. Any adjacent couples must be overlapped by a hazard cover to eliminate the possibility of a static hazard. Although the system is no longer a minimal system, it can now operate reliably when static hazards are significant.

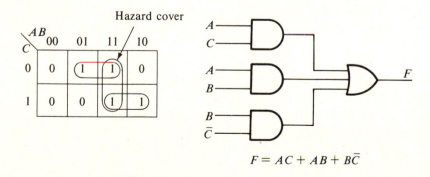

$$F = AC + AB + B\bar{C}$$

Figure 2.41 Use of a hazard cover.

A dynamic hazard differs from a static hazard in that a transition is made from 1 to 0 or 0 to 1 accompanied by two or more unwanted transitions. Figure 2.42 demonstrates two dynamic hazards. If the circuit of Fig. 2.38 is followed by a gate, a dynamic hazard might result as indicated in Fig. 2.43.

Figure 2.42 Dynamic hazards.

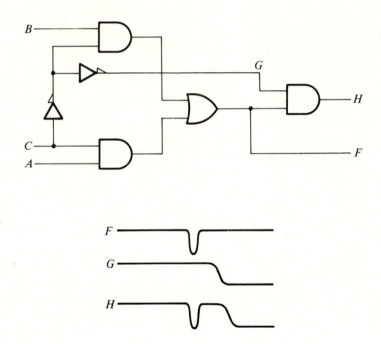

Figure 2.43 Circuit with dynamic hazard.

It can be shown that a dynamic hazard can only exist as a result of static hazards. Therefore, if hazard covers are used to eliminate all static hazards, we need not be concerned with dynamic problems.

Although we will not be concerned further with hazards until we consider asynchronous state machines, it is mentioned here to emphasize that minimal circuits may not be hazardfree. When hazards adversely affect system operation, we must sacrifice minimization for reliability. A few extra gates used for hazard covers will be worth far more to reliable circuit operation than the small increase in cost.

Drill Problems: Sec. 2.2E

1. With the aid of a timing chart determine the transition of A that will produce a static hazard on X.

Figure DP2.2E-1

2.3 VARIABLE-ENTERED MAPS

Section overview: The variable-entered map is closely related to the K-map but is more useful in some cases. The construction of the map and its use in minimization is covered in this section.

The K-map is a very useful tool in logic function minimization if less than six variables are involved. When the number of variables exceeds six, computer programs can be used to solve this problem. The variable-entered map can be used to plot an n-variable problem on an $n - 1$ variable map. It is actually possible to reduce the map dimension by two or three in some cases, but a reduction of one is a more typical situation.

A second advantage of the variable-entered map occurs in design problems involving multiplexers which can be solved more readily by this approach than by K-maps.

A. Plotting the Variable-Entered Map

We will introduce the map-entered variable approach by considering the three-variable function

$$X = \overline{AB}\overline{C} + AB\overline{C} + A\overline{B}\overline{C} + ABC$$

Of course, this function could be reduced by means of a three-variable K-map. On the other hand, we could plot the function on a two-variable map if we consider the value of X to be a function of map location and the variable C. This possibility is demonstrated in Fig. 2.44.

The first map lists the conditions required of C at each location of A and B to result in $X = 1$. The only exception to this is location $\overline{A}B$, which yields $X = 0$ regardless of the value of C. The second map conveys the same information in a more efficient manner. This map is called a variable-entered map and contains the same information as a three-variable K-map. In this case, C is called the map-entered variable.

Figure 2.45 demonstrates the procedure used to plot the variable-entered map (VEM). For $A = 0$, $B = 0$, and $C = 0$, we see from the table that X follows the value of D. In the corresponding map location a D is

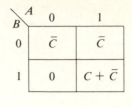

	A 0	1
B 0	$X = 1$ if $\overline{C} = 1$	$X = 1$ if $\overline{C} = 1$
1	$X = 0$	$X = 1$ if $\overline{C} = 1$ or if $C = 1$

	A 0	1
B 0	$\overline{C}$	$\overline{C}$
1	0	$C + \overline{C}$

Figure 2.44 Variable-entered maps.

entered. This location then generates a 1 in the map when $D = 1$ and a 0 when $D = 0$. For $A = 0$, $B = 0$, and $C = 1$, the output X is 0 regardless of the value of D. When $A = 0$, $B = 1$, and $C = 0$, X is 1 for either value of D; thus a 1 is entered in this location of the map. We note that a 1 is equivalent to $D + \overline{D}$. For $A = 0$, $B = 1$, and $C = 1$, the table shows that when $D = 0$, $X = 1$ and when $D = 1$, $X = 0$. The corresponding map location then contains a $\overline{D}$. When $A = 1$, $B = 0$, and $C = 0$, we see that when $D = 0$, we do not care what value X takes on. This implies that this particular combination of variables will never occur. However, for the same values of A, B, and C, $X = 1$ when $D = 1$. The expression reflecting this information, $D + \overline{D}\Theta$ is then entered into the map. Location 110 on the map contains $D\Theta + \overline{D}$ as indicated by the truth table. Location 111 contains a Θ since both values of D are "don't care" conditions. A three-variable map now contains the same information as does a four-variable K-map, but each individual location in this map contains more information than the K-map.

B. Reducing Expressions with the Variable-Entered Map

In order to reduce an expression using a VEM, we again choose groups of similar terms to cover all nonzero terms appearing in the map, except "don't care" terms that are taken as 0s. Figure 2.46 shows some examples of VEMs with the appropriate groups. In these maps 1s are written as the MEV plus its complement. The first map of Fig. 2.46 reduces to

$$F_R = \overline{A}\overline{C}D + B\overline{C}D + A\overline{B}C$$

All three groups in this map are couples including the one in location 101. This term involves $D + \overline{D}$ and consequently will not contain D when reduced. The second map reduces to

$$F_R = \overline{A}\overline{C}D + \overline{A}B\overline{C} + AC\overline{D} + A\overline{B}\overline{D} + A\overline{B}C$$

We see in this map that a D which is adjacent to a 1 or $D + \overline{D}$ can form a couple with the D of this term. This couple does not cover the $\overline{D}$ term, but

Inputs A B C D (MEV)	Output X
0 0 0 0	0
0 0 0 1	1
0 0 1 0	0
0 0 1 1	0
0 1 0 0	1
0 1 0 1	1
0 1 1 0	1
0 1 1 1	0
1 0 0 0	Θ
1 0 0 1	1
1 0 1 0	0
1 0 1 1	0
1 1 0 0	1
1 1 0 1	Θ
1 1 1 0	Θ
1 1 1 1	Θ

(a)

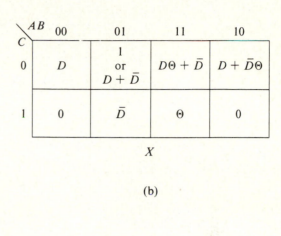

(b)

Figure 2.45 (a) Truth table for X. (b) Variable-entered map for X.

this variable can be covered by grouping it with the D term in the same map location.

The last map can be expressed as

$$F_R = A\overline{B}\overline{C} + \overline{B}\overline{C}E + \overline{A}BC\overline{E} + \overline{A}BC\overline{D}$$

Here we see that when a term such as $E + \overline{E}$ is adjacent to an identical term, a quad can be formed.

As another example of the use of the VEM, consider the expression of Fig. 2.36. Using E as the MEV, we plot the map of Fig. 2.47. The reduced expression is

$$F_R = B\overline{D}\overline{E} + BC\overline{D} + \overline{A}\overline{B}C D + \overline{A}C\overline{D}E$$

We always want the largest groups possible to get maximum reduction. In Fig. 2.47 we see a quad involving $\overline{E}$ and one involving $E + \overline{E}$.

"Don't care" conditions serve the same purpose in reducing VEMs as in K-map reduction. The map of Fig. 2.48 shows several "don't care" conditions. There are two types of "don't care" situations possible. The first is a location containing only a Θ. This location can then be taken as a 0, 1, E, or $\overline{E}$, whichever is appropriate. The second type of "don't care" condition is

$$F = A\bar{B}CD + \bar{A}B\bar{C}D + A\bar{B}C\bar{D} + AB\bar{C}D + \bar{A}\bar{B}CD$$

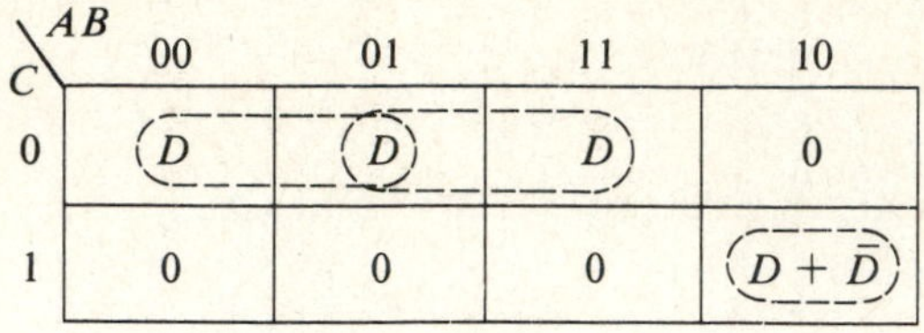

$$F = \bar{A}\bar{B}\bar{C}D + \bar{A}B\bar{C}D + \bar{A}B\bar{C}\bar{D} + A\bar{B}C\bar{D} + A\bar{B}C\bar{D} + A\bar{B}CD + ABC\bar{D}$$

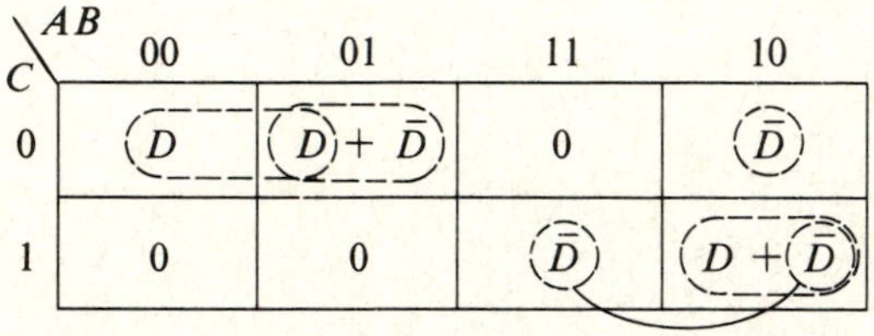

$$F = \bar{A}\bar{B}\bar{C}\bar{D}E + \bar{A}\bar{B}CDE + \bar{A}\bar{B}C\bar{D}\bar{E} + \bar{A}B\bar{C}D\bar{E} + A\bar{B}\bar{C}D\bar{E} + A\bar{B}\bar{C}DE + A\bar{B}\bar{C}D + \bar{A}B\bar{C}D\bar{E}$$

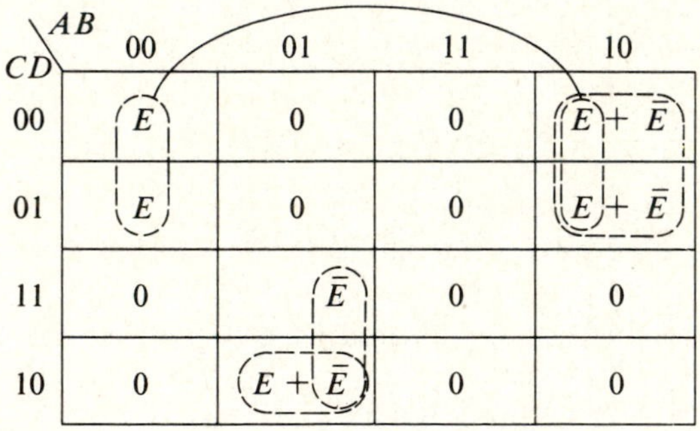

Figure 2.46 Variable-entered maps.

exemplified by $E\Theta + \bar{E}$. In this case, the "don't care" can be taken as a 0 or 1 to give either $\bar{E}$ or $E + \bar{E}$. In the map of Fig. 2.48, the Θs in the second column are taken as an E, the Θs in column 3 are taken as $E + \bar{E}$, and each Θ in column 4 is taken as a 1. This allows two quads to cover all terms, giving

$$F_R = B\bar{C}E + AD$$

■ **Example 2.4** Construct a VEM from the truth table of Fig. 2.49 using E as the MEV. Write the reduced expression for X.

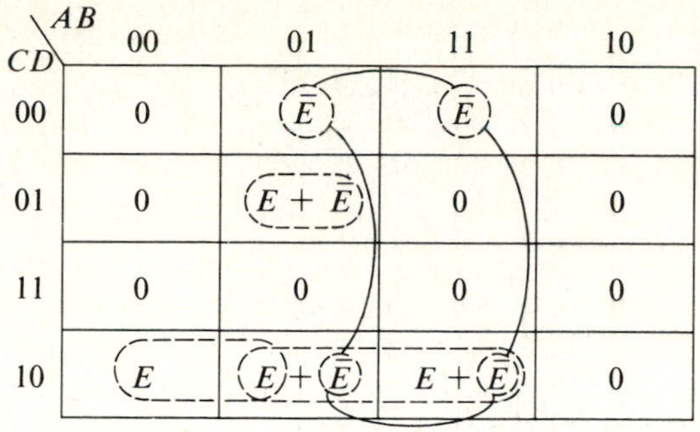

Figure 2.47 A five-variable VEM.

☐ ***Solution:*** A five-variable problem with one MEV requires a four-variable map to implement. Figure 2.50 shows the map for this example.

CD \ AB	00	01	11	10
00	0	Θ	E	0
01	0	Θ	Θ	$E\Theta + \bar{E}$
11	0	0	Θ	$E + \bar{E}\Theta$
10	0	0	0	0

Figure 2.48 "Don't care" conditions.

As we group terms we attempt to create the largest nonredundant groups possible. The map shown in Fig. 2.50 takes all Θs within the groups as 1s and all exterior Θs as 0s. The expression reduces to

$$X = \bar{A}\bar{B}\bar{D} + \bar{B}DE + AB\bar{D} \quad \blacksquare$$

One disadvantage of the VEM is the difficulty in realizing a logic function by first finding the minimum form of $\bar{F}$ and then complementing this result. The K-map is much easier to use for this purpose.

A	B	C	D	E	X
0	0	0	0	0	Θ
0	0	0	0	1	1
0	0	0	1	0	0
0	0	0	1	1	1
0	0	1	0	0	1
0	0	1	0	1	1
0	0	1	1	0	0
0	0	1	1	1	Θ
0	1	0	0	0	Θ
0	1	0	0	1	Θ
0	1	0	1	0	0
0	1	0	1	1	0
0	1	1	0	0	0
0	1	1	0	1	0
0	1	1	1	0	0
0	1	1	1	1	0

A	B	C	D	E	X
1	0	0	0	0	0
1	0	0	0	1	0
1	0	0	1	0	0
1	0	0	1	1	Θ
1	0	1	0	0	Θ
1	0	1	0	1	0
1	0	1	1	0	0
1	0	1	1	1	1
1	1	0	0	0	1
1	1	0	0	1	Θ
1	1	0	1	0	0
1	1	0	1	1	0
1	1	1	0	0	1
1	1	1	0	1	1
1	1	1	1	0	Θ
1	1	1	1	1	Θ

Figure 2.49 Truth table for Example 2.4.

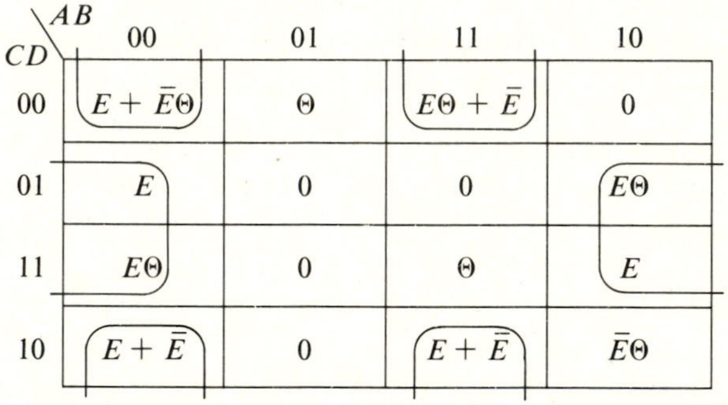

Figure 2.50 VEM for Example 2.4.

Drill Problems: Sec. 2.3A and 2.3B

1. Using C as the MEV, reduce $X = \bar{A}B\bar{C} + \bar{A}BC + AB\bar{C} + A\bar{B}\bar{C}$.

2. Using D as the MEV, reduce $Y = \bar{A}\bar{B}\bar{C}\bar{D} + \bar{A}BC\bar{D} + A\bar{B}C\bar{D} + A\bar{B}\bar{C}\bar{D} + A\bar{B}CD + A\bar{B}C\bar{D}$.

3. Repeat drill problem 2 from Sec. 2.2D with D as the MEV.

C. The Variable-Entered Map with Two Map-Entered Variables

There are occasions when it is useful to reduce the dimension of the K-map by 2. This can be done by plotting two MEVs rather than one. Let us plot the VEM for $X = \overline{AB}\overline{C}D + \overline{AB}C\overline{D} + ABCD + A\overline{B}CD + A\overline{B}\overline{C}\overline{D}$ using C and D as MEVs. We first factor the expression into terms involving the map locations $\overline{A}\overline{B}$, $\overline{A}B$, $A\overline{B}$, and AB. This results in

$$X = \overline{A}\overline{B}(\overline{C}\overline{D} + C\overline{D}) + A\overline{B}(C\overline{D} + CD) + ABCD$$

The map is shown in Fig. 2.51. After plotting the map of Fig. 2.51(a), we note that the terms in the upper two locations can be simplified, leading to the map of Fig. 2.51(b). This type of map is useful when combining gates with multiplexers to realize logic functions. We will consider this combination of elements in the next chapter.

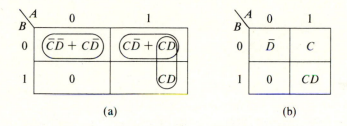

Figure 2.51 A VEM with two MEVs: (a) original map; (b) simplified map.

The reduction of the map is more complex than the single MEV case. Figure 2.51(a) can be used for this reduction. With the groups shown, the resulting expression is

$$X_R = ACD + A\overline{B}C + \overline{A}\overline{B}D$$

■ **Example 2.5** Plot the VEM for the expression

$$Y = \overline{A}\overline{B}\overline{C}\overline{D} + \overline{A}\overline{B}\overline{C}D + \overline{A}\overline{B}C\overline{D} + \overline{A}BCD + ABCD + A\overline{B}CD$$

using C and D as MEVs. Write the reduced expression for Y.

□ **Solution:** The expression is first written in factored form as

$$Y = \overline{A}\overline{B}(\overline{C}\overline{D} + \overline{C}D + C\overline{D}) + \overline{A}BCD + A\overline{B}CD + ABCD$$

The VEM is then plotted in Fig. 2.52. We note that $\overline{C}\overline{D} + \overline{C}D + C\overline{D}$ is also equal to NOT CD. When either $\overline{C}\overline{D}$, $\overline{C}D$, or $C\overline{D}$ occurs, the term CD cannot be true; thus NOT CD is equivalent to the sum of these three terms.

The reduced expression is

$$Y_R = \overline{A}\overline{B}\overline{C} + \overline{A}\overline{B}\overline{D} + BCD + ACD \, ■$$

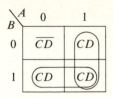

Figure 2.52 The VEM for Example 2.5.

Drill Problems: Sec. 2.3C

1. Plot the function $F = A\overline{B}\overline{C}DE + A\overline{B}C\overline{D}E + \overline{A}B\overline{C}DE + \overline{A}B\overline{C}D\overline{E} + AB\overline{C}D\overline{E} + ABCD\overline{E} + \overline{A}\overline{B}CDE + \overline{A}B\overline{C}DE + A\overline{B}C\overline{D}E + \overline{A}BC\overline{D}\overline{E}$ on a VEM with D and E as MEVs.

2. Plot and reduce $X = A\overline{B}CD + AB\overline{C}D + ABC\overline{D} + \overline{A}BCD + A\overline{B}C\overline{D} + ABCD + \overline{A}\overline{B}CD + AB\overline{C}\overline{D}$ using a VEM with C and D as MEVs.

2.4 REALIZING LOGIC FUNCTIONS WITH GATES

Section overview: This section considers some practical aspects of using gates to implement logic functions. The exclusive OR and coincidence gates are also discussed.

A. Gates

During the 1950s and 1960s, gates were used exclusively to implement logic functions. In some cases, these realizations took the form of diode matrices, which were essentially equivalent to gate circuits. In the late 1960s and early 1970s, as MSI circuits became available, different forms of function realization were developed. The multiplexer, the decoder, the read-only memory, and the programmable logic array now allow alternative forms of function implementation. These devices will be considered in Chap. 3.

Although newer methods are available, gate realization is still used. Minimization techniques along with gate implementations generally lead to minimal component cost of a finished circuit. This section notes a few practical aspects involved in using gate circuits.

When SSI circuits are used to construct a system, IC chips with 1, 2, 3, or 4 gates will be utilized. There may be several logic functions to implement in a system; consequently when one function is realized with chips, unused gates often result. These gates can then be used in constructing some other logic function of the system.

Furthermore, it is often appropriate to convert to NOR or NAND gates to fully utilize the several gates of the same type contained on each IC chip. For example, the expression $X = AB + CD$ could be realized with two 2-input AND gates and a 2-input OR requiring two chips. A single quad 2-input NAND could implement this expression as shown in Fig. 2.53. The application of DeMorgan's laws is quite useful in converting functions to implementations involving a single type of gate, resulting in a lower chip count.

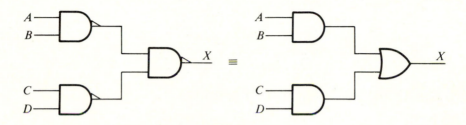

Figure 2.53 NAND gate realization.

Another practical problem often encountered is that of unused inputs. Suppose we have an expression $Y = \overline{A\bar{B}C}$ to implement and have a 4-input NAND available. If we connect these three variables to the gate inputs, we must determine how to connect the fourth input. Remembering that $A\bar{B}C \cdot 1 = A\bar{B}C$ dictates that the fourth input be tied high in this case. In TTL circuits, a floating input is interpreted as a logic one; hence we might be tempted to allow the fourth input to float. An open input exhibits a reasonably high impedance and becomes less immune to noise. We should always tie unused inputs to either ground or the $+5$ V supply, whichever is appropriate, rather than allowing the input to float and possibly introduce noise problems. If an input is tied to $+5$ V, a series resistance of 1 to 10 kΩ should be used between the source and gate input to limit current.

A 4-input OR gate can be used to implement a three-variable OR expression, but in this case we note that $A + B + C + 0 = A + B + C$. Thus, here we want the fourth input connected to ground rather than $+5$ V.

It might also be convenient to construct functions requiring a larger number of inputs with 2- or 3-input gates. Figure 2.54 shows two such examples. Four-input OR gates or AND gates can be based on 2-input devices. If unused gates are available, this method may conserve chip count even though 4-input gates are included in the TTL line.

A last point that should be emphasized is that a NAND or NOR gate can be used as an inverter if needed. Utilization of a spare gate may avoid the addition of an inverter chip. All inputs can be tied together to convert a NAND or NOR to an inverter, but the driving circuit is loaded more heavily than necessary in this configuration. Instead of driving all inputs in parallel,

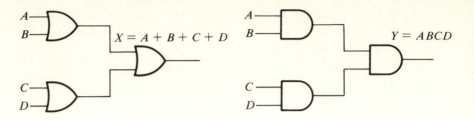

Figure 2.54 Expansion of inputs.

the practical method shown in Fig. 1.9(b) is generally preferred if minimal current loading of the driving gate is appropriate.

B. The EXOR Gate in Function Realization

The exclusive OR function obeys the truth table of Fig. 2.55. The symbol for the EXOR gate and the equation it satisfies are also shown in the figure. A 1 is generated at the gate output only if one input equals 1 while the other input is 0. If both inputs are equal, the output is 0. The K-map for the EXOR function is indicated in Fig. 2.56. In writing the expression for this function we see that three 2-input gates are required since

$$X = A \oplus B = A\bar{B} + \bar{A}B$$

Fortunately, the function is implemented on a chip. The 7486 is a quad 2-input EXOR gate chip.

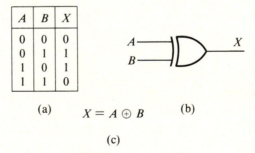

A	B	X
0	0	0
0	1	1
1	0	1
1	1	0

(a) $X = A \oplus B$ (b)

(c)

Figure 2.55 The exclusive OR gate: (a) truth table; (b) symbol; (c) equation.

If the output of the EXOR gate is inverted, it becomes a coincidence circuit or a NEXOR gate. In this case, two equal inputs result in a 1 output while two different inputs lead to a 0 output. The expression for the NEXOR is $X = AB + \bar{A}\bar{B}$.

Figure 2.57 shows the schematic and truth table for the 74135 Quad EXOR/NEXOR gate which represents a rather clever circuit arrangement. A and B are the inputs to an EXOR gate. The output of the first EXOR

	A	0	1
B			
0		0	1
1		1	0

Figure 2.56 *K*-map for the EXOR.

gate connects to the input of a second EXOR gate. If the C input to this gate is low, the output equals the signal applied to the other input. Therefore, when C is low, the circuit output Y represents the EXOR of inputs A and B.

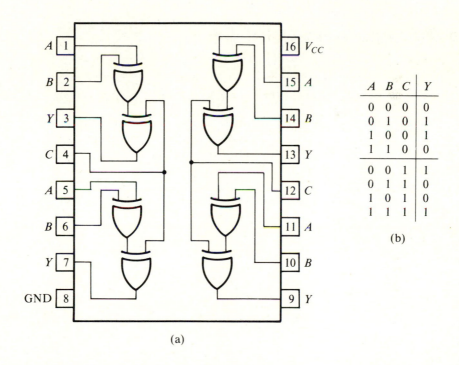

(a)

A	B	C	Y
0	0	0	0
0	1	0	1
1	0	0	1
1	1	0	0
0	0	1	1
0	1	1	0
1	0	1	0
1	1	1	1

(b)

Figure 2.57 The 74135 EXOR/NEXOR gate: (a) schematic; (b) truth table.

In order to produce a NEXOR gate, input C is taken high. For this condition, the EXOR function of gate 1 is inverted by gate 2 resulting in a NEXOR function for Y. Each input C drives two gate circuits in order to conserve pins on the chip.

Since EXOR and NEXOR gates are available on chips, their use in logic function realization should be considered. In certain cases, these gates can implement functions more efficiently than conventional gates. The map of

Fig. 2.58 reflects one such function. There is no reduction possible, and the resulting expression is

$$X = \overline{A}B\overline{C} + \overline{A}\,\overline{B}C + ABC + A\overline{B}\,\overline{C}$$

A total of 5 gates and 17 inputs are required to implement this expression.

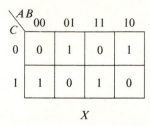

Figure 2.58 A *K*-map.

The expression for X can be factored to yield

$$X = \overline{A}(B\overline{C} + \overline{B}C) + A(BC + \overline{B}\,\overline{C}) = \overline{A}(B\oplus C) + A(\overline{B\oplus C})$$

This function could be implemented with two AND gates, an OR gate, an EXOR gate, and a NEXOR gate requiring 10 inputs. On the other hand, we can note that X can also be expressed as

$$X = A\oplus(B\oplus C)$$

We can implement this expression with two 2-input EXOR gates as indicated in Fig. 2.59.

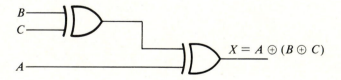

Figure 2.59 Implementation of the map of Fig. 2.58.

The map of Fig. 2.58 represents a special case that allows the EXOR to be applied much more efficiently than conventional gates. This map contains four 1s with no adjacencies. Any map with more nonadjacent 1s than adjacent 1s is a candidate for realization using EXOR circuitry. We should note in the map of Fig. 2.58 that the diagonal 1s represent a nonadjacent pair as also do the pairs of 1s separated by a single 0.

A theory can be developed for absolute minimization using EXOR circuits, but a rather simple method can be used to decrease gate count. This

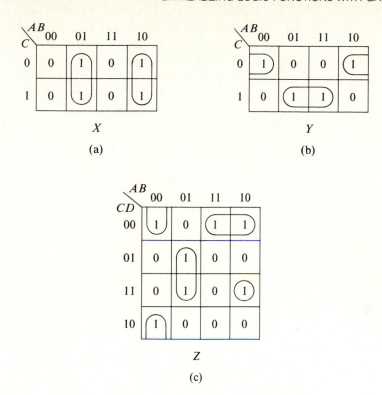

Figure 2.60 *K*-map examples implemented with EXOR gates.

method consists simply of reducing the circuit by conventional methods, then looking for EXOR or NEXOR relations within the reduced expression. We can demonstrate this method with the maps of Fig. 2.60. The expression for X in Fig. 2.60(a) is reduced to

$$X = \overline{A}B + A\overline{B}$$

This expression obviously reduces to

$$X = A \oplus B$$

The expression for Y in Fig. 2.60(b) is reduced to

$$Y = \overline{B}\overline{C} + BC$$

which can also be written as

$$Y = \overline{B \oplus C}$$

The map of Fig. 2.60(c) gives an expression

$$Z = \overline{A}\overline{B}\overline{D} + \overline{A}BD + A\overline{C}\overline{D} + A\overline{B}CD$$

Although the first two terms could be combined in a NEXOR relation to reduce total inputs, the number of gates would not be reduced. Unless an unused NEXOR gate happens to be available in the system, conventional gate realization would probably lead to a lower chip count.

Drill Problems: Sec. 2.4B

1. Minimize the number of gates used to implement the expression $X = \overline{A}\overline{B}\overline{C} + \overline{A}\overline{B}C + AB\overline{C} + ABC$ using conventional gates only. Repeat if EXOR/NEXOR gates are allowed.

2. Repeat problem 1 if $X = \overline{A}\overline{B}\overline{C} + \overline{A}\overline{B}C + \overline{A}BC + AB\overline{C} + ABC + A\overline{B}C$.

SUMMARY OF SIGNIFICANT POINTS IN CHAP. 2

1. Boolean algebra can be used to concisely express combinational logic problems.

2. Boolean algebra can be used to reduce the number of gates required to implement a logic function and can be used to convert the implementation to a particular type of gate.

3. The K-map is based on Boolean algebra and can be used to minimize logic functions.

4. Hazards may result in a minimized circuit. If important to eliminate these transient signals, hazard covers can be used.

5. The variable-entered map can also be used to minimize logic circuits. An advantage of this map over the K-map is its smaller size. We will later see that the VEM approach is useful in multiplexer design also.

6. The traditional method of function realization uses gates. This method minimizes component cost.

7. The EXOR or NEXOR gate can sometimes be used to reduce chip count over conventional gate implementation.

REFERENCES AND SUGGESTED READING

1. D.J. Comer, *Electronic Design with Integrated Circuits*. Reading, Mass.: Addison-Wesley, 1981, chaps. 3 and 4.
2. W.I. Fletcher, *An Engineering Approach to Digital Design*. Englewood Cliffs, N.J.: Prentice-Hall, 1980, chaps. 2 and 3.
3. Engineering Staff, *The TTL Data Book for Design Engineers*. Dallas: Texas Instruments, 1983.

4. National Semiconductor Staff, *Memory Databook*. Santa Clara, Calif.: National Semiconductor, 1982.
5. C.H. Roth, Jr., *Fundamentals of Logic Design*. St. Paul, Minn.: West, 1975.

CHAP. 2 PROBLEMS

Secs. 2.1A to 2.1D

For Probs. 2.1 through 2.9 use Boolean algebra to determine whether the given equations are true or false.

*2.1 $\overline{AB} + C\overline{A} + \overline{B}C = \overline{AB}$

2.2 $\overline{A}\overline{B}\overline{C} + \overline{A}BC = \overline{A}$

2.3 $\overline{A}\overline{B}\overline{C}\overline{D} + D = \overline{A + B + C} + D$

2.4 $AB\overline{C} + A\overline{B}C + \overline{A}BC = \overline{\overline{A}\overline{B} + \overline{A}\overline{C} + \overline{B}\overline{C} + ABC}$

*2.5 $AB + CD + \overline{A}\overline{C} = \overline{A} + B + \overline{C} + D$

2.6 $\overline{A} + ABC = \overline{A\overline{B} + A\overline{C}}$

2.7 $CFG + C\overline{D}\overline{E} + EFG + DFG = DFG + EFG + C(\overline{D + E})$

2.8 $AB + \overline{A}\overline{C}\overline{D} + \overline{B}\overline{C}\overline{D} = AB + \overline{C}\overline{D}$

*2.9 $AB + BC + \overline{A}C + \overline{A}\overline{B}C = AB\overline{C} + AB + \overline{A}C$

2.10 Realize, using NAND gates only, $X = \overline{\overline{A} + \overline{B} + C + \overline{B} + D}$.

2.11 Realize, using NOR gates only, $X = A(B + \overline{C})(A + D)$.

2.12 Realize the expression of Prob. 2.10 using only NOR gates.

2.13 Realize the expression of Prob. 2.11 using only NAND gates.

*2.14 Realize using NAND gates only, $Y = \overline{A}C + AB$.

2.15 Use algebraic methods to express F as a function only of variables and single-variable complements:

$$F = \overline{\overline{\overline{(A + \overline{B})C} + D + \overline{E}}}$$

Sec. 2.1E

2.16 Write the equation for F in Fig. P2.16. Then convert the gates to reflect proper assertion level and write the expression for F directly from the circuit.

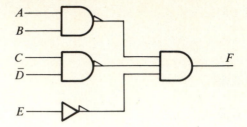

Figure P2.16

2.17 Repeat Prob. 2.16 for the circuit shown in Fig. P2.17.

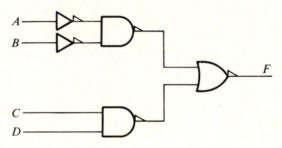

Figure P2.17

2.18 Repeat Prob. 2.16 for the circuit shown in Fig. P2.18.

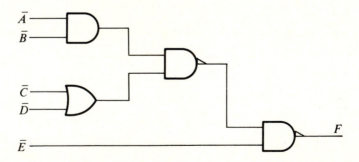

Figure P2.18

Sec. 2.2A

*2.19 Write the expression for F in Prob. 2.17 and put in SSOP form.

2.20 Write the expression for F in Prob. 2.16 and put in SSOP form.

Secs. 2.2B to 2.2D

Minimize the following expressions using K-maps.

*2.21 $F = ABC + \overline{A}BC + \overline{A}\overline{B}C$

2.22 $F = \overline{A}B\overline{C}\overline{D} + \overline{A}B\overline{C}D + \overline{A}BC\overline{D} + A\overline{B}CD + AB\overline{C}D$

2.23 $F = (AB + CD)(A\overline{C} + BD)$

*2.24 $F = (A + BC)(A + B + CD + \overline{A}C)$

2.25 $F = ABC + \overline{A}BCDE + \overline{\overline{A}BCD} + A\overline{B}C\overline{D}E + A\overline{B}C\overline{D}E$

2.26 Build a minimal hardware system to realize a function F that equals 1 when a 4-bit input code equals 1, 2, 5, 6, 8, 11, 12, and 14. F should equal 0 for input codes of 4, 7, 9, and 10. The remaining possible codes will never occur.

2.27 Build a code converter using gates and minimal design techniques to convert the binary code to the fictitious W-code shown.

Binary Input			W-Code Output		
A	B	C	X	Y	Z
0	0	0	1	1	1
0	0	1	1	Θ	1
0	1	0	1	1	0
0	1	1	0	1	1
1	0	0	1	0	0
1	0	1	0	0	1
1	1	0	0	1	0
1	1	1	0	Θ	0

*2.28 Realize the expression represented by the K-map with two gates (Fig. P2.28). *Hint*: Consider factoring the reduced expression.

C	AB 00	01	11	10
0	0	1	0	0
1	0	Θ	1	0

Figure P2.28

Sec. 2.2E

2.29 If X is realized as shown in Fig. P2.29, use a hazard cover to eliminate any static hazards. Explain hazards with a timing chart. Under what conditions of R, S, and X would you expect a glitch?

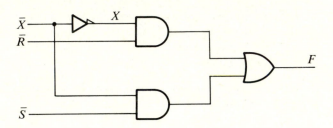

Figure P2.29

2.30 Instead of using a hazard cover, can you add a noninverting delay circuit at an appropriate point to eliminate the hazard in Prob. 2.29?

Secs. 2.3A and 2.3B

2.31 Work Prob. 2.21 using C as a map-entered variable.

2.32 Work Prob. 2.21 using B as a map-entered variable.

*__2.33__ Work Prob. 2.22 with D as a map-entered variable.

2.34 Work Prob. 2.24 with A as a map-entered variable.

2.35 Work Prob. 2.25 with E as a map-entered variable.

2.36 A seven-segment LED display is shown in Fig. P2.36. Using VEMs, design the gate circuitry necessary to produce the display shown. Note that the inputs to the display are asserted low, that is, if the voltage is 5 V, the diode will not conduct current and is not asserted. To turn a segment on, the input must be taken to 0 V to cause current flow through the diode.

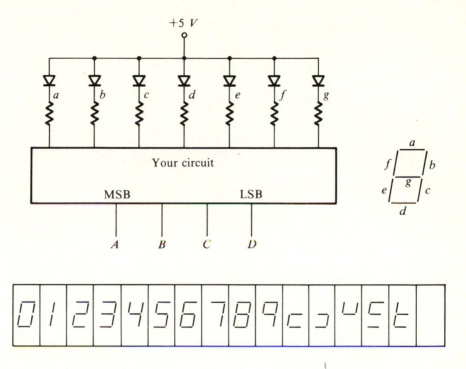

Figure P2.36

2.37 Repeat Prob. 2.36 if the input codes for 10, 11, 12, 13, 14, or 15 will never occur.

2.38 Design a code converter using gates to convert 4-bit binary code to 5-bit *BCD*, that is, 1 bit for the tens column and 4 bits for the ones column.

Sec. 2.4

* **2.39** Write the expression X realized by the circuit shown in Fig. P2.39. Use a K-map to implement this same function in a minimal way with no EXOR gates.

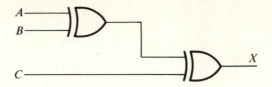

Figure P2.39

2.40 Implement the function represented by the K-map with conventional gates (Fig. P2.40); then repeat if EXOR and NEXOR gates are also allowed. Which method requires fewest gates?

C \ AB	00	01	11	10
0	1	0	1	1
1	0	1	1	0

Figure P2.40

CHAPTER

3

Logic Function Realization with MSI Circuits

The traditional method of using gates to implement combinational logic functions has been applied for over three decades. Since this method often results in minimum component cost for many combinational systems, it continues to represent a popular approach. There are several newer circuits now available for function realization that lead to other advantages over gate implementations. This chapter will consider digital systems that use multiplexers, decoders, read-only memories, and programmable logic arrays along with applicable design methods. Advantages of each approach will also be considered.

The latter part of the chapter will cover MSI circuits that implement standard logic functions such as the parity checker/generator, the 7-segment LED driver, the encoder, the comparator, and the adder. A discussion of addition and subtraction methods using MSI adders concludes the chapter.

3.1 COMBINATIONAL LOGIC WITH MSI CIRCUITS

Section overview: This section considers several MSI circuits that can be used to implement combinational logic functions. The multiplexer, decoder, read-only memory, and programmable logic array are each discussed and design methods are developed. Advantages of each device are covered.

A. Multiplexers

The multiplexer (MUX) is also called a data selector and was originally designed to time-division multiplex several lines of parallel information onto a single line. More recently the MUX has become popular in combinational logic circuit applications. The MUX has n select lines, 2^n input lines, and a single output line. It can be visualized as a switching circuit with the binary code on the select lines determining which input data line is connected to the output, hence the name data selector. Figure 3.1 shows the symbol for a 4:1 MUX and the switching circuit used to explain its operation.

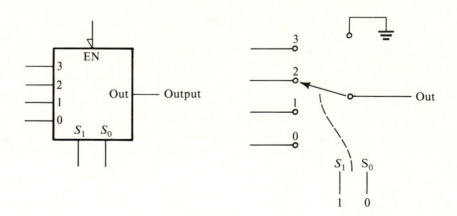

Figure 3.1 The multiplexer.

The EN input enables the circuit when low. When high, the EN input either forces the output to remain at the low level or puts the output into the high-impedance state depending on the MUX used. For example, the 74151 is a MUX that forces the output low while the 74251 presents a high output impedance when the EN input is high. If the MUX is enabled, values of $S_1(\text{MSB}) = 1$ and $S_0 = 1$ cause input line 3 to be connected to the output. If

$S_1 = 1$ and $S_0 = 0$, input line 2 is connected to the output, and so on. Of course, the actual MUX does not use a mechanical switch to selectively connect input to output. Instead AND gates are used as shown in Fig. 3.2. A particular line is gated to the output only when EN is asserted and S_1 and S_0 contain the code to open the corresponding AND gate.

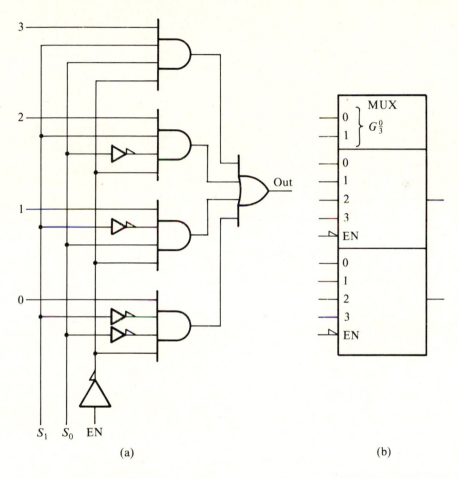

(a)

(b)

Figure 3.2 Implementation of a 4:1 MUX: (a) actual circuit; (b) IEEE/IEC logic symbol for dual 4:1 MUX chip.

The IEEE/IEC symbol requires some explanation. The upper block is a common control block. It is understood that an input to this block represents an input to all other common blocks of the symbol. The G indicates an AND dependency, and $G\frac{0}{3}$ indicates there are four AND dependencies of lines 0 through 3 in the circuit. The two control inputs select a particular AND gate to allow data to pass through this gate to the output.

The MUX realizes logic functions by setting up a one-to-one correspondence between input line number and minterm number of a logic expression. This is demonstrated in Fig. 3.3 where a two-variable K-map is shown with the minterm locations numbered.

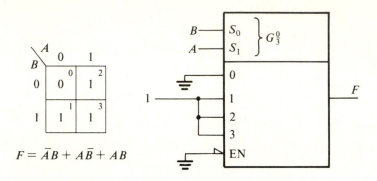

$$F = \bar{A}B + A\bar{B} + AB$$

Figure 3.3 Correspondence between K-map and MUX.

From the K-map we see that a 1 should appear at the output when either minterm 1, 2, or 3 is present. A zero should result when minterm zero occurs. By connecting lines A and B to the select lines of the MUX, and then applying logic 1 to lines 1, 2, and 3, the correct function appears on the output line of the MUX. Logic zero is connected to the zero input line.

It is quite easy to extend the 4:1 MUX to a three-variable problem by using the map-entered variable approach. Let us assume we wish to synthesize the circuit to result in

$$X = AB\bar{C} + A\bar{B}C + \bar{A}B$$

The VEM and MUX implementation is shown in Fig. 3.4.

When $AB = 00$, the VEM shows that a 0 output should result. The MUX implements this requirement by tying input line 0 to ground. When $AB = 01$, the VEM indicates that X should equal 1. The MUX produces a 1 at the output for this combination of A and B by tying input line 1 to logic 1 ($+5$ V). For $AB = 10$, X should equal 1 only if $C = 1$. Connecting C to input line 2 generates the correct value at the MUX output. Connecting $\bar{C}$ to input line 3 completes the implementation of X by producing a 1 at the output only if $A = 1$, $B = 1$, and $\bar{C} = 1$.

In general, an n-variable function requires a $2^{n-1}:1$ MUX to implement. For example, a four-variable design can be implemented by an 8:1 MUX having three select lines. Figure 3.5 demonstrates such a design.

It is possible to implement a system involving more variables without increasing the inputs to the MUX under certain conditions. One instance is

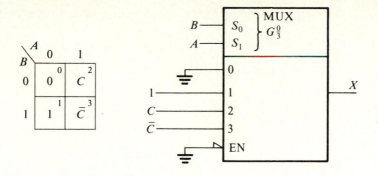

Figure 3.4 A three-variable realization.

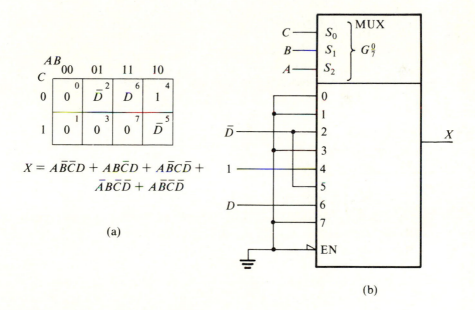

$$X = A\bar{B}\bar{C}D + AB\bar{C}\bar{D} + A\bar{B}C\bar{D} + \bar{A}BC\bar{D} + A\bar{B}C\bar{D}$$

(a)

Figure 3.5 A four-variable realization: (a) VEM; (b) 8:1 MUX realization.

that of seldom-used variables, a case that occurs often in sequential system design. A seldom-used variable is one that appears only occasionally in a logic expression. Each term of the expression

$$Y = A\bar{B}C + A\bar{B}\bar{C}D + \bar{A}BC\bar{D} + \bar{A}B\bar{C}E + \bar{A}B\bar{C}$$

contains the variables A, B, and C while D and E appear fewer times and do not appear simultaneously in the same term. Again the variables A, B, and C can be used to select the minterms or input lines while both D and E can be used as inputs. This function is implemented in Fig. 3.6.

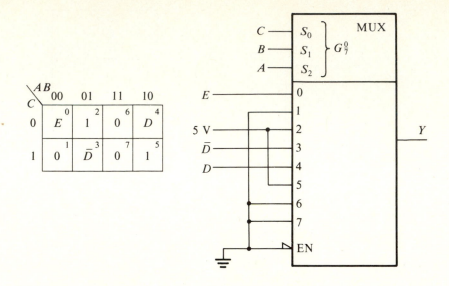

Figure 3.6 A seldom-used variable implementation (74LS138).

Another method of increasing the number of variables without using a larger MUX is to use gating at the inputs. In order to demonstrate this method, let us consider the expression

$$X = A\bar{B}C\bar{D}\bar{E} + AB\bar{C}\bar{D}\bar{E} + A\bar{B}C\bar{D}\bar{E} + \bar{A}\bar{B}CDE$$

$$+ \bar{A}B\bar{C}\bar{D}\bar{E} + \bar{A}B\bar{C}D\bar{E} + ABCDE$$

This five-variable expression can be realized with a three-select line MUX and additional gates. If variables A, B, and C are connected to the select lines, these same variables should be used to create a VEM. The remaining two variables, D and E, both become MEVs in this map. The original and simplified VEMs are shown in Fig. 3.7. Since the variables A, B, and C act as minterm as well as MUX line selectors, the expression at each minterm location is implemented on the corresponding MUX line using gates. Figure 3.8 shows the finished circuit which requires only two gates in addition to the 8:1 MUX.

There is a pin limitation on IC chips that generally restricts the size of the circuit that can be fabricated on the chip. A very complex circuit could be fabricated on a small chip, but the number of pins to the outside world ultimately determines what can be practically included on the chip. Most SSI or MSI circuits use 14 to 24 pins. Thus, a 16:1 is the largest available MUX on a chip. It is possible however, to use several MUXs to expand to a larger multiplexing circuit. Figure 3.9 shows a 64:1 circuit. Six variables, designated A, B, C, D, E, and F, are required to address 64 input lines. The four least significant bits are connected in parallel to the select lines of

C \ AB	00	01	11	10
0	DE ⁰	$D\bar{E} + \bar{D}\bar{E}$ ²	$\bar{D}\bar{E}$ ⁶	0 ⁴
1	0 ¹	0 ³	DE ⁷	$\bar{D}\bar{E} + D\bar{E}$ ⁵

C \ AB	00	01	11	10
0	DE	$\bar{E}$	$\bar{D}\bar{E}$	0
1	0	0	DE	$\bar{D}$

(a)	(b)

Figure 3.7 (a) A VEM. (b) Simplified VEM.

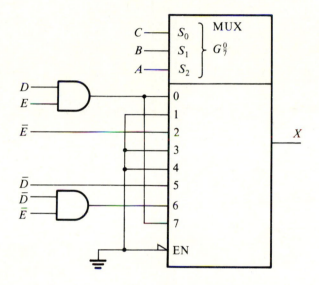

Figure 3.8 Implementation of a five-variable function with an 8:1 MUX (74LS138).

four 16:1 MUXs. The two most significant bits connect to the select lines of a 4:1 MUX. A given input code will address one line on each of the 16:1 MUXs, but the 4:1 MUX will gate only one of the outputs of the larger MUXs to the circuit output. If the code happens to be $A\bar{B}CD\bar{E}F$, which corresponds to 101001_2 or 41_{10}, each 16:1 MUX connects the ninth input line ($CD\bar{E}F$) to its output. These four outputs connect to the input lines of the 4:1 MUX which addresses input line 2 for this case ($A\bar{B}$). Thus, the input line which then connects to the circuit output is the ninth input line of MUX 3. This line corresponds to the forty-first circuit input line.

In chips containing more than one MUX, the select lines will normally appear in parallel for all MUXs. The 74LS153 is a dual 4:1 MUX chip, but select lines S_1 for both MUXs are connected to a single pin as also are lines

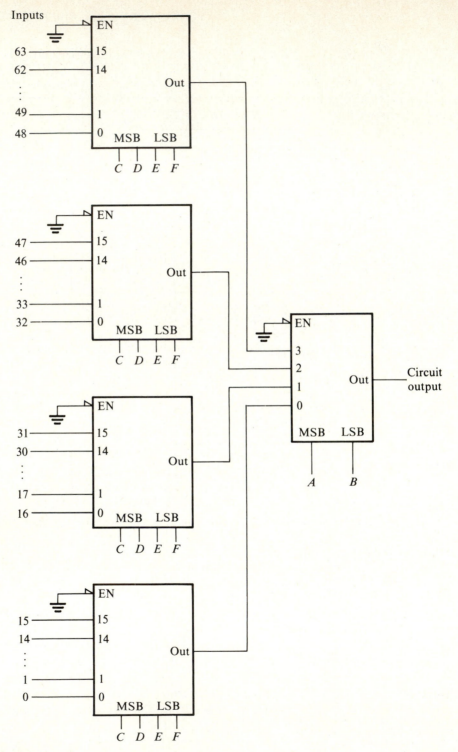

Figure 3.9 A 64:1 multiplexing circuit.

S_0. Consequently, the MUXs cannot be addressed independently. The same is true of the 74LS157 quad 2:1 MUX chip. The select line S_0 is common to all 4 MUXs. Although this arrangement sacrifices some flexibility, the IC pin number requirement is minimized. Both the 74LS153 and the 74LS157 are 16 pin chips. The 74150 is a 24 pin 16:1 MUX.

In comparing MUX design of logic circuits to gate design, we note that map minimization techniques are not applicable to the MUX circuit. Less design time is then required for the MUX design. Another practical advantage of this circuit is that debugging and trouble-shooting is easier than it is for gate circuits. If a MUX circuit malfunctions, the values of the variables connected to the select lines are measured. The corresponding input line and the output line are then checked to see if correct levels are present. In general, the problem can be readily isolated in this way. This is not the case for a complex gate circuit with several levels of gates. Pinpointing a malfunctioning gate may take considerably longer.

A third advantage is that chip count can often be minimized when using MUXs for function realization. Figure 3.10 compares MUX and gate design for the function

$$X = \bar{A}\bar{B}\bar{C}\bar{D} + \bar{A}B\bar{C}\bar{D} + \bar{A}B\bar{C}D + AB\bar{C}D + \bar{A}\bar{B}C\bar{D} + ABC\bar{D}$$

The gate realization, after minimization, requires three 3-input AND gates, one 4-input AND gate, and a 4-input OR gate. Converting to NAND gates allows the implementation to be accomplished with one triple 3-input NAND chip and one dual 4-input NAND chip. The MUX realization requires only one 8:1 MUX chip.

Although fewer chips may be needed for the MUX design, the chip cost will usually be higher than that of the gate design. An 8:1 MUX chip (74LS151) may cost 89 cents compared with a cost of 35 cents for each gate chip (74LS12 and 74LS21). The smaller number of inputs that must be wired may lead to a lower in-circuit cost than the gate circuit.

Drill Problems: Sec. 3.1A

1. Realize the expression $X = \bar{A}\bar{B}C + A\bar{B}\bar{C} + \bar{A}B\bar{C} + ABC + \bar{A}\bar{B}\bar{C}$ with an 8:1 MUX.

2. Realize X from problem 1 with a 4:1 MUX.

3. Realize X from problem 1 with a 2:1 MUX and a minimal number of gates preceding the MUX inputs.

B. Decoders

A binary decoder has n input lines and 2^n output lines. The output lines are numbered in accordance with the decimal equivalent of the input binary code. When the decoder is enabled, a code on the input asserts the corresponding output line while all other outputs remain unasserted. Figure 3.11

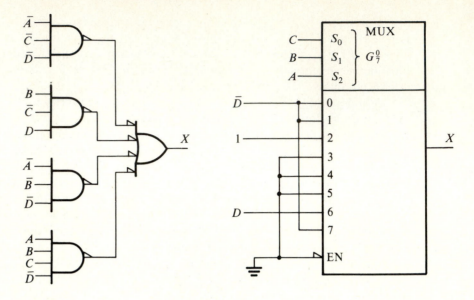

Figure 3.10 Gate and MUX realization of X.

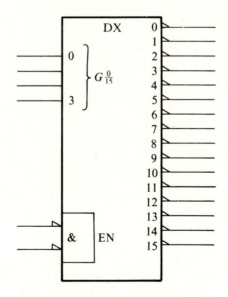

Figure 3.11 A 4-line to 16-line decoder (74LS154).

shows a 4-line to 16-line decoder. This device decodes all possible input codes with the outputs being asserted low. In order to implement a logic function, the outputs corresponding to all minterms that equal 1 in the function are ORed.

We will demonstrate the use of the decoder by implementing the function of Fig. 3.5. This is shown in Fig. 3.12. In decimal notation, the function can be written as

$$X = 4 + 8 + 9 + 10 + 13$$

These output lines are ORed to form the desired function. Since the decoder outputs are asserted low, a low-assertion OR gate (NAND gate) is required. When either of the codes 4, 8, 9, 10, or 13 is present at the input, X will be asserted.

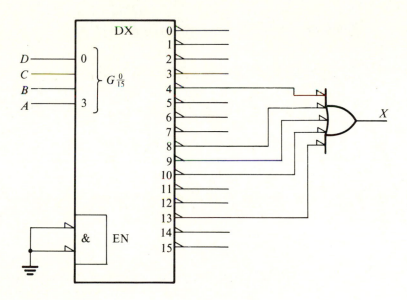

Figure 3.12 Decoder implementation of logic function.

From the standpoint of efficiency, the decoder is somewhat wasteful in that all possible input combinations are decoded, but not all outputs are used. Chip count can be quite low since a single chip and a NAND gate can implement a rather complex function. The decoder is also quite efficient in the realization of multiple logic functions as the following example will demonstrate.

■ *Example 3.1* Use a decoder to convert the binary code to Gray code.

Binary	Gray	Binary	Gray
A B C D	W X Y Z	A B C D	W X Y Z
0 0 0 0	0 0 0 0	1 0 0 0	1 1 0 0
0 0 0 1	0 0 0 1	1 0 0 1	1 1 0 1
0 0 1 0	0 0 1 1	1 0 1 0	1 1 1 1
0 0 1 1	0 0 1 0	1 0 1 1	1 1 1 0
0 1 0 0	0 1 1 0	1 1 0 0	1 0 1 0
0 1 0 1	0 1 1 1	1 1 0 1	1 0 1 1
0 1 1 0	0 1 0 1	1 1 1 0	1 0 0 1
0 1 1 1	0 1 0 0	1 1 1 1	1 0 0 0

□ **Solution:** We first write the decimal equivalent of each bit of the Gray code:

$$W = 8 + 9 + 10 + 11 + 12 + 13 + 14 + 15$$

$$X = 4 + 5 + 6 + 7 + 8 + 9 + 10 + 11$$

$$Y = 2 + 3 + 4 + 5 + 10 + 11 + 12 + 13$$

$$Z = 1 + 2 + 5 + 6 + 9 + 10 + 13 + 14$$

With a single 4-line to 16-line decoder and four NAND gates, the code converter of Fig. 3.13 is constructed. Actually, to minimize the circuit, we could note that $W = A$ and eliminate one gate. ■

It is possible to use several chips to expand the number of possible input codes to a decoding circuit. Let us assume that a 6-bit binary code is to be decoded. This would result in 64 possible input combinations requiring 64 output lines. Figure 3.14 shows a method of implementing this circuit using one 2-line to 4-line decoder and four 4-line to 16-line decoders.

The two most significant bits of the input are used to enable the proper 4-line to 16-line decoder through the 2-line to 4-line device. If $AB = 00$, for example, the upper decoder is active. The four least significant bits, $CDEF$, determine which of the 16 lines of this decoder is asserted. The upper decoder has one output line asserted for each input code between 000000 and 001111. The next lower decoder is active for input codes of 010000 to 011111. The third decoder is active for input codes of 100000 to 101111, and the last is active for input codes in the range of 110000 to 111111.

The 74154 chip is a 4-line to 16-line decoder packaged in a 24-pin circuit. The 74155 is a 16-pin chip that contains two 2-line to 4-line decoders with common input address lines. By using the separate enable lines, a 3-

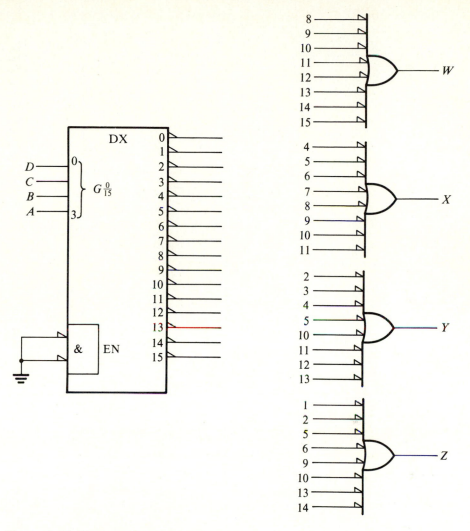

Figure 3.13 Binary to Gray code converter.

line to 8-line decoder can be constructed with the 74155 chip as shown in Fig. 3.15.

The MSB of the input code is used to enable decoder 1 when low and decoder 2 when high. If decoder 1 is enabled while decoder 2 is disabled, all outputs of decoder 2 are high and a single output of decoder 1 will be asserted low. An input code of 010 will assert circuit output line 2, while the code 110 will assert output line 6.

As mentioned earlier, the decoder may not be as efficient as the MUX in realizing a single function since a NAND chip is required. The real advantage of the decoder occurs in realizing several functions simultaneously. For

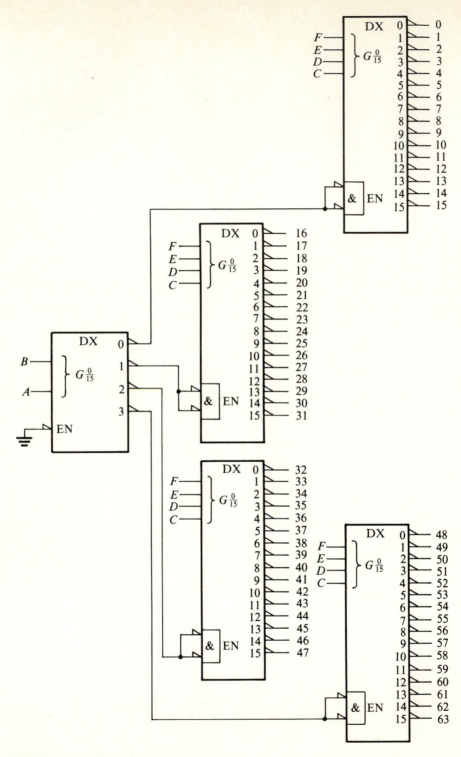

Figure 3.14 Expanded decoding circuit.

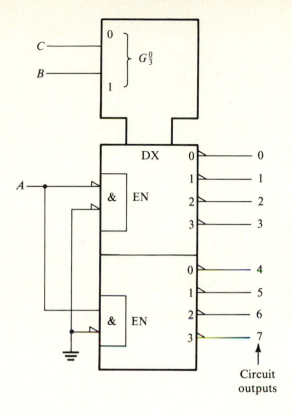

Figure 3.15 Dual 2-line to 4-line decoders used for a 3-line to 8-line decoder.

comparison purposes, we will implement the same function realized by the gates and the MUX of Fig. 3.10. In terms of decimal notation, this function is

$$X = 0 + 2 + 4 + 5 + 13 + 14$$

Figure 3.16 shows the decoder implementation of this function. The 74LS154 cost is $1.75, while an 8-input NAND gate is 35 cents (74LS30). This component cost is more than twice that of the MUX realization and exactly three times that of the gate realization.

Next we will compare the costs of the decoder and MUX designs for the multifunction realization,

$$F1 = \overline{A}BCD + A\overline{B}\overline{C}D + AB\overline{C}\overline{D} + \overline{A}B\overline{C}D + \overline{A}BC\overline{D}$$
$$+ \overline{A}\overline{B}CD + ABCD$$

$$F2 = \overline{A}\overline{B}\overline{C}D + A\overline{B}\overline{C}\overline{D} + A\overline{B}CD + ABC\overline{D} + ABCD + \overline{A}\overline{B}\overline{C}\overline{D}$$

$$F3 = A\overline{B}\overline{C}\overline{D} + \overline{A}\overline{B}CD + \overline{A}B\overline{C}D + A\overline{B}\overline{C}D + \overline{A}\overline{B}C\overline{D} + \overline{A}\overline{B}\overline{C}D + AB\overline{C}\overline{D}$$
$$+ A\overline{B}CD + \overline{A}\overline{B}C\overline{D}$$

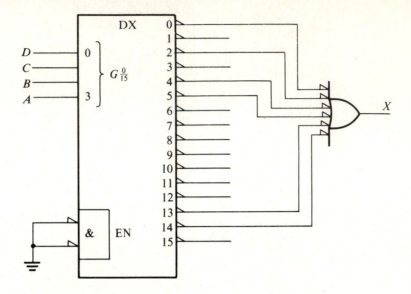

Figure 3.16 Decoder implementation of X.

$$F4 = A\overline{B}\overline{C}D + \overline{A}B\overline{C}D + \overline{A}BCD + AB\overline{C}\overline{D} + A\overline{B}C\overline{D}$$

The decoder design is given in Fig. 3.17, while the MUX design is shown in Fig. 3.18. One decoder (74LS154-$1.75), four 8-input NAND chips (74LS30-$0.35), and one quad 2-input NAND chip (74LS03-$0.29) are required for the decoder implementation resulting in a component cost of $3.44. The four MUXs (74LS151-$0.89) cost $3.56.

Drill Problems: Sec. 3.1B

1. Realize the expression $X = \overline{A}B\overline{C} + A\overline{B}\overline{C} + \overline{A}BC + ABC + \overline{A}\overline{B}\overline{C}$ with a 3-line to 8-line decoder and an 8-input NAND gate.

2. Realize the expressions $F1 = A\overline{B}C + AB\overline{C} + \overline{A}$, $F2 = \overline{A}B\overline{C} + \overline{A}B\overline{C} + A\overline{B}C$, and $F3 = \overline{A}B\overline{C} + ABC$ with a decoder followed by gates.

3. Show the pin connections necessary to implement X from problem 1 using a 74155 decoder chip along with an 8-input NAND gate.

C. Read-Only Memory

The read-only memory or ROM finds its major use as a storage unit for fixed programs in a computer. Generally, these ROMs are fairly large, storing several thousands of binary bits. Smaller ROMs are available that can

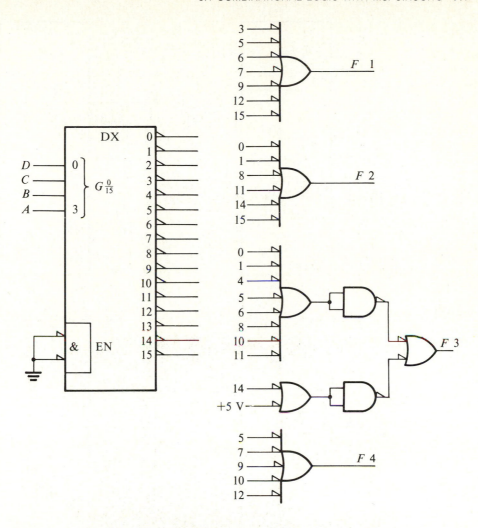

Figure 3.17 Multifunction realization with decoder.

be used to realize logic functions. These devices lead to more expensive function implementations than those devices previously discussed.

The advantage of ROM implementation occurs in systems requiring changeable functions. If function X is to be generated on some line during a certain time period for a given input combination, and if function Y is to be generated during some other time period for the same input combination, then a ROM system is very useful. When function X is required, a ROM producing this function can be plugged into the circuit. A second ROM for function Y can be plugged into the circuit when appropriate. No wiring changes are required for this change of logic function.

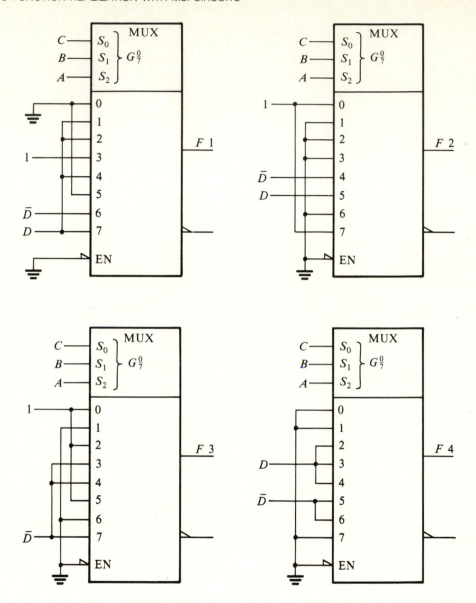

Figure 3.18 Multifunction realization with MUXs.

A block diagram of a ROM is shown in Fig. 3.19. The address lines point to locations within the ROM that store words of n bits. Since 2^m unique binary codes can be set up on the m input lines, 2^m different locations can be addressed. The ROM size is defined by the number of locations and the word size. A 256×4 ROM indicates that the device has 256 storage locations each holding a 4-bit word. This ROM would require eight address lines to access 256 locations. When a binary code is applied to the address lines, the contents of the location specified appear on the output lines. A smaller ROM size might be 32×8, while a size of 8192×8 represents a large ROM.

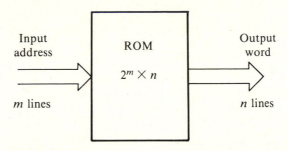

Figure 3.19 Block diagram of ROM.

There are several different methods of entering the words into the ROM locations. This process is referred to as programming the ROM. The first method allows the system designer to indicate to the manufacturer the desired contents of the ROM. The manufacturer then makes the necessary connections of gates to produce these contents. This type of ROM is called a mask-programmable ROM. The process involves creation of photographic metal masks that, after reduction, are used to determine the areas on which vaporized metal conductors are deposited on the chip.

A second method of programming can be performed by the circuit designer. This type of device is called a field-programmable ROM. The contents of all locations are set to 1s during the manufacturing process. A 1 bit can be changed to a 0 by removing a metal-conducting link. These metal conductors, called fusible links, are designed to carry normal currents, but will evaporate if large amounts of current flow. By applying specified voltages along with appropriate addresses, the fusible links can be selectively evaporated to program the ROM with the desired contents. Once programmed, the 0 bits can never be restored to 1 bits.

There are other types of ROM such as the ultraviolet erasable ROM (EPROM) and the electrically alterable ROM (EAROM) which are important in computer applications. These devices are not significant in logic function realization and will not be discussed further here.

The ROM can be visualized in terms of a minterm decoder followed by a set of OR gates. For m address lines, a total of 2^m AND gates, each having m inputs, would be required. For an n-bit word, n output OR gates with up to 2^m inputs would be necessary. Figure 3.20 shows this configuration for an 8×4 ROM. Three input lines lead to 2^3 or 8 AND gates to decode the 8 minterms. When a given minterm or location is accessed by the address lines, one AND gate output will be asserted. The output word generated depends on the connections from this gate output to the OR gate inputs. Connected as shown, this ROM would generate the following outputs for the given inputs.

Input		Output X_0	X_1	X_2	X_3
$\bar{A}\bar{B}\bar{C}$	(000)	0	0	1	0
$\bar{A}\bar{B}C$	(001)	1	0	1	1
$\bar{A}B\bar{C}$	(010)	0	1	0	1
$\bar{A}BC$	(011)	1	0	0	1
$A\bar{B}\bar{C}$	(100)	1	1	0	1
$A\bar{B}C$	(101)	0	1	1	0
$AB\bar{C}$	(110)	1	0	0	0
ABC	(111)	0	1	1	0

Normally the ROM would be used for function realization only in applications requiring multifunction generation. The code converter of Example 3.1 could be realized by a 16 × 4 ROM. Each address given by the inputs A B C D would access a word that would correspond to the desired outputs W X Y Z. For example, the contents of location 0010 would be 0011, and the contents of location 0111 would be 0100.

Drill Problems: Sec. 3.1C

1. Specify the contents of a 32 × 8 ROM to implement the functions $F1$, $F2$, $F3$, and $F4$ realized by the decoder circuit of Fig. 3.17.

2. Realize the expressions $F1 = A\bar{B}C + AB\bar{C} + \bar{A}BC$, $F2 = \bar{A}\bar{B}C + \bar{A}B\bar{C} + A\bar{B}\bar{C}$, and $F3 = \bar{A}\bar{B}\bar{C} + ABC$ with an 8 × 4 ROM. Show the contents of all locations.

D. Programmable Logic Arrays

The programmable logic array or PLA can be thought of as a ROM that has had a large percentage of its locations deleted. A 2^{16} × 8 ROM would contain 2^{16}, or 65,536, storage locations of 8-bit width. There is a PLA manufactured with 16 address lines and 8-bit word size, but instead of storing 65,536 words, it stores only 96. From a hardware standpoint, the ROM must decode every possible minterm of the input variables, while the PLA

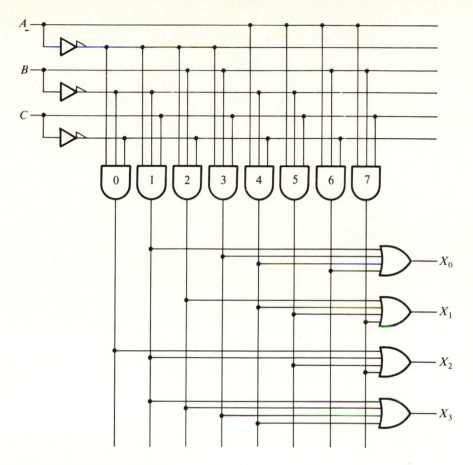

Figure 3.20 An 8 × 4 ROM.

decodes some very small percentage of the minterms. In the actual PLA just mentioned, less than 0.2 percent of the input variable minterms are decoded.

Since the number of storage locations or minterms decoded in the PLA is significant, the size of this device includes this number in the specifications. A PLA is specified by the number of input lines, the number of minterms that can be realized, and the number of output lines. A $16 \times 96 \times 8$ PLA describes the device of the preceding paragraph.

The major applications of the ROM is that of word storage in programmable systems. Logic function realization is an area in which the ROM can be applied, but it may be inefficient in terms of unused storage locations. The PLA is used almost exclusively for logic function realization and therefore does not require a large storage capacity. While the ROM can perform any function performed by the PLA, in logic function realization the ROM is more expensive and often requires much more programming. The PLA can be mask-programmed or field-programmed (FPLA) in the same way a ROM is programmed.

The PLA is sometimes used to produce a system with a small number of chips in a minimum time. The keen competition among digital system manufacturers often makes it desirable for a company to move from a prototype to the marketplace as quickly as possible. Chip count of the system is also important and often must be minimized. The most effective means of decreasing chip count is to integrate the system on a single chip or small number of chips. Unfortunately, the process of integrating a prototype circuit may require three or four weeks. If the finished circuit does not operate up to specifications, the design is modified and several more weeks are required for a second fabrication. Rather than implement the logic circuit by IC fabrication, many companies use the PLA. The field-programmable unit can be used to verify system operation before the mask-programmable units are used. A single metal deposition is then performed to produce the finished circuits. This process reduces turnaround time from weeks to days in creating a low-chip-count logic system.

The PLA contains an AND gate to decode each minterm followed by an OR gate to produce each bit of the output word. The inputs of the AND gates are connected to decode the desired minterms while the outputs of these gates become inputs to the OR gates to form a sum of products.

As an example of the use of the FPLA, consider the logic functions

$$F1 = A\bar{B}C\bar{D}EF + A\bar{B}\bar{C}DEF + \bar{A}B\bar{C}\bar{D}EF + AB\bar{C}D\bar{E}F + \bar{A}B\bar{C}D\bar{E}F$$
$$+ A\bar{B}C\bar{D}E\bar{F} + \bar{A}B\bar{C}DEF + \bar{A}B\bar{C}\bar{D}EF$$

$$F2 = \bar{A}B\bar{C}\bar{D}EF + \bar{A}\bar{B}\bar{C}DE + A\bar{B}\bar{C}DE$$

$$F3 = \bar{A}B\bar{C}DE\bar{F} + \bar{A}B\bar{C}DEF + AB\bar{C}\bar{D}EF + \bar{A}\bar{B}\bar{C}\bar{D}E\bar{F}$$
$$+ \bar{A}BCDE\bar{F} + ABCD$$

Those terms with fewer than six variables imply "don't care" conditions relative to the missing variable or variables. These variables can be disconnected from the AND gates that decode the minterms during the programming procedure. For example, the term $\bar{A}\bar{B}\bar{C}DE$ is not a function of variable F. Thus, both the F and $\bar{F}$ inputs are disconnected from the gate that implements this term.

Each of the three functions are now examined for common terms to avoid duplication. We note that $\bar{A}B\bar{C}DEF$ appears in both F_1 and F_3. This minterm need only be formed with one AND gate which can then drive both output OR gates.

These functions are realized by the PLA of Fig. 3.21 which requires at

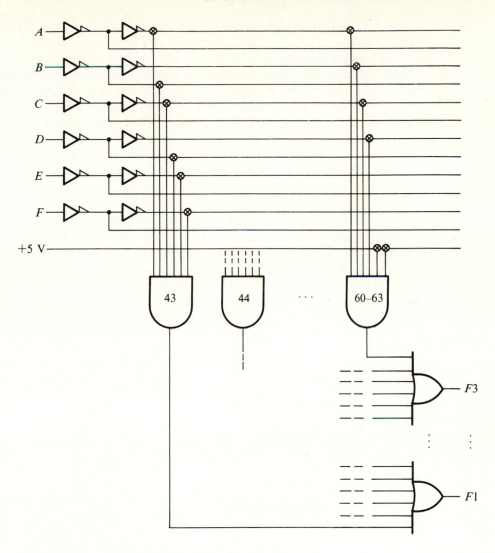

Figure 3.21 Realization of functions by a PLA.

least six input lines, at least sixteen AND gates, and at least three output OR gates each having eight or more inputs.

E. MUX, Decoder, ROM, and PLA Comparison

Chapter 2 considered the use of logic gates to implement logic functions. These same gates are used as building blocks for each of the MSI circuits considered earlier in this section. Some flexibility is lost since the gates in MUXs or decoders have fixed configurations and interconnections.

For single-function realization, gates lead to the lowest component cost while a MUX can realize any function up to a five-variable expression with only one chip. For realization of several functions simultaneously, the decoder becomes more cost effective. The ROM is also useful for this purpose, but it costs more than the decoder. When the functions that are realized must be modified at various times, ROM replacement becomes appropriate. Replacement of the ROM chip in a hardwired socket leads to a new set of logic functions. The PLA is less expensive than the ROM and also has the advantage of programmability. This chip can be used to create a minimal system in a very short time.

There is a physical similarity between the decoder, ROM, and PLA. The decoder can be visualized as a group of AND gates, each of which drives an output line. Each gate has the same number of inputs as does the decoder and each realizes a minterm. A binary decoder uses all possible input combinations. If there are n input lines, there will be 2^n output lines. There will also be 2^n gates each having n inputs. The pin limitation on IC chips restricts the maximum decoder size. A 4-line to 16-line is the largest decoder available as a single MSI chip. Custom integrated circuits may include larger decoders, which generally use several levels of gating to minimize required internal connections. In MSI circuits, a BCD to decimal decoder is also available. This device is a 4-line to 10-line decoder. The outputs represent the decimal numbers 0 through 9 while the four inputs represent the binary codes of these numbers.

The ROM is a simple extension of the decoder. One form of ROM adds a set of n OR gates after the AND gates to form an n-bit output word each time an AND gate is asserted. Figure 3.20 shows this arrangement.

The PLA is very similar to the ROM, also consisting of minterm decoders followed by OR gates. There are much fewer decoder gates in the PLA than contained in the ROM.

3.2 STANDARD LOGIC FUNCTIONS WITH MSI CIRCUITS

Section overview: This section will consider several MSI chips that implement well-used logic functions. The 7-segment LED driver, the parity generator/checker, the encoder, and the comparator are chips that perform important standard functions. The operation of these chips are covered in this section.

The combinational logic functions considered up to this point are often called random logic functions. Each function depends on the application and any arbitrary function can be implemented as needed. There are certain applications encountered frequently enough that the required functions are implemented on single chips. These standard functions can then be used in appropriate applications without the necessity of redesigning functions.

A. The 7-Segment LED Driver

The light-emitting diode (LED) is very useful in logic circuits serving as a visual indication of output voltage levels. This device also is used for the popular 7-segment display that forms numeric and other symbols when driven by binary input signals. Because of its importance in these applications, we will discuss the LED and the 7-segment display.

The LED is a semiconductor diode constructed from gallium arsenide or gallium phosphide. When current flows through the diode, electrons recombine with holes in the semiconductor material yielding energy in the process. This energy is in the form of both heat and light with a significant amount of energy appearing in the latter form for the materials used in the LED. A current ranging from 10 to 40 mA will cause a reasonable light output. A typical value of current for practical LEDs is 15 mA. The voltage drop across a forward-biased LED ranges from 1.6 V to over 3 V depending on the material used.

Although the LED circuit can be arranged as a low assertion or a high assertion circuit, a low assertion arrangement is generally used when TTL is the driving circuit. Figure 3.22 shows both arrangements. The resistor is selected to limit the current through the diode when asserted. This value is typically 180 Ω.

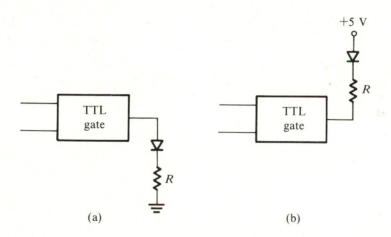

Figure 3.22 LED circuit: (a) high-assertion input; (b) low-assertion input.

The LED is used to give a visual indication of the state of the circuit output. In Fig. 3.22(a), when the gate output is at the low level, no current will flow. To activate the LED, the output must switch to the high level creating a 4 to 5 V drop across the resistor-diode combination. The second circuit is asserted when the gate output goes low, while no current flows when this output is high. The resistor is calculated by noting that a total voltage drop of 5 V will appear across the LED-resistor combination when the diode is asserted. Approximately 2.2 V will drop across the LED, leaving 2.8 V to drop across the resistor. If 15 mA must flow through the LED, the resistor value is given by $R = 2.8 \text{ V}/15 \text{ mA} = 187 \text{ }\Omega$.

Another application of the LED is in the 7-segment display that is used to form numbers or characters. This display is popular in calculators, digital watches and clocks, computer readouts, digital meters, and other digital instrumentation. The 7-segment display uses either a common cathode or a common anode arrangement as shown in Fig. 3.23. The common anode configuration of Fig. 3.23(a) is a low assertion circuit while the common cathode is a high assertion circuit. A current-limiting resistor is placed in series with each LED. TTL gates can sink considerably more current at the low output voltage level than they can produce at the high voltage level; thus the common anode configuration is generally used with TTL circuits. If a common cathode circuit is used, current amplifiers or buffers must be inserted between the TTL gates and the LEDs. A light diffuser is placed over each LED to create a uniformly illuminated line for each segment.

In order to create each number, the correct segments must be activated. For example, a 1 is created by activating segments b and c. A 3 requires segments a, b, c, d, and g. An 8 requires that all segments be activated.

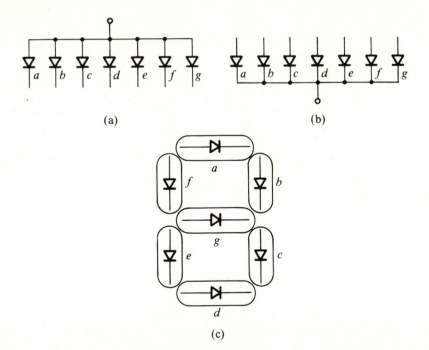

(a)

(b)

(c)

Figure 3.23 7-segment display: (a) common anode; (b) common cathode; (c) actual physical arrangement.

A 4-output digital circuit can represent the binary code for any number from 0 to 15. In order to use the LED to indicate the number contained on the four lines, a code-converting circuit must be used to drive the correct segments. The 7447A and 7448 are chips that contain a BCD to 7-segment decoder. The 7447A is an open-collector circuit and is the more popular of the two chips because of its low asserted outputs. The 7448 has high asserted outputs. The numbers 10 through 15 do not cause displays of corresponding decimal numbers, but display the symbols indicated in Fig. 3.24. There are hex displays available that produce the symbols A, B, C, D, E, and F for the numbers 10 through 15.

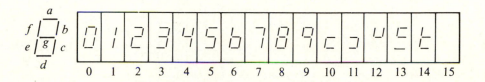

Figure 3.24 7-segment display for 16 inputs.

The schematic for the BCD to 7-segment decoder/driver is shown in Fig. 3.25. This circuit drives the common anode display of Fig. 3.23(a) with a current limiting resistor of approximately 180 Ω in series with each LED.

This code converter was designed by considering the Boolean expression that activates each segment and reducing each expression to lowest terms. Segment a must be asserted low when 0000, 0010, 0011, 0101, 0111, 1000, 1001, or 1101 is present. If A is the MSB and D is the LSB, the K-map for a is shown in Fig. 3.26.

The expression can be reduced by implementing the NOT a function. Since a must be asserted low, this function actually represents the desired expression. Hence, from the groupings shown on the map,

$$a = B\overline{D} + AC + \overline{A}\overline{B}\overline{C}D$$

A careful examination of Fig. 3.25 will indicate that this expression is implemented for segment a on the 7447A chip. The expressions for all other segments were reduced in the same way before implementing with gates.

There are two blanking control pins of special interest in the construction of multidigit LED displays. These are the $\overline{RBI}$, ripple-blanking input, and $\overline{BI/RBO}$, blanking input or ripple-blanking output. If the $\overline{BI/RBO}$ pin is held low, the display is blank; that is, no segment will be illuminated regardless of the input. If this pin is held high or floated, there is another possibility for blanking to occur. This requires that the $\overline{RBI}$ pin be held low and $A = B = C = D = 0$. For this combination of binary zero at the input and

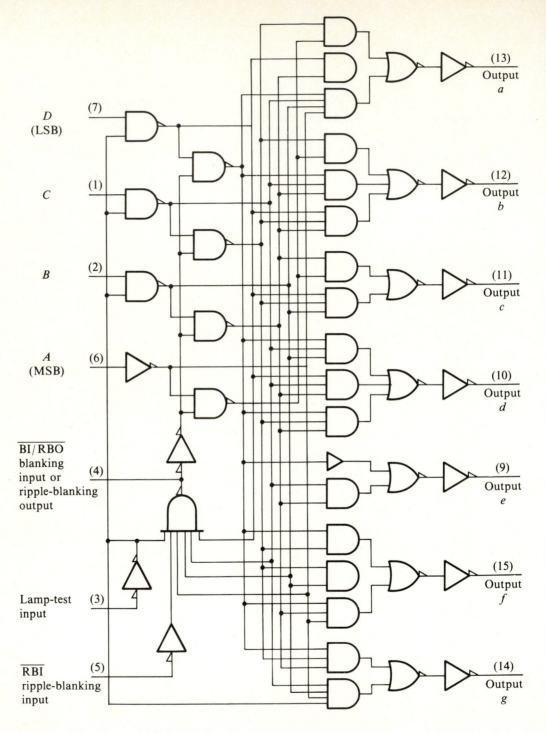

Figure 3.25 BCD-to-7-segment decoder driver.

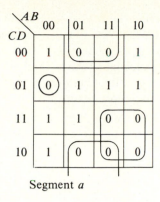

Figure 3.26 *K*-map for segment *a*.

$\overline{\text{RBI}}$ low, the pin $\overline{\text{BI/RBO}}$ is pulled low internally and the display is blanked. If both $\overline{\text{RBI}}$ and $\overline{\text{BI/RBO}}$ are high, no blanking occurs.

One of the major purposes of these control pins is to blank leading or trailing zeros of a multidigit display. For example, if the number 0070.1600 is to be displayed as 70.16, the two leading zeros and the two trailing zeros must be blanked. This can be done by tying the $\overline{\text{RBI}}$ pin of the MSB low, which will blank the MSB zero. The $\overline{\text{BI/RBO}}$ pin of the MSB should be tied to the $\overline{\text{RBI}}$ pin of the next MSB. This arrangement should continue down to the LSB integer display. The two leading zeros would be blanked, but the zero next to the decimal (70.16) would not be blanked. This is because a nonzero input (7) to the adjacent driver will drive $\overline{\text{RBI}}$ of the least significant integer high to disable blanking of this driver. The decimal part is treated in the same manner except that the least significant driver has the $\overline{\text{RBI}}$ pin tied low and the $\overline{\text{BI/RBO}}$ pin connected to the $\overline{\text{RBI}}$ pin of the more significant decimal driver. Each decimal driver is connected in this manner up to the most significant decimal bit. All trailing zeros and all leading zeros are blanked by this configuration.

Drill Problem: Sec. 3.2A

Design a 3-bit binary to 7-segment driver using low asserted LEDs. Numbers 0 through 7 can be represented by the display.

B. The Parity Generator/Checker

The concept of parity is important in many digital systems. As a result of its importance we will discuss this general topic before proceeding to a discussion of the parity circuit.

Chapter 1 discussed the use of a number of binary bits to represent code words. For example, the ASCII code consists of 7 bits. Each 7-bit binary code word represents a character of the ASCII set such as an "A", "+", or "2". The formation of binary words is the basic method of coding information in digital systems. Numbers may be coded in straight binary, in BCD, in a signed magnitude representation, or in other appropriate ways. The code may be peculiar to the digital system being used, or it may be an accepted standard.

As a digital system carries out its function, it moves binary words from one section of the system to another or it may exchange words with other digital systems. A computer, for example, stores binary words in memory and retrieves them as needed by the overall system. A computer terminal transmits binary words to a digital computer and receives other binary words back from the computer.

As this movement of binary information takes place, an occasional error may be introduced into the information. A bit of a word may be changed due to noise in the system. Extraneous signals or unwanted signals are classified as noise and represent a common, but serious, problem in all digital systems. Although it is beyond the scope of this book to discuss the sources of noise, for our purposes we will state that this problem can change the value of an occasional binary bit from its correct value. When a bit of a word is changed, an incorrect word results.

There are two methods of dealing with this problem. One simply identifies words containing errors while the second actually corrects the errors. These two methods are called error detection and error correction. We will briefly discuss the notion of error detection.

Typically noise will introduce errors that are rather widely separated in time. Perhaps a terminal transmitting information to a computer may have one bit identified incorrectly for each 1000 words transmitted. Thus, we assume that no word will ever contain more than one error. Based on this assumption the set of words are now chosen to have a minimum distance of two. This means that in order to confuse one word of the set with any other, at least 2 bits must be changed. Figure 3.27 demonstrates the concept of distance.

If the minimum distance of a set is two, a single error introduced into a word will result in a word that is not a member of the set. This word then can be identified to the system as an erroneous word and proper steps can be taken to recover the correct word.

Although it may appear complex to implement the ideas of the preceding paragraphs, in fact it is surprisingly simple. If all words of a set contain an even number of 1 bits, the minimum distance is two. Similarly, the same minimum distance results if all words of a set contain an odd number of 1 bits. When a word contains an even number of 1 bits, the word is said to have even parity. Odd parity requires an odd number of 1 bits in the word.

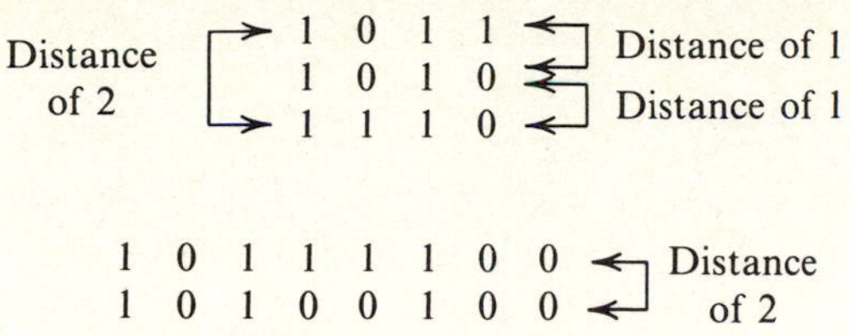

```
Distance      ┌─→ 1  0  1  1  ┐  Distance of 1
 of 2         │   1  0  1  0  ┤  Distance of 1
              └─→ 1  1  1  0  ┘
```

```
1  0  1  1  1  1  0  0  ┐  Distance
1  0  1  0  0  1  0  0  ┘    of 2
```

Figure 3.27 Distance between code words.

Transmission of words all having the same parity, allows us to detect single errors by checking parity at the receiving end. If correct parity does not exist for the received word, it indicates that this is an erroneous word. The words of a given set do not normally exhibit the same parity, but an additional bit position can be used to create the same parity for each word. This extra bit position is called the parity bit. In Chap. 1, we saw that the ASCII code requires 7 bits to represent each character. To use error detection, an eighth bit is added to create correct parity for each word. Figure 3.28 shows the creation of even parity for several different 7-bit words.

Original 7-bit word	Transmitted 8-bit word with even parity
0 1 1 0 0 1 0	0 1 1 0 0 1 0 1
0 1 0 0 1 0 0	0 1 0 0 1 0 0 0
0 1 1 1 1 0 0	0 1 1 1 1 0 0 0
0 0 0 1 0 0 1	0 0 0 1 0 0 1 0
1 0 0 1 0 1 0	1 0 0 1 0 1 0 1

parity bit ↑

Figure 3.28 Creation of even parity words.

When a word is received, a parity checker circuit is used to determine if an error has occurred. This parity checking circuit can be composed of EXOR gates as shown in Fig. 3.29. Each EXOR will check a 2-bit group for parity, generating a 1 if odd parity exists and a 0 for even parity. If an even number of 2-bit groups generate odd parity, the overall word is an even parity word. If all four groups exhibit odd parity, the second level EXOR gates will generate 0 outputs. If two groups generate odd parity and 1s are applied to different EXOR gates at level 2, the level 3 EXOR will output a 0. The inverter produces a high assertion output for even parity. Odd parity will result in an output of 1 by the third-level EXOR gate.

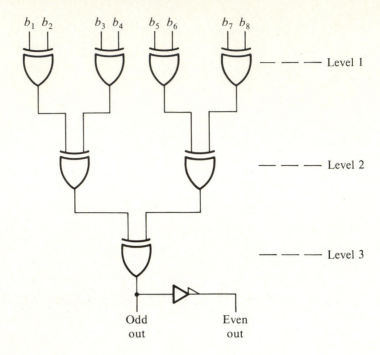

Figure 3.29 A parity checker.

The parity checker can also be used as a parity generator. Before transmitting a 7-bit ASCII code character, the parity is checked. The output of the parity checker, corresponding to the opposite type parity desired, will generate the correct parity bit. This function is demonstrated in Fig. 3.30 for an even parity system.

Because 7- and 8-bit codes commonly occur in data transmission (ASCII and EBCDIC), a standard MSI chip has been designed to generate or check parity for words up to 9 bits. This chip is the 74180 which is shown in Fig. 3.31. This circuit is similar to that of Fig. 3.29 with two exceptions. We note that NEXOR gates are used at the first and third levels of gating and additional control gates appear at the output.

Each input gate produces a high output if even parity exists for the pair of bits this gate checks. An even parity input will result in either zero, two, or four of these gates having a high output. Point X will then be at a high level if an even parity word is applied to the inputs. If our system requires that an even parity input result in a high asserted output on the Even Out line, the control inputs P_E and P_O are high and low, respectively. The high level on P_E opens output gates 2 and 3 resulting in Even Out being high and Odd Out being low for an even parity word. For an odd parity input word, Even Out and Odd Out would be low and high, respectively. Reversing the

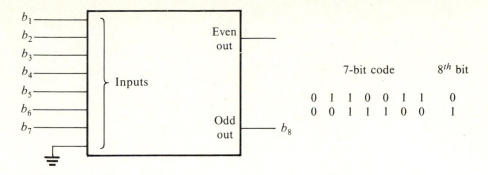

Figure 3.30 Even parity generation.

levels of the inputs P_E and P_O results in a reversal of output levels as indicated by the table of Fig. 3.31. This discussion assumes that the 1 bit is asserted high at the input.

Two items of information must be known in order to generate parity for a word. We must know whether the generated parity bit is to be low or high asserted and we must know whether odd or even parity is to be produced. Knowing these specifications allows us to select the correct levels on P_E and P_O. In order to demonstrate, let us assume we want to produce a high assertion parity bit that results in odd parity for a 9-bit word. The 8 code bits would be applied as inputs, and P_E could be driven low while P_O is driven high. This will result in Odd Out being high whenever the 8-bit code word has even parity. Hence, Odd Out can be used as the ninth parity bit of the word.

If parity checking of a 9-bit input word is desired, 8 bits of the word are applied to the normal inputs while the ninth bit connects to P_E or P_O. For true high asserted parity checks, the ninth bit connects to P_O with an inverter connected from P_O to P_E. The Even Out will then be asserted high if even parity exists, while the Odd Out line will be asserted high if odd parity exists.

■ *Example 3.2* Show how to connect the 74180 to generate even parity for a 9-bit word that is to be transmitted. The circuit should also check parity on a 9-bit received word. Assume that the words have high asserted 1 bits and that a control signal called GENCHK is high when generating parity and low when checking.

□ *Solution:* This problem requires the production of two signals: an error signal when the received word has odd parity and a parity bit that must be added to each word that will be transmitted. We will produce these signals on two separate output lines. We will arbitrarily assign the Even Out line to

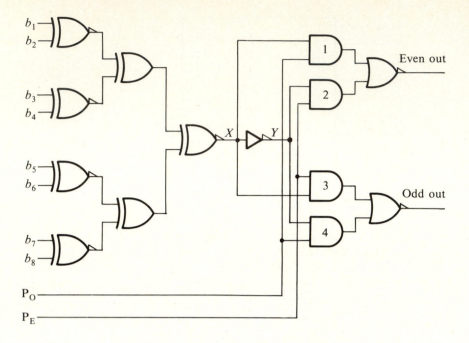

Inputs			Outputs	
Number of high data inputs b_1–b_8	P_E	P_O	Even out	Odd out
Even	H	L	H	L
Odd	H	L	L	H
Even	L	H	L	H
Odd	L	H	H	L
X	H	H	L	L
X	L	L	H	H

Figure 3.31 The 74180 parity generator/checker and function table.

generate the ninth bit of a transmitted word, while Odd Out will produce the error signal when odd parity of the received word is detected.

In checking parity for a 9-bit word, the ninth bit can be connected to P_O with an inverter connected from P_O to P_E. This will result in Odd Out being asserted high if odd parity is detected; thus this output will become the error indicator.

In generating parity, if P_E is low and P_O high, Even Out will be asserted high when the 8-bit code has odd parity. This output then contains the correct parity bit to be inserted as the ninth bit.

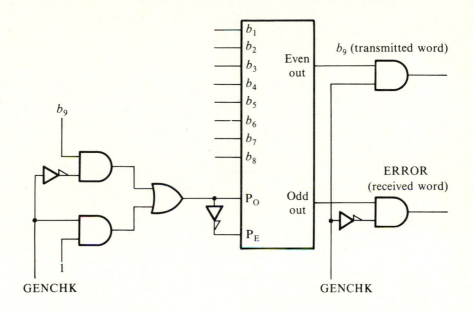

Figure 3.32 A parity generator/checker system.

Figure 3.32 shows the implementation of these ideas. For parity checking of a received word, GENCHK is taken low. When the GENCHK line is low, the ninth bit of the received word is gated to P_O and Odd Out is gated to the error output. This assumes that the other 8 bits of the received word appear at the bit inputs of the 74180. An odd parity input word will lead to the high assertion of the ERROR line. When GENCHK is taken high, a parity bit is to be generated. The 8 bits of the word to be transmitted must now appear at the bit inputs of the 74180. The pin P_O is now forced high and P_E is low. If odd parity of the word exists, the Even Out line will be asserted high, producing the necessary ninth parity bit. ∎

The 74180 is expandable in 8-bit increments. The Even Out and Odd Out pins of the first chip connect to the P_E and P_O input pins, respectively, of the second chip.

Drill Problems: Sec. 3.2B

1. Show how to connect the 74180 to check parity of a 7-bit word. When odd parity is not true, a high asserted output should occur.

2. Repeat problem 1 for a 12-bit input word.

C. The Encoder

A binary encoder has n output lines and 2^n input lines. The input lines are numbered consecutively from 0 to $2^n - 1$. When a single input line is asserted, the binary code corresponding to the line number appears at the output.

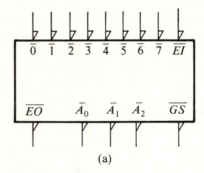

(a)

	Inputs									Outputs				
$\overline{EI}$	$\overline{0}$	$\overline{1}$	$\overline{2}$	$\overline{3}$	$\overline{4}$	$\overline{5}$	$\overline{6}$	$\overline{7}$		$\overline{GS}$	$\overline{A}_0$	$\overline{A}_1$	$\overline{A}_2$	$\overline{EO}$
H	θ	θ	θ	θ	θ	θ	θ	θ		H	H	H	H	H
L	H	H	H	H	H	H	H	H		H	H	H	H	L
L	θ	θ	θ	θ	θ	θ	θ	L		L	L	L	L	H
L	θ	θ	θ	θ	θ	θ	L	H		L	H	L	L	H
L	θ	θ	θ	θ	θ	L	H	H		L	L	H	L	H
L	θ	θ	θ	θ	L	H	H	H		L	H	H	L	H
L	θ	θ	θ	L	H	H	H	H		L	L	L	H	H
L	θ	θ	L	H	H	H	H	H		L	H	L	H	H
L	θ	L	H	H	H	H	H	H		L	L	H	H	H
L	L	H	H	H	H	H	H	H		L	H	H	H	H

Figure 3.33 (a) Logic symbol for 74148 encoder; (b) function table. (Although the IEEE/IEC standard would not use overbars for low asserted inputs, we will do so throughout the remainder of the text.)

The priority encoder is an extension of the binary encoder. This device allows several input lines to be asserted simultaneously while the output presents the binary code of the highest-numbered, asserted input line. For example, if input lines 3, 5, and 8 are asserted, the output code generated is 1000. The priority encoder was specifically designed to be used in conjunction with priority interrupt systems for computers. We note that if only one input line is asserted at a time, the priority encoder will function as a standard encoder.

The 74148 is an 8-input priority encoder that can be used in several applications. This particular device has low assertion input and output signals. The 74148 chip is shown in Fig. 3.33 along with its function table. When $\overline{EI}$, the enable input, is high, the chip is disabled and all outputs are at the high level (deasserted). Asserting $\overline{EI}$ by taking it low enables the operation of the chip. If $\overline{EI}$ is low, but all inputs are high, $\overline{EO}$ is low. This is the only condition for which this enable output is low. As long as the chip is enabled, the output binary code will indicate the number of the highest numbered line asserted. The output $\overline{GS}$ is asserted when any input line is active.

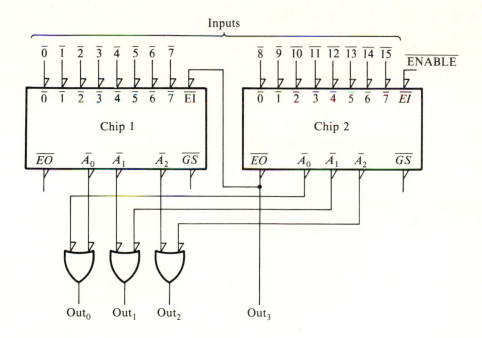

Figure 3.34 A 16-line to 4-line encoder.

With these additional outputs the 74148 can be used to create larger encoders. For example, the 16-line to 4-line encoder with high asserted outputs can be constructed as shown in Fig. 3.34. When an input line numbered from 0 to 7 is asserted, chip 1 will present the corresponding binary code to the OR gates. This will drive Out_0, Out_1, and Out_2 while $\overline{EO}$ remains low since no input line is asserted on this chip. If an input line from 8 to 15 is asserted, the $\overline{EO}$ of chip 2 goes high to disable chip 1. This output also provides the MSB of the output code. Thus, if line 13 is asserted, chip 2 develops the code 101 and drives the OR gates to produce $Out_0 = 1$, $Out_1 = 0$, and $Out_2 = 1$. The value of Out_3 is provided by $\overline{EO}$ which is now high, generating a 1 for the MSB.

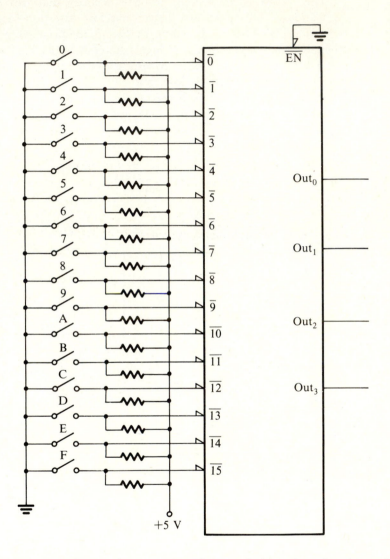

Figure 3.35 An encoder used to generate binary-coded hex for a hex keypad.

One application of a 16-line to 4-line encoder is in generating the proper code for a hex keypad. The 16 keys on the pad each represent a hex number. When depressed, each key closes a corresponding set of switch contacts. When the switch contacts close, the encoder input is driven low. The encoder then generates the binary code corresponding to the depressed key. Figure 3.35 shows this arrangement.

Drill Problem: Sec. 3.2C

Show how to connect the 74147, 10-line-to-4-line priority encoder, to a decimal keypad.

D. The Comparator

It is often necessary to compare two binary numbers to determine which is larger. The basic method of comparison starts by comparing the MSB of input A to the MSB of input B. If one of these bits is 1 and the other 0, the process is completed and the number containing 1 as the MSB is identified as the largest number. If the MSB of A equals the MSB of B, then the next most significant bits are compared. This process continues until a bit of one number differs from the corresponding bit of the other. Rather than make these bit comparisons successively, the comparator is implemented so that all comparisons are made after only two gate delays.

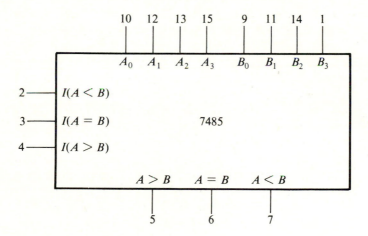

Figure 3.36 The 7485 4-bit comparator.

There are three outputs for a comparator. One is asserted if $A = B$, one is asserted if $A > B$, and the third is asserted when $A < B$. In addition to these outputs and the inputs, other terminals are often provided on an MSI chip to allow expansion to larger word comparison. The 74LS85 4-bit comparator is shown in Fig. 3.36. The inputs $I(A < B)$, $I(A = B)$, and $I(A > B)$ are considered inputs from a comparison of bits of lesser significance. In a 4-bit word comparison, $I(A < B)$ and $I(A > B)$ should be tied low while the $I(A = B)$ input should be tied high. Figure 3.37 shows an 8-bit comparator based on the 7485.

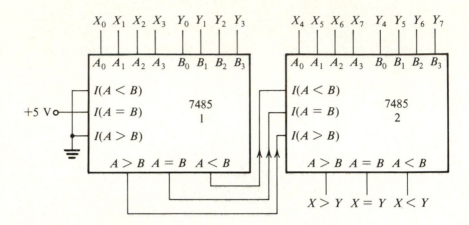

Figure 3.37 An 8-bit comparator.

IC number 2 compares the most significant 4 bits, or nibble. If this nibble shows an unequal value between X and Y, the correct output is immediately determined. If X and Y are identical in the four most significant bits, the input from the four least significant bit comparison determines the output. When large words are compared, the method of Fig. 3.37 suffers from a time-delay problem. Four comparators can only present a valid result after an overall propagation delay equal to four times the delay of an individual comparator. This figure is a "worst case" value based on the three higher-order nibbles being equal while the nibble of lowest significance determines the result. One IC delay time is required before comparator 2 receives an input from comparator 1. Two delay times are completed before the inputs to comparator 3 are stable. Three delay times are spent before comparator 4 receives stable inputs. The output of comparator 4 cannot be considered valid until four propagation delay times have been expended.

A parallel expansion scheme is recommended for comparison of larger words when high speed is required. Comparison of two 24-bit code words can be accomplished by the configuration of Fig. 3.38 [3]. The output comparison is completed in this case after two IC propagation delay times.

The output comparator is driven by the five parallel comparators. The output of the upper comparator is the most important result and consequently drives the most significant bits of the output comparator. If $A_{23}A_{22}A_{21}A_{20}A_{19} < B_{23}B_{22}B_{21}B_{20}B_{19}$, then the $A < B$ output of comparator 5 will drive the B_3 input of comparator 6, resulting in an overall output of $A < B$. If the A_3 input of comparator 6 is driven, the overall output is $A > B$. When comparator 5 produces an output of $A = B$, neither A_3 nor B_3 is asserted at the input of comparator 6. In this case comparator 4 will

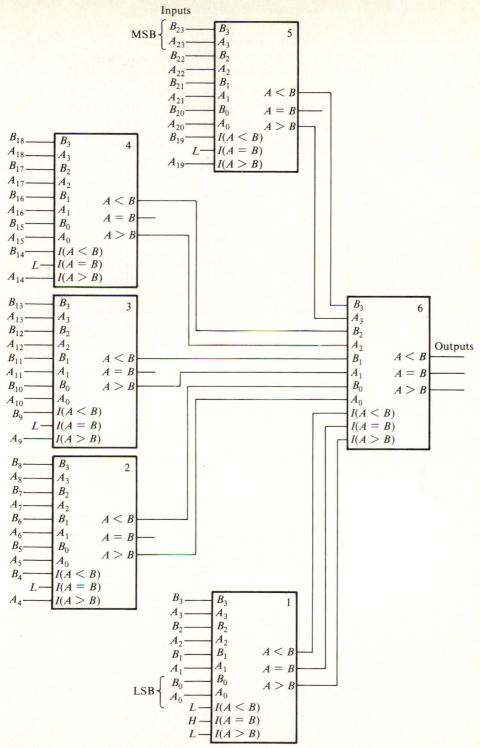

Figure 3.38 A 24-bit comparator.

determine the circuit output assuming $A_{18}A_{17}A_{16}A_{15}A_{14}$ is not equal to $B_{18}B_{17}B_{16}B_{15}B_{14}$. We note that the cascading inputs $I(A < B)$, $I(A = B)$, and $I(A > B)$ are used as normal inputs for each comparator except comparator 1. These inputs determine the output in exactly the same way a fifth bit comparison would, that is, they do not affect the output unless the higher four A and B inputs are equal.

There are several applications of comparators in digital systems. One common use is that of determining if a digital code or word falls within a certain range of values. The system of Fig. 3.39 determines if an input word is within the range 00100000 to 10000000.

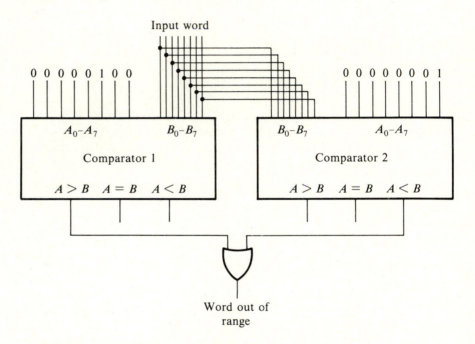

Figure 3.39 Circuit for determining if input word is out of range.

If the input word is below the boundary 00100000, the $A > B$ output of comparator 1 asserts the system output. When the input word is above the boundary 10000000, the $A < B$ output of comparator 2 drives the system output. For input words equal to or between the boundary values, the system output is not asserted.

Drill Problem: Sec. 3.2D

Show how to connect two 7485 chips to construct a 6-bit comparator.

3.3 ARITHMETIC CIRCUITS

Section overview: This section considers methods and circuits for adding or subtracting two binary numbers. The most popular methods for subtraction with digital systems use complementing schemes; thus complement formation is discussed. The implementation of 8-bit adding/subtracting units is covered, and the section concludes with a brief coverage of a standard MSI arithmetic/logic unit chip.

The original development of the digital computer was sparked by the desire to make rapid arithmetic calculations. Although computers can perform many other types of operation, the very name of the device indicates its mathematical capability. Since the computer deals with binary or binary-related codes, the arithmetic performed is also done with binary circuits. This section will first consider the binary adder and then proceed to a discussion of methods of subtraction. After developing these topics, the arithmetic logic unit or ALU will be considered.

A. The Adder

In adding two single-digit binary numbers, we are concerned with four possible combinations. This information is expressed in the truth table of Fig. 3.40. When both A and B are equal to 0, the sum output S and carry output C are both equal to 0. If either A or B is equal to 1, but not both, $S = 1$ and $C = 0$. When $A = B = 1$, the sum output is $S = 0$ while $C = 1$. A binary circuit that produces these outputs is called a half-adder and is shown in Fig. 3.41.

Inputs		Outputs	
A	B	S	C
0	0	0	0
0	1	1	0
1	0	1	0
1	1	0	1

Figure 3.40 Truth table for a single-digit adder.

Of course a single digit adder has little practical value by itself, but it can be used to construct a several-digit adder. In adding a multibit number, only the LSB can be added with a half-adder. All other columns must consider the carry from the column of next least significance in addition to the

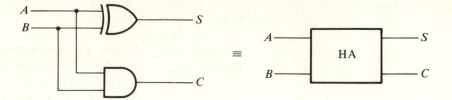

Figure 3.41 A half-adder.

two inputs. A circuit to consider these three inputs and produce the sum and carry outputs is called a full adder. The truth table for the full adder is shown in Fig. 3.42. The full adder can be implemented with two half-adders and an OR gate as demonstrated in Fig. 3.43.

A	B	C_{in}	S	C_{out}
0	0	0	0	0
0	0	1	1	0
0	1	0	1	0
0	1	1	0	1
1	0	0	1	0
1	0	1	0	1
1	1	0	0	1
1	1	1	1	1

Figure 3.42 Truth table for full adder.

The truth table yields the two equations

$$S = \bar{A}\bar{B}C_{in} + \bar{A}B\bar{C}_{in} + A\bar{B}\bar{C}_{in} + ABC_{in}$$

and

$$C_{out} = \bar{A}BC_{in} + A\bar{B}C_{in} + AB\bar{C}_{in} + ABC_{in}$$

Representing each half-adder by its equivalent gates as shown in Fig. 3.43 allows us to verify that these equations are implemented by this circuit. The equation for S can be written as

$$S = C_{in} \oplus (A \oplus B) = C_{in} \oplus (A\bar{B} + \bar{A}B)$$

$$= (A\bar{B} + \bar{A}B)\bar{C}_{in} + (\overline{A\bar{B} + \bar{A}B})C_{in}$$

$$= A\bar{B}\bar{C}_{in} + \bar{A}B\bar{C}_{in} + \bar{A}\bar{B}C_{in} + ABC_{in}$$

This circuit can also be used to find that

$$C_{out} = C_{in}(A \oplus B) + AB = A\bar{B}C_{in} + \bar{A}BC_{in} + AB\bar{C}_{in} + ABC_{in}$$

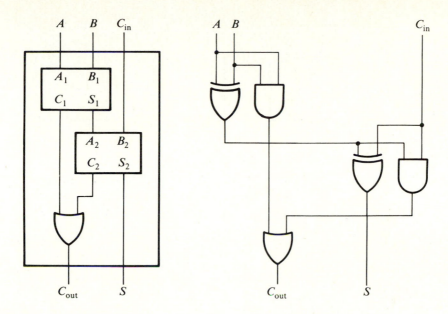

Figure 3.43 The full adder.

Thus, two half-adders and an OR gate can be used to construct the full adder.

For slower speed adders an n-bit adder can be constructed from n full adders. For higher speed systems the propagation delay of the circuit can become excessive. For the "worst case" of two numbers containing all 1s, the carry propagates from the LSB through the MSB. This requires one propagation delay time for each bit of the numbers being added. This is called the ripple carry method and is demonstrated in Fig. 3.44 for a 4-bit adder.

The problem of carry propagation time can be minimized by using a look-ahead-carry circuit. In this method the carry signals are generated by gates that look for those combinations of inputs that result in carrys. For example, the carry from the second column C_2 of a binary number should be zero if the first two columns add to 0, 1, 2, or 3. If the sum is 4, 5, or 6, a carry C_2 should be generated. The sums 4, 5, and 6 can occur in several different ways. These possibilities are tabulated in Fig. 3.45. We see that C_2 can be generated by a set of gates rather than being propagated through the first two adder stages.

The 7483 is a 4-bit full adder using look-ahead-carry on C_1, C_2, C_3, and C_4. In order to generate the carry for higher significance columns (C_3 and C_4) more gates are required. This does not present a serious problem for MSI circuits since many gates can be fabricated on each chip. The 7483 chip can be used as a building block to create larger adders such as 8-, 12-,

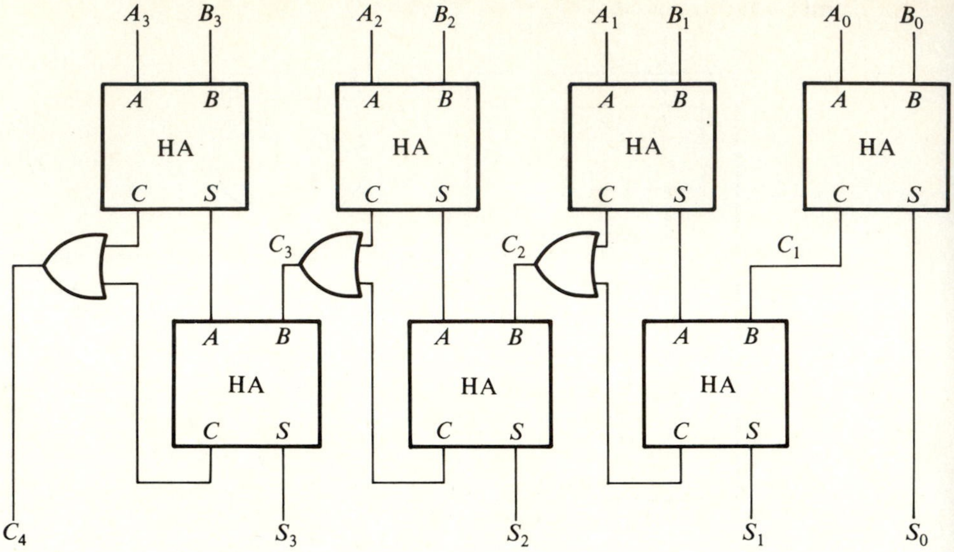

Figure 3.44 A 4-bit binary adder.

A_1	A_0	B_1	B_0	Decimal result	C_2
0	1	1	1	4	1
1	0	1	0	4	1
1	1	0	1	4	1
1	0	1	1	5	1
1	1	1	0	5	1
1	1	1	1	6	1

(a)

B_1B_0 \ A_1A_0	00	01	11	10
00	0	0	0	0
01	0	0	1	0
11	0	1	1	1
10	0	0	1	1

(b)

Figure 3.45 (a) Inputs that lead to $C_2 = 1$. (c) Truth table for C_2.

or 16-bit units. We note that for expansion purposes a carry input to the LSB is required.

Drill Problems: Sec. 3.3A

1. Show how to use two 74LS83 chips to construct a 6-bit adder.

2. Repeat problem 1 for a 12-bit adder.

B. Number Complements

Although subtracting circuits can be constructed they are rarely utilized in computers. In order to conserve circuitry, adders are used to create both sums and differences of binary numbers. Before we discuss how differences are formed using an adder, we must first consider the formation of complements in general terms.

There are two types of complement of concern to the logic designer. One is called the radix complement; the other is the diminished radix complement. The radix refers to the base of the number system being used. If we are dealing with decimal numbers, the radix is 10 and we speak of the 10's complement. The diminished radix complement for decimal numbers is the $(10-1)$'s or 9's complement. In binary the radix complement is the 2's complement while the diminished radix complement is the 1's complement.

By definition a complement is a quantity needed to make a thing complete. In number systems the "thing" we are attempting to complete depends on the number of columns with which we are dealing. For the 10's complement, a 4-column number leads to a "thing" equal to 10^4 or 10,000. Given a decimal number, for example, 6784, the 10's complement equals that number that adds to 6784 resulting in 10,000. In this case, the 10's complement is 3216. This information can be expressed numerically as

$$N_n + \overline{N}_n(10) = 10^n$$

where n is the maximum number of columns dealt with. Generally we know the number N_n along with the value of n and our problem is to find the complement $\overline{N}_n(10)$. Thus, the more useful form of the previous equation is

$$\overline{N}_n(10) = 10^n - N_n \quad \text{(10's complement)} \tag{3.1}$$

The 9's complement of a decimal number is formed by finding the number that adds to the given number to result in a sum of $10^n - 1$. The 9's complement of 6784 is 3215. Here the equation is

$$N_n + \overline{N}_n(9) = 10^n - 1$$

Solving for $\overline{N}_n(9)$ gives

$$\overline{N}_n(9) = 10^n - 1 - N_n \quad \text{(9's complement)} \tag{3.2}$$

There is another way to form the 9's complement of a number that is quite easy. Equation (3.2) suggests that we first calculate 10^n, subtract 1 from it, then subtract N_n. We can form this same value by considering each column individually, subtracting each number from 9. The 9's complement of 6784 can be formed by subtracting 4 from 9, 8 from 9, 7 from 9, and 6 from 9, giving 3215. If we were dealing with a 6-column number and wanted the complement of 6784, we simply add leading zeros to this number to result in 6 total columns. The 9's complement here would be $\overline{N}_n(9) = 993215$.

The 10's complement can be formed in the same manner as the 9's complement, by subtracting each single column digit from 9. After this is done, the result, which is the 9's complement, has 1 added to it to form the 10's complement. The 10's complement of 006784 is $\overline{N}_n(10) = 993216$.

The 2's and 1's complements are formed in much the same way as the 10's and 9's complements. For 2's complement the number and its complement must add to give 2^n, where n is the number of columns used. The equation is

$$N_n + \overline{N}_n(2) = 2^n$$

Arranging this equation to solve for the complement gives

$$\overline{N}_n(2) = 2^n - N_n \quad \text{(2's complement)} \tag{3.3}$$

The 2's complement of 1101 is 0011. The sum of these numbers is 10000 which equals 2^n.

The 1's complement differs from the 2's complement by 1, giving the equation

$$\overline{N}_n(1) = 2^n - 1 - N_n \quad \text{(1's complement)} \tag{3.4}$$

The 1's complement of 1101 is 0010. In this case the sum of N_n and $\overline{N}_n(1)$ equals 1111, which is equivalent to 10000 − 1.

The general equations for radix complement and diminished radix complement are

$$\overline{N}_n(r) = r^n - N_n \quad \text{(radix complement)} \tag{3.5}$$

and

$$\overline{N}_n(r - 1) = r^n - 1 - N_n \quad \text{(diminished radix complement)} \tag{3.6}$$

The 1's complement can be formed very easily by inverting each bit of the given number. The 1's complement of $N_8 = 10110010$ is formed by replacing each 0 with a 1 and each 1 with a 0. This gives $\overline{N}_8(1) = 01001101$. From a physical standpoint, a set of n output lines containing the binary code for N_n can be followed by a set of inverters to form $\overline{N}_n(1)$.

Another method of forming the 2's complement is simply to form the 1's complement and then add 1 to this result. The 2's complement of $N_8 = 10110010$ is found by adding 1 to the 1's complement, that is, $\overline{N}_8(2) = 01001101 + 1 = 01001110$.

■ *Example 3.3* Form the 10's, 9's, 2's, and 1's complements of the decimal numbers 213, 146, 37, and 13.

□ **Solution:** From Eq. (3.1) we form the 10's complement:

$$\overline{N}_3(10) = 10^3 - 213 = 787 \qquad \overline{N}_3(10) = 10^3 - 146 = 854$$

$$\overline{N}_2(10) = 10^2 - 37 = 63 \qquad \overline{N}_2(10) = 10^2 - 13 = 87$$

The 9's complement uses Eq. (3.2) or simply subtracts 1 from the 10's complement giving $\overline{N}_3(9) = 786$, $\overline{N}_3(9) = 853$, $\overline{N}_2(9) = 62$, and $\overline{N}_2(9) = 86$. Note that $213 + 786 = 999$, as does $146 + 853$. Note also that $37 + 62 = 99$ and $13 + 86 = 99$.

In order to form the 2's and 1's complements, we must convert the decimal numbers to binary giving $213_{10} = 11010101_2$, $146_{10} = 10010010_2$, $37_{10} = 100101_2$, and $13_{10} = 1101_2$. We will form the 1's complement first since this can be done by inspection. The four complements are formed by replacing each 0 with 1 and each 1 with 0 to give $\overline{N}(1) = 00101010$, 01101101, 011010, and 0010. The 2's complement simply adds 1 to the 1's complement to result in $\overline{N}(2) = 00101011$, 01101110, 011011, and 0011. ■

Drill Problems: Sec. 3.3B

1. Form the 1's and 2's complements of the decimal number 193.

2. Repeat problem 1 for the binary number 11010001.

3. Form the 9's complement of 2639.

4. Form the 10's complement of 2639.

C. Subtracting Positive Binary Numbers with Adders

Now that we have examined methods of complementing numbers, we move to the subject of using complements and adders to form differences. We will deal with binary numbers exclusively here since we are interested in methods of digital system subtraction.

The first method we will consider is used only for positive binary numbers. The basic idea here is to form the difference $A - B$ by forming $\overline{A} + B$. We can see that $A - B = \overline{\overline{A} + B}$ if we carry out the complementing operations. This method can be applied to either 1's or 2's complement systems.

2's Complement

Since this discussion is limited to 2's complement in the following paragraphs, we will drop the parentheses containing the radix for convenience.

The 2's complement of A is

$$\overline{A} = 2^n - A$$

Adding B results in

$$\overline{A} + B = 2^n - A + B$$

Complementing $\overline{A} + B$ gives

$$\overline{\overline{A} + B} = 2^n - (2^n - A + B) = A - B$$

This last operation yields the correct result if A is greater than or equal to B. If B is greater than A, we must replace the last step of the procedure. In this case, the result would be negative, but we can only deal with a positive result. Thus, when B is larger than A, we must detect this fact and form $B - A$ rather than $A - B$.

We note in the procedure that $\overline{A} + B = 2^n - A + B$. When B is greater than A, this result can be written as $\overline{A} + B = 2^n + B - A$. Since this number is greater than 2^n, it will contain a carry bit, that is, a 1 will be present in the $(n + 1)$st column. This amounts to a carry bit from the nth column. If A were larger than B, the result would not contain a 1 in this column since the result would be less than 2^n. We can therefore examine the carry bit to determine if B is larger than A by the presence of a 1. If this bit is 0, then A is larger than B.

When B is larger than A, we do not want to complement $\overline{A} + B$. Instead we want to simply subtract 2^n from this result to leave us with

$$\overline{A} + B - 2^n = B - A$$

We can effectively subtract 2^n simply by ignoring the $(n + 1)$st bit after it is used to determine that B is larger than A.

The final procedure to form $A - B$ then is as follows:

1. Complement A.

2. Add B to this complement.

3. If the carry bit is 0, complement the result.

4. If the carry bit is 1, the result is already present if we do not consider the carry bit as part of the result.

We will demonstrate this procedure for 8-bit words by forming $A - B$ for two cases: $A = 01011011$, $B = 00111011$ and $A = 01101100$, $B = 11001101$. For the first case using 2's complement,

$$\overline{A} = 10100101$$

and

$$\overline{A} + B = 11100000$$

Since there is no carry bit, we complete the procedure by forming

$$\overline{\overline{A} + B} = A - B = 00100000$$

In the second case,

$$\overline{A} = 10010100$$

and

$$\overline{A} + B = 101100001$$

The carry bit is 1; therefore we drop the carry bit which is equivalent to subtracting 2^n. The result is negative with a magnitude of 01100001.

Using the 1's complement system requires some modification of the procedure. If A is greater than B, then

$$\overline{A} + B = (2^n - 1) - (2^n - 1 - A + B) = A - B$$

When B is larger than A, this result would be negative. Instead of forming the complement of $\overline{A} + B$, this expression must be slightly modified. We see that

$$\overline{A} + B = 2^n - 1 - A + B = B - A + 2^n - 1$$

We need to drop the 2^n and add 1 to this expression to yield the correct result. This can be done by using the carry bit to add to the LSB. This is referred to as end-around-carry.

1's Complement

The procedure for 1's complement subtraction of one positive binary number B from another positive binary number A is as follows:

1. Complement A.

2. Add B to this complement.

3. If the carry bit is zero, complement the result.

4. If the carry bit is 1, add this bit to the LSB of the existing result.

We will again demonstrate the procedure for the two cases $A = 01011011$, $B = 00111011$ and $A = 01101100$, $B = 11001101$. For the first case

$$\overline{A} = 10100100$$

and

$$\overline{A} + B = 11011111$$

The absence of a carry bit dictates that this result be complemented to form the difference. This gives

$$\overline{\overline{A} + B} = A - B = 00100000$$

In the second case

$$\overline{A} = 10010011$$

and

$$\overline{A} + B = 101100000$$

The carry bit is 1 and must be added to the LSB recognizing that the result is the magnitude of a negative number. When this is done, the answer is found to be 01100001. A practical implementation of the 1's complement system is shown in Fig. 3.46. The LSB must have a carry input to utilize this scheme.

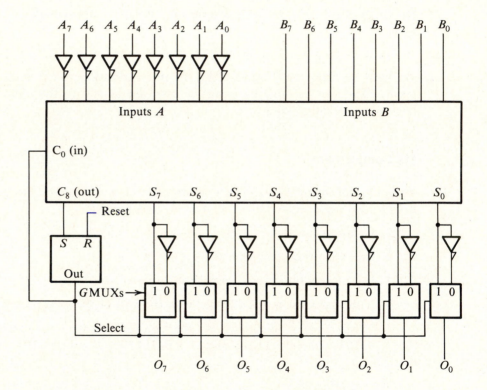

Figure 3.46 A subtracting circuit.

The unit labeled storage register may be an SR flip-flop which will be studied in detail in Chap. 4. A carry on C_8 will cause the storage register output to equal 1. This value will remain even if C_8 returns to a 0 value. The flip-flop will only return to a 0 output value when the reset input is asserted.

The process starts when the numbers A and B are presented to the input lines. The input inverters form the complement of A. The storage register must be reset prior to this time. If no carry is present, G is not asserted and the 8 inverted sum bits are gated through MUXs to the output. This produces the complement of the adder output. When a carry bit is generated from the process, C_8 sets the storage register to a 1. The output of this device is applied to the C_0 input to add 1 to the result. The signal G is now high to present the uncomplemented output values as the final results. The 1 in the storage register is also used to indicate that the result represents the magnitude of a negative number.

After we consider registers we will see that storage registers can be used to hold results and to easily form the 1's complement without requiring inverters. Since results are normally stored in registers anyway, a complementing register can be used to eliminate the output inverters and gating in Fig. 3.46.

Drill Problems: Sec. 3.3C

1. If $A = 10010110$ and $B = 01110011$, determine the values of G, S_0 to S_7, and O_0 to O_7 for the circuit of Fig. 3.46.

2. Repeat problem 1 if $A = 01110011$ and $B = 10010110$.

D. Adding and Subtracting Signed Numbers

A plus or minus sign can be included in a binary number by expressing this sign as the leftmost bit. A 1 is used to represent a minus while a 0 represents a plus. Using this method allows us to express $+13$, as 01101 and -13 as 11101. If the leftmost bit indicates the sign and the remaining bits express the magnitude of the number, the method is called signed magnitude. There are two other methods that are of more value in calculating applications. These are the 1's complement representation and the 2's complement representation of signed numbers. In both these schemes positive numbers are expressed in the same way as are signed magnitude numbers. Negative numbers continue to use a 1 for the sign bit, but the number is expressed in complement form. Some examples of 1's complement numbers are $+5 = 0101$, $-5 = 1010$, $+37 = 0100101$, $-37 = 1011010$, $+127 = 01111111$, and $-127 = 10000000$. The same numbers in 2's complement form are $+5 = 0101$, $-5 = 1011$, $+37 = 0100101$, $-37 = 1011011$, $+127 = 01111111$, and $-127 = 10000001$.

The 1's complement form leads to some ambiguity since zero can be represented as either $+0$ or -0, leading to 00000 or 11111, for 5-bit words. Any circuit using 1's complement arithmetic must account for both possible forms. In 2's complement form, only 00000 applies as a representation of zero since a negative 0 cannot be represented.

Microprocessors use either 1's or 2's complement arithmetic. Because of the popularity of the 8-bit microprocessor, we will limit the following discussion to 8-bit words.

2's Complement Arithmetic

We will mention that the system used for both 1's and 2's complement arithmetic actually adds the sign bit as if it were part of the number. Thus, as we form the following sums, note that the leftmost bit is added even though it represents the sign.

If B is to be subtracted from A, this number is first complemented and then added to A. This results in

$$A + \bar{B} = A + 2^n - B = 2^n + A - B$$

When both A and B are positive and the magnitude of A is greater than that of B, $A - B$ will be positive. In this case, the result will exceed 2^n and therefore will carry a bit into the sign column (the eighth bit). The sign column will contain a 0 and a 1 since A is positive and B is negative. Adding the carry bit results in a 0 for the sign bit and carries a 1 to the ninth column. If we ignore this 1, it is equivalent to subtracting 2^n from the result, leading to the difference $A - B$.

If B has a larger magnitude than that of A, the sum $A + \bar{B} = 2^n - (B - A)$. This result is the 2's complement form of $B - A$ which again is the correct result. Note here that a 1 will not carry into the eighth column.

The adder simply adds the 8 bits and ignores the carry to the ninth column. If the eighth bit contains a 0, the result is positive. A 1 in this position means the result is negative and therefore appearing in 2's complement form. The only exception to this is when we add two positive numbers or two negative numbers and the result is out of range. We will discuss this case as an example in the following paragraphs.

In order to develop a set of rules or an algorithm for doing 2's complement arithmetic, we must consider six different situations. We will add the following six pairs of numbers using 2's complement numbers: $+51$ and $+32$, -51 and -32, $+51$ and -32, -51 and $+32$, $+68$ and $+86$, and -68 and -86.

The first pair gives

```
+51    00110011
+32    00100000
       01010011    = +83
```

When two positive numbers are added, the result is given directly if the resulting sum is less than 128.

The two negative numbers give

$$
\begin{array}{rcl}
-51 & = & 11001101 \\
-32 & = & 11100000 \\
\hline
& & 110101101
\end{array}
$$

If the carry to the ninth column is ignored, the result is the correct number, expressed in 2's complement form. This result is converted as follows:

$$10101101 \ = \ -1010011 \ = \ -83$$

The next example gives

$$
\begin{array}{rcl}
+51 & = & 00110011 \\
-32 & = & 11100000 \\
\hline
& & 100010011
\end{array}
$$

Ignoring the ninth bit gives a result of

$$00010011 \ = \ +19$$

The next pair of numbers are

$$
\begin{array}{rcl}
-51 & = & 11001101 \\
+32 & = & 00100000 \\
\hline
& & 11101101
\end{array}
$$

This result is the correct negative number in 2's complement form since

$$11101101 \ = \ -0010011 \ = \ -19$$

The last two cases are numbers that yield out-of-range results. The two sums are

$$
\begin{array}{rcl}
+68 & = & 01000100 \\
+86 & = & 01010110 \\
\hline
& & 10011010
\end{array}
$$

and

$$
\begin{array}{rcl}
-68 & = & 10111100 \\
-86 & = & 10101010 \\
\hline
& & 101100110
\end{array}
$$

The actual numbers resulting are unimportant. The significant point is that when two numbers of similar sign are added and the sum exceeds $2^7 - 1$, the resulting sign bit is opposite to that of the two numbers.

Assuming each negative operand is presented in 2's complement form, the algorithm for addition is:

1. Add the two operands.

2. Check the two sign bits of the operands and compare to the sign bit of the result. If the two operand signs are the same, but the resulting sign is different, the result is out of range. In this case an overflow flag should be asserted. If the two operand signs are different, the result is valid. If the two operand signs and the sign of the result are the same, the result is also valid.

A system to implement the ideas discussed above is shown in Fig. 3.47. The NEXOR and EXOR check for out-of-range answers, generating an overflow flag if such occurs.

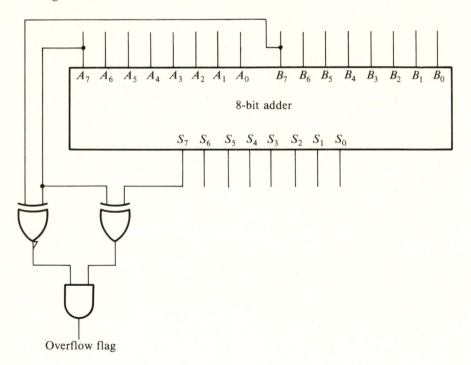

Figure 3.47 A 2's complement arithmetic circuit.

1's Complement Arithmetic

Use of 1's complement arithmetic is very similar to the use of 2's complement. To form the difference $A - B$, the number B is complemented and added to A, giving $A + \bar{B} = A + (2^n - 1) - B$. Just as in the 2's complement case, when the magnitude of A is greater than B, a carry bit propagates through the sign column to the ninth bit. If this bit is added to the least significant bit (end-around-carry), it is equivalent to subtracting 2^n and adding 1. With these modifications, the result is $A - B$.

When the magnitude of B is greater than that of A, the result is $A + \bar{B} = A + (2^n - 1) - B = (2^n - 1) - (B - A)$. This is the 1's complement of $B - A$ which is the correct result. Since 1's complement arithmetic is so similar to 2's complement, we will not discuss its implementation.

The significant points relative to complement arithmetic are:

1. Differences of numbers can be formed by complementing and then adding.

2. The adder always performs the same function whether a sum or difference results.

3. The sign bit is treated by the adder in the same manner as all other bits, thus allowing signed or unsigned numbers to be added with this unit.

Drill Problems: Sec. 3.3D

1. If $A = 10010001$ and $B = 00011011$ in the circuit of Fig. 3.47, indicate the values of S_0 to S_7 and the overflow flag.

2. Repeat problem 1 for $A = 00011011$ and $B = 10010001$.

3. Repeat problem 1 for $A = 10010001$ and $B = 11111000$.

E. The Arithmetic Logic Unit

MSI circuits are available that combine many arithmetic and logic functions on a single chip. These devices incorporate principles discussed in previous sections resulting in a multipurpose circuit at a very low cost. The 74181 is a 4-bit arithmetic logic circuit or ALU that can be used separately for 4-bit word applications or as a building block in larger word systems.

This particular chip performs the basic arithmetic operations plus several other occasionally helpful operations. A set of select lines is provided to choose which function the chip is to perform. Addition, subtraction, comparison, complementation, and doubling are included in the set of arithmetic operations available. ORing, NORing, ANDing, NANDing, EXCLUSIVE ORing, and logical comparing are included in the set of logical operations performed by the chip.

When used as a component in larger word systems, a look ahead carry generator chip, the 74182, can be used with the 74181 to speed up operation. Typical chip propagation delays for conventional TTL range from 20 to 50 ns depending on the operation performed.

SUMMARY OF SIGNIFICANT POINTS IN CHAP. 3

1. Logic functions can be realized by MSI circuits such as MUXs, decoders, ROMs, and PLAs. Although gate circuits may lead to a lower component cost, MSI circuits can yield a lower chip count, require less wiring, or allow simpler function changes.

2. The number of inputs and outputs of MUX and decoder circuits are limited by the pins available on MSI circuit chips. These smaller elements can be used as building blocks to create larger MUXs and decoders.

3. Several standard functions are implemented on IC chips. Among these are the 7-segment LED driver chip, the parity generator/checker, the encoder, and the comparator. It is important to understand how to use these standard chips.

4. Some digital systems can add and subtract binary numbers. Subtraction can be performed by an adder if complements are used.

REFERENCES AND SUGGESTED READING

1. A.M. Abd-alla and A. C. Meltzer, *Principles of Digital Computer Design, vol. 1.* Englewood Cliffs, N.J.: Prentice-Hall, 1976, chap. 2.
2. W.I. Fletcher, *An Engineering Approach to Digital Design.* Englewood Cliffs, N.J.: Prentice-Hall, 1980, chap. 1.
3. Signetics Staff, *Signetics Logic-TTL Data Manual.* Sunnyvale, Calif.: Signetics, 1982.

CHAP. 3 PROBLEMS

Sec. 3.1A

* **3.1** Realize the expression $Y = \overline{A}\overline{B}C\overline{D} + \overline{A}BCD + \overline{A}BC\overline{D} + \overline{A}BC\overline{D} + A\overline{B}C\overline{D} + A\overline{B}C\overline{D} + AB\overline{C}\overline{D} + ABC\overline{D}$ using a 16:1 MUX.

3.2 Realize the expression in Prob. 3.1 using an 8:1 MUX.

3.3 Use a 4:1 MUX preceded by a minimal number of gates to realize Y of Prob. 3.1.

3.4 Realize the expression $X = \overline{A}\overline{B}\overline{C} + \overline{A}B\overline{C} + A\overline{B}\overline{C} + A\overline{B}C$ with three 2:1 MUXs.

* **3.5** On a quad 2:1 MUX chip, all four select lines are connected to a common pin to minimize the number of pins on the chip. Using this type of chip and a minimal number of gates, realize the function X of Prob. 3.4.

3.6 Use seven 8:1 MUXs to realize the 7-segment LED driver of Prob. 2.36.

3.7 Use seven 4:1 MUXs preceded by gates to realize the 7-segment LED driver of Prob. 2.36.

Sec. 3.1B

* **3.8** Realize the expression Y of Prob. 3.1 using a 4-line to 16-line decoder.

3.9 Realize the expression Y of Prob. 3.1 and the expression X of Prob. 3.4 using a 4-line to 16-line decoder.

3.10 Show how to construct a 5-line to 32-line decoder from standard MSI decoders.

3.11 Use a 4-line to 16-line decoder and gates to realize the 7-segment driver of Prob. 2.36.

Sec. 3.1C

3.12 Show the contents of all locations of a 16×4 ROM that will accomplish the code conversion of Example 3.1.

* **3.13** Show the contents of all locations of a 16×4 ROM that will produce $F1$, $F2$, $F3$, and $F4$ of Fig. 3.17.

Sec. 3.2

3.14 Show the pin diagram of four 7447A chips driving 7-segment displays, used to display numbers from .01 to 99.99. Connect to blank leading and trailing zeros.

3.15 Show how to connect the 74180 to check even parity of a 16-bit word. When even parity is not true, a high asserted output should occur.

* **3.16** Design an 8-line to 3-line encoder using gates.

3.17 Design an 8-line to 3-line priority encoder using gates.

3.18 Using 74148 chips, design a 10-line to 4-line BCD encoder.

Sec. 3.3

3.19 Write the expression for C_3 of a 4-bit look-ahead-carry adder. C_3 is the carry from the third to the fourth column.

* **3.20** Find the 10's complement of 1967.

3.21 Form the 1's complement of 33_{10} using an 8-bit number.

* **3.22** Form the 2's complement of 33_{10} using an 8-bit number.

3.23 If $A = 10010110$ and $B = 00110010$, use 2's complement arithmetic to form $A - B$. Assume that A and B are positive, unsigned numbers.

3.24 Repeat Prob. 3.23 using 1's complement arithmetic.

* **3.25** Repeat Prob. 3.23 if A and B are both signed numbers.

3.26 Repeat Prob. 3.23 using 1's complement arithmetic if A and B are both signed numbers.

CHAPTER
4
Multivibrators, Counters, and Registers *

The circuits considered in the first three chapters are combinational logic circuits. As explained earlier, these systems have outputs that depend only on the applied input signals. For a given set of inputs, the outputs will always be the same. There are many applications of combinational circuits as indicated in the last two chapters.

---

* Some material in this chapter comes from the text by David J. Comer, *Electronic Design with Integrated Circuits*, © 1981, Addison-Wesley, Reading, Mass. Used with permission.

There is another class of circuits, called sequential circuits, that behave quite differently from combinational circuits. A sequential circuit depends not only on the present inputs but also on past history of the inputs and time. This chapter will consider an important class of sequential circuits called multivibrators. Multivibrators are rather simple sequential circuits designed for specific applications. Certain types of multivibrator, particularly flip-flops, are used as building blocks to construct larger sequential circuits called state machines. In Chaps. 5 to 7 we will discuss state machine design after considering the simple, but useful multivibrator in this chapter.

4.1 TYPES OF MULTIVIBRATOR

Section overview: Three types of multivibrator are treated in this section: the monostable, the astable, and the bistable. The monostable and astable are popular in timing applications and are discussed with this application in mind. The bistable multivibrator contains a very important circuit subclass called the flip-flop. This component is very important in state machine design and is considered in detail in this section.

There are three types of multivibrator: the monostable, the bistable, and the astable. Monostable multivibrators are also referred to as one-shot or single-shot multivibrators. The astable is often called a free-running multivibrator. The bistable multivibrator has several subclasses of circuit such as the flip-flop and the Schmitt trigger.

The difference among the three types of multivibrator is in the number of stable states of each circuit. The output of the monostable exists in only one stable state, the bistable output has two stable states, and the astable has none. Figure 4.1 shows input and output waveforms for each of the three types.

The one-shot multivibrator requires an input to initiate an output. Once the output has moved to the unstable state (high in this case), it will remain there for some time T before returning to the stable state. The circuit will remain in the stable state until another input signal is applied. The time T is referred to as the period of the one-shot. This value can be determined by certain components of the circuit. The bistable flip-flop moves from one stable state to another upon application of an input signal. The output remains in this second stable state until another input signal is applied at which time the output resumes the original stable state. The astable requires no input and continually moves from one quasistable state to another at fixed intervals of time. These half-periods are also determined by circuit components. We note that the outputs of the first two multivibrators are functions of input, whereas the astable multivibrator output is a function of time only.

Multivibrators play an important role in digital systems. One-shot and astable circuits are important in waveform generation while bistable flip-flops

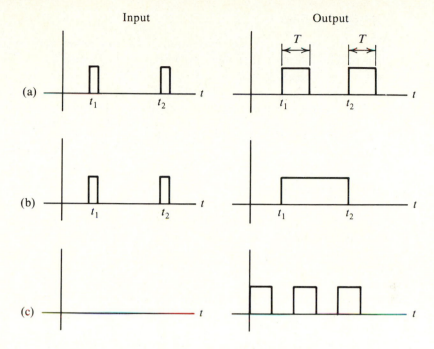

Figure 4.1 Multivibrator waveforms: (a) monostable; (b) bistable flip-flop; (c) astable.

are very important in state machines, counters, shift registers, and other digital systems. The flip-flop is perhaps the most important single class of circuit in the entire digital field.

A. The Bistable Multivibrator

The flip-flop is used much more than any other type of bistable circuit. This element will be considered in various forms before we proceed to the Schmitt trigger circuit, a bistable of importance in a limited number of applications.

Flip-flops operate in one of two modes: direct or clocked. Direct-mode flip-flops respond directly to applied inputs. The outputs change as a direct result of the inputs. In clocked flip-flops, a change of input has no effect on the output until a clock signal is applied. When a clock transition from one voltage level to another occurs, the output changes to a value dictated by the input levels. The SR flip-flop and gated D flip-flop are examples of direct-mode operation devices. The toggle, JK, and clocked D flip-flops are clocked devices. Some clocked flip-flops also have direct inputs, allowing operation in either mode.

The *SR* Flip-Flop

This form of flip-flop has two input lines and either one or two outputs. The output line that is always present is often labeled Q. Generally, a second output called $\overline{Q}$ will also be present. One input line is used to set the device to the $Q = 1$ state while the other input sets the device to the $Q = 0$ state. These inputs are called the SET or S input and the RESET or R input; hence the name *SR* flip-flop. The outputs Q and $\overline{Q}$ make up what is called a double-rail output; that is, $\overline{Q}$ always equals the complement of Q except when both S and R are asserted simultaneously. This condition is normally avoided in actual circuit operation. The *SR* flip-flop requires high assertion inputs to operate properly. The low assertion input device is called the $\overline{SR}$ flip-flop. The symbols for both active high input and active low input flip-flop are shown in Fig. 4.2 along with the function and characteristic tables for each.

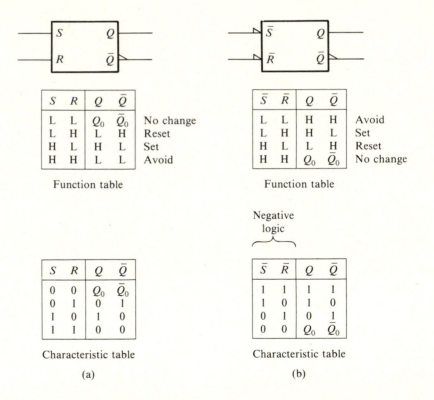

S	R	Q	$\overline{Q}$	
L	L	Q_0	$\overline{Q}_0$	No change
L	H	L	H	Reset
H	L	H	L	Set
H	H	L	L	Avoid

Function table

$\overline{S}$	$\overline{R}$	Q	$\overline{Q}$	
L	L	H	H	Avoid
L	H	H	L	Set
H	L	L	H	Reset
H	H	Q_0	$\overline{Q}_0$	No change

Function table

Negative logic

S	R	Q	$\overline{Q}$
0	0	Q_0	$\overline{Q}_0$
0	1	0	1
1	0	1	0
1	1	0	0

Characteristic table

(a)

$\overline{S}$	$\overline{R}$	Q	$\overline{Q}$
1	1	1	1
1	0	1	0
0	1	0	1
0	0	Q_0	$\overline{Q}_0$

Characteristic table

(b)

Figure 4.2 The set-reset flip-flop and function tables: (a) active high inputs (*SR*); (b) active low inputs ($\overline{SR}$).

The function table is similar in purpose to the function table used for combinational circuits. The flip-flop table implies a time dependence, however, while the function table for a gate does not. Outputs Q and $\overline{Q}$

refer to conditions applying after the inputs S and R take on particular values. Q_0 and $\overline{Q}_0$ refer to conditions before the input conditions are applied. For example, row 1 of the SR function table tells us that the state of Q and $\overline{Q}$ prior to applying $S = R = L$ will remain after these input conditions are applied.

For the SR flip-flop, inputs are normally kept low until a change in output state is desired. To guarantee that a 1 is set into the device ($Q = 1$), S is taken high while R remains low. The input S can then be returned to a low value and Q will continue at the high level. In order to reset the circuit, R is taken high while S remains low. When this occurs, $Q = 0$ and this condition will persist even after R returns to a low level. Because the flip-flop remains in a particular state after the inputs move to the no-change condition, it is also called a latch.

From the characteristic table we see that Q may equal 0 or 1 for the input condition of $S = R = 0$, depending on the input conditions prior to setting both inputs low. This is a distinct difference from combinational circuits which always produce a unique output for a given set of inputs.

When both S and R are asserted, Q and $\overline{Q}$ are driven low. If both S and R are returned simultaneously to the low level, it is impossible to determine if Q will assume the 1 state or the 0 state. Generally we avoid the input condition of $S = R = 1$ unless the following input condition is either $S = 1$ and $R = 0$ or $S = 0$ and $R = 1$. In either of these cases, the output will be uniquely determined.

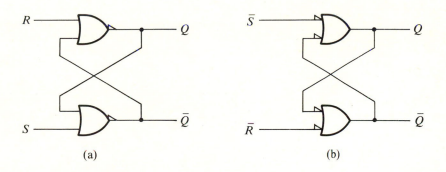

Figure 4.3 (a) An SR latch; (b) An $\overline{S}\,\overline{R}$ latch.

The SR flip-flop can be constructed from NOR gates as shown in Fig. 4.3. This figure also shows the $\overline{S}\,\overline{R}$ device constructed from NAND gates. For the SR latch, applying a 1 on S (high level) and a 0 on R results in $\overline{Q}$ being asserted low. Since $\overline{Q}$ drives an input to the other NOR gate, both inputs are low, resulting in a high value for Q. When S returns to a low value, the high level of Q continues to drive the lower NOR gate to keep $\overline{Q}$ at a low level. If R is now asserted, the upper NOR gate drives Q low which now allows $\bar{Q}$ to return to a high level. The $\bar{S}\bar{R}$ flip-flop behaves in a similar way except the inputs must be asserted low.

An equation describing the SR latch is

$$Q = S + \bar{R}Q_0$$

This tells us that Q will be asserted if we assert S or if R is not asserted and Q has previously been asserted.

A popular SSI latch is the 74279 which is a quad $\bar{S}\bar{R}$ latch. Two of the latches have an extra set input on them. This circuit is shown in Fig. 4.4.

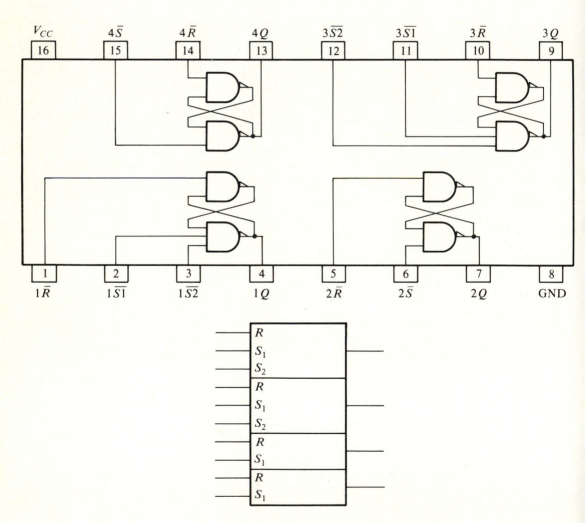

Figure 4.4 The 74279 TTL quad $\bar{S}\bar{R}$ latch and IEEE/IEC symbol.

The *SR* flip-flop can function as a 1-bit memory. To store a 1 in this device, the *S* input is asserted. This input can then be deasserted, and the flip-flop remains in the *Q* = 1 state as long as the two inputs are not asserted. The 0 bit can be stored by asserting *R*. Several *SR* flip-flops can be combined to form a storage register for several bits as we shall see later.

A very common application of the *SR* or $\overline{SR}$ flip-flop is that of debouncing switch contacts in digital systems. Toggle or push-button switches do not make sudden, firm, mechanical contact as a switch is closed or opened. If the output of the switch of Fig. 4.5(a) were observed on an oscilloscope with proper triggering and time-scale setting, the waveform would typically appear as shown in Fig. 4.5(b). This contact bounce lasts 10 to 20 ms for the small switches used in electronic circuits.

If the switch output were connected to a gate or inverter input, the output would change state several times as the electrical signal bounced through the trigger level of the gate. In some systems this multiple switching at the output of the gate cannot be tolerated. If the signal were applied to the *S* input of an *SR* latch, the latch would set the first time the switch output exceeded the 1-state level. If the switch output then dropped low again, the latch would remain set in that state until the *S* signal returned to the low level and the *R* input went high. Thus, a clean signal with only one transition would result at the latch output. A latch can be used to debounce a switch for both positive and negative transitions as shown in Fig. 4.6. In this instance the $\overline{SR}$ latch is used. When the switch is in the reset position, $\overline{R}$ is at the 0 V level and $\overline{S}$ is at $+5$ V due to the pull-up resistor. As the switch changes to the set position, $\overline{R}$ moves to a voltage of $+5$ V and $\overline{S}$ is pulled toward ground. Although this voltage on $\overline{S}$ may fluctuate rapidly, once the $\overline{S}$ input drops to the low logic level, the latch switches to Q = 1 creating a single transition. When the switch moves from the set to the reset position, a single transition to Q = 0 also occurs. The waveform is shown in Fig. 4.7.

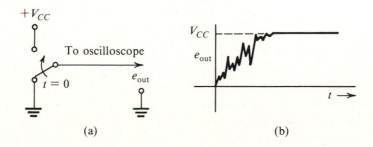

Figure 4.5 (a) Switch as state is changed. (b) Typical output.

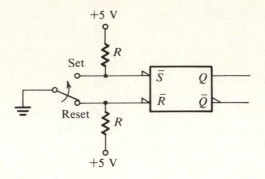

Figure 4.6 Switch debouncer.

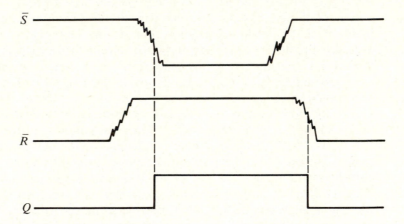

Figure 4.7 Waveform of switch debouncer.

A useful variation of the SR latch is the gated SR latch of Fig. 4.8. This circuit will not allow the S or R inputs to affect the output when G is low; when G is high, the circuit behaves like a normal SR latch.

The *D*-Gated Latch

In some applications the S and R inputs will always be complementary, that is, $S = 0$ when $R = 1$, and when $S = 1$, $R = 0$; this can be expressed as $S = \overline{R}$. Because pin connections on an IC chip can be minimized for this situation, this circuit has become a popular device known as the D-gated latch. Figure 4.9 shows the schematic for the D-gated latch. If $D = 1$ and $G = 1$, then the latch will be set to the $Q = 1$ state. If $D = 0$ and $G =$

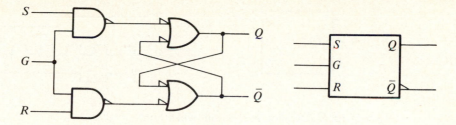

Figure 4.8 A gated *SR* latch and symbol.

1, the latch will return to the $Q = 0$ state. If $G = 0$, the state cannot be changed by the D input.

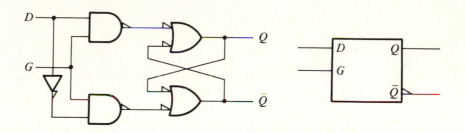

Figure 4.9 *D*-gated latch and symbol.

The *T* Flip-Flop

A multivibrator that changes state or toggles with each successive input is called a T flip-flop. The symbol for this device is shown in Fig. 4.10 along with its typical input and output waveforms. If the T input is high, positive transitions of the CK or clock input will cause the output to change state. The wedge-shaped symbol on the CK input indicates that the output change will occur on the positive transition. If T goes low, then the CK input has no effect on the flip-flop output. This circuit can be used as a frequency divider for rectangular waveforms and is useful in counting applications.

The *JK* Flip-Flop

This bistable circuit has two gating inputs along with a clock input. The voltage level of the gates determine the output state to which the clock input will shift the flip-flop. Often direct-set and reset inputs are provided that can override the clock input. For clocked operation the R and S inputs are set to the inactive level; but if a certain output condition is to be preset into

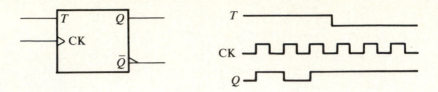

Figure 4.10 A *T* flip-flop and output waveform.

the flip-flop before clocked operation takes place, these inputs can be used. Figure 4.11 shows the symbol for the *JK* flip-flop and its corresponding function table.

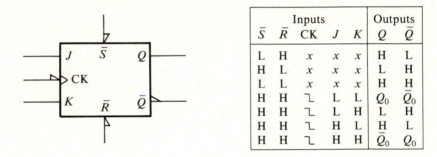

	Inputs				Outputs	
$\bar{S}$	$\bar{R}$	CK	J	K	Q	$\bar{Q}$
L	H	x	x	x	H	L
H	L	x	x	x	L	H
L	L	x	x	x	H	H
H	H	⌐	L	L	Q_0	$\bar{Q}_0$
H	H	⌐	L	H	L	H
H	H	⌐	H	L	H	L
H	H	⌐	H	H	$\bar{Q}_0$	Q_0

Figure 4.11 A *JK* flip-flop and function table.

The x in the function table means that regardless of what state this input has, the output will not be affected. When the direct inputs are high, clocked operation results. The notation Q_0 means the state of Q prior to the occurrence of the negative clock transition. If J and K are both high, the clock input toggles the flip-flop indicated by Q taking on a value $\bar{Q}_0$ and by $\bar{Q}$ taking on a value of Q_0. When J and K are both low, the clock input has no effect on the output. To set a 1 into the flip-flop with the clock input, J must be high and K low prior to the negative clock transition. There is generally some minimum time before the clock pulse occurs that the J and K inputs must be stable. This time is referred to as gate set-up time and, for TTL flip-flops, it is typically in the nanosecond range. The 74112 chip contains two *JK* devices with operation similar to that indicated in Fig. 4.11. The small triangle preceding the wedge on this circuit indicates the fact that output changes take place on the negative-going clock transition.

The *D* Flip-Flop

The clocked *D* flip-flop or simply *D* flip-flop is closely related to the *JK* flip-flop and has become quite useful since the development of the integrated circuit. This device eliminates the *K* gate external input connection by including an on-chip inverter from the *J* input to the *K*. This always forces *K* to equal $\bar{J}$. The input is then labeled *D* rather than *J*. If *D* = 1, a negative clock transition will result in *Q* = 1. If *D* = 0, then *Q* = 0 after the transition. The symbol and function table for the *D* flip-flop are shown in Fig. 4.12. This table reflects only the clocked operation of the device. The 7474 is a typical example of a *D* flip-flop.

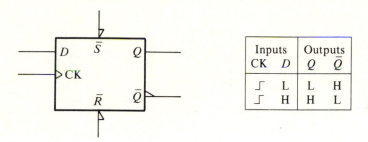

Inputs		Outputs	
CK	*D*	*Q*	*Q̄*
⌐	L	L	H
⌐	H	H	L

Figure 4.12 A *D* flip-flop and function table.

Constructing a *T* Flip-Flop

Manufacturers do not produce *T* flip-flops. Instead they are created from *JK* or *D* flip-flops. The *JK* flip-flop of Fig. 4.11 will toggle on each negative clock transition if *J* = *K* = 1. Tying *J* and *K* together creates the *T* input of the toggle flip-flop. When this input is dropped low, the toggle function is disabled. A *D* flip-flop can be connected to toggle by connecting *Q̄* to *D*. Once connected, the toggling function can only be disabled by overriding clocked operation with the direct set or reset input. For this reason, the *JK* flip-flop is used when a controlled toggle is required.

Edge-Triggered vs Level-Triggered

Clocked flip-flops are designed to cause an output change either when the clock signal makes a transition or when this signal reaches some particular level. A given flip-flop is referred to as an edge-triggered device or as a level-triggered device. In order to understand the differences between these triggering methods, let us consider some basic design problems in the *JK* flip-flop. The circuit of Fig. 4.13 demonstrates a *JK* flip-flop that exhibits an oscillation problem. The coupling from outputs to input gates is done to cause the circuit to toggle when both *J* and *K* are high and the CK input is applied. If $Q = 1$, the K gate is enabled while the J gate is disabled. A CK input will cause a high level to reach R with a low level remaining on S. The SR flip-flop will then change to the $Q = 0$ state. At this point, the J gate opens and the K gate closes. Herein lies the problem. This transition from $Q = 1$ to $Q = 0$ takes place very shortly after the CK pulse goes high. Typically, the propagation delay through a TTL gate and RS flip-flop is in the nanosecond range. Since the clock pulse is still high when the J gate opens and the K gate closes, a high level reaches S while a low level is applied to R. The circuit toggles again and instead of toggling just once, it will continue toggling until CK goes low to disable both gates.

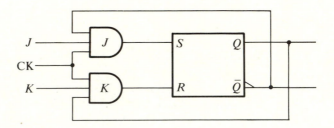

Figure 4.13 An impractical *JK* flip-flop.

There are two popular means of overcoming this problem to create a circuit that toggles only once for each CK pulse applied. The first method uses edge-triggering while the second uses a master-slave arrangement. Edge-triggering implies that the flip-flop is sensitive to the rising or falling edge of the CK input. In a positive edge-triggered circuit, the flip-flop can only change state in response to the positive edge of the CK signal. This resolves the continuous toggling problem of the circuit of Fig. 4.13. The circuit can only toggle once each time the CK input goes positive. The question remains of how to produce a circuit that responds only to the edge of the CK waveform. The answer is to use a pulse-narrowing circuit for the CK signal before applying it to the gates. Figure 4.14 shows this modification. The pulse-narrowing circuit could be an RC differentiating circuit as it was in the era of discrete flip-flops. For integrated circuits, a delaying circuit is used to generate the narrow pulse width. Although the circuit of Fig. 4.15 is not precisely the one applied in IC flip-flops, it demonstrates the basic operating principle of a pulse-narrowing circuit [4].

When the CK pulse is low, point A will be high and B will be low. The output of the AND gate will then be low. When CK rises, point B rises to the high level while A remains high. The AND condition will be satisfied until the delaying inverter allows point A to drop to its low level. This time will be approximately equal to the propagation delay time of the inverter.

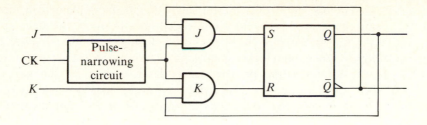

Figure 4.14 A practical *JK* flip-flop.

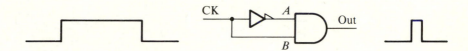

Figure 4.15 A pulse-narrowing circuit.

The AND gate output will be high during the period of time that both *A* and *B* are high and drops to the low level as *A* goes low.

When the pulse-narrowing circuit is used for the *JK* flip-flop, the pulse width must be less than the total propagation delay through the flip-flop. This allows the narrow pulse to initiate a change of state of the *SR* flip-flop when either the *J* or *K* gate is enabled, but then the pulse disables both gates before the flip-flop output changes. The edge-triggered circuit can be designed to respond to either the positive-going or negative-going transition of CK.

The master-slave flip-flop is a level triggered-circuit that solves the oscillation problem in a different way. The master-slave flip-flop consists of two latches, the first of which accepts the input information on one clock transition and transfers this information to the second latch on the alternate clock transition. The latch that accepts the input information is called the master latch; the output latch is called the slave. Figure 4.16 is a timing chart of a master-slave unit.

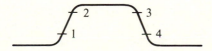

1. Isolate slave from master.
2. Enable data inputs to master.
3. Disable data inputs.
4. Transfer data from master to slave.

Figure 4.16 Master-slave timing chart relative to clock input.

The operation of a master-slave *JK* can be explained in terms of the circuit of Fig. 4.17. As the clock signal goes positive, the slave gates are closed by CK. Information is then shifted through the master gates to the master latch. When the clock signal goes negative, the master gates are first closed so that information on the master latch cannot be changed, and then the slave gates open to transfer data from the master latch to the slave latch. The cross-coupling of outputs to gates guarantees that the flip-flop will toggle when *J* = *K* = 1.

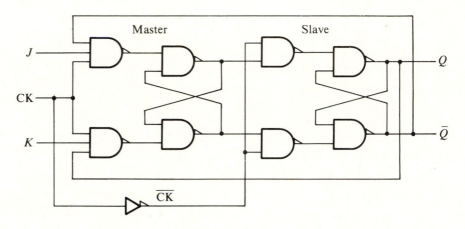

Figure 4.17 A master-slave *JK* flip-flop.

There are some very subtle differences between the edge-triggered and master-slave flip-flops that must be understood by the digital system designer. The symbol for each device is generally the same in the manufacturers' handbooks, and the specifications must be examined to determine the triggering mode of a given flip-flop.

Before discussing these differences, we will return to a consideration of the meaning of the wedge-shaped clock symbol used with clocked flip-flops. Figure 4.18 shows this symbol. The symbol in Fig. 4.18(a) indicates that the output will only change as the CK input goes positive while the symbol of (b) indicates that output changes can only occur as CK goes negative. It should be emphasized that these symbols are used for both the edge-triggered and the master-slave circuit. The major difference between the two types of flip-flop is in the time that input information must be presented to result in an output change. A negative-transition edge-triggered *JK* will shift to that output state determined by the data on the *J* and *K* inputs at the time of the negative clock transition. Until the next negative transition, the *J* and *K* inputs have no effect on the flip-flop output.

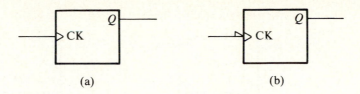

(a)   (b)

Figure 4.18 Symbols for flip-flops that change on: (a) positive-going clock;
(b) negative-going clock.

For the master-slave device, the two preceding sentences do not apply.
When the clock goes positive on a negative-transition device such as that of
Fig. 4.17, the master flip-flop will accept information. If J and K change
while the clock is positive, the master flip-flop state will change also. Let us
assume that CK goes positive while $Q = 0$ and $J = K = 0$, and the master
flip-flop contains a 0. If J goes positive during positive clock time, the master flip-flop will switch to the 1 state. Even if J goes back to the zero level
prior to the negative transition of CK, a 1 remains in the master. When the
negative clock transition occurs, the output switches to $Q = 1$, although both
J and K equal zero at this point in time. Figure 4.19 demonstrates these differences by showing the results of driving an edge-triggered and a master-
slave flip-flop with the same inputs. We note that the short pulses on J and
K during the positive half-cycle of the clock signal have no effect on the
edge-triggered flip-flop but determine the output of the master-slave device.
For this reason, the master-slave circuit is called a "ones-catching" flip-flop.
That is, it will register the fact that a one occurred on an input even if that
input has dropped to zero before the clock transition. In some cases this
"ones-catching" feature is useful, but in state machine design we will see that
the master-slave device is generally avoided to simplify design. We will also
see in a later section that both the master-slave and edge-triggered flip-flops
are useful in counter and register applications.

The Schmitt Trigger

The Schmitt trigger is a bistable circuit, but its function is different from that
of the flip-flop. This circuit is used to detect levels of an input signal.
When a particular level is exceeded by the input signal, the Schmitt trigger
asserts the output voltage. This input level that must be exceeded is called
the upper trip point (UTP). As the input signal decreases below this UTP,
the output remains asserted until the input drops to a value called the lower
trip point (LTP). As the signal drops below the LTP, the output is
deasserted.

A normal trigger-level detector has a single trip point. When the input
exceeds this value, the output is asserted; when the input drops below the

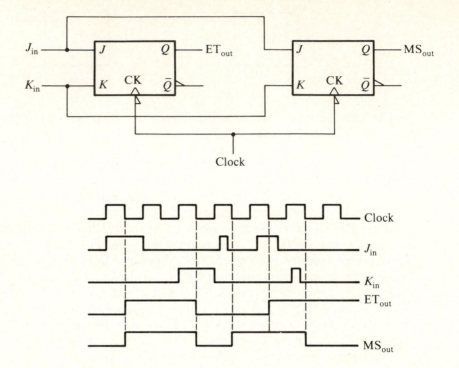

Figure 4.19 Circuit and waveforms for edge-triggered and master-slave devices.

trip point, the output is deasserted. The Schmitt trigger is said to have hysteresis which refers to the two trip points. It is this difference in trigger level that allows the Schmitt trigger to discriminate against noise as a trigger level is detected. The noisy waveform of Fig. 4.20 is used to compare a conventional trigger-level detector to the Schmitt trigger. Assuming a trigger level for the conventional circuit equal to the UTP of the Schmitt, the waveform of Fig. 4.20(b) results. The noise causes the input signal to cross the UTP three times. The trigger-level detector exhibits a transition each time the input level is crossed. This rapid switching of the output is called "chattering" and can have an adverse effect on following circuits in many instances. The Schmitt trigger eliminates chattering since the output will become asserted when the UTP is exceeded, but will remain asserted when the noise drops the input below this level. It is not until the input waveform drops below the LTP that the Schmitt trigger output is deasserted. Although the noise may again cause the input to exceed the LTP, the output will not change unless the UTP is exceeded. For the Schmitt trigger to discriminate

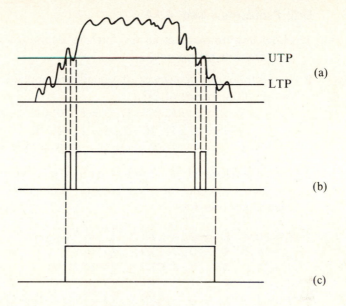

Figure 4.20 (a) Input signal with noise. (b) Output of conventional trigger-level detector. (c) Output of Schmitt trigger.

against noise effectively, the maximum noise amplitude must be less than the difference UTP–LTP.

We may note some similarity in the noise discrimination property of the Schmitt trigger and the switch-debouncing ability of the *SR* flip-flop. Both circuits output a single transition when a trigger level is crossed, but suppress further transitions when the level is crossed repeatedly. The flip-flop requires a second input signal to resume its original state, while the Schmitt trigger will move to its original state when the single input signal decreases below the LTP.

Although it is possible to construct Schmitt trigger circuits which allow both the UTP and LTP to be determined by external resistors, most logic circuits involving the Schmitt trigger have fixed trigger levels. The 7414 is a hex inverter with a Schmitt trigger input. The UTP is typically 1.7 V while the LTP is 0.9 V. This would effectively discriminate against noise with a maximum amplitude of less than 800 mV. The 7413 is a dual 4-input NAND with Schmitt trigger inputs. The UTP and LTP are equal to those of the inverter. The 74244 chip contains 8 three-state buffers with Schmitt trigger inputs. The difference between UTP and LTP with these circuits is approximately 400 mV.

Drill Problems: Sec. 4.1A

1. Show how to connect an *SR* latch to debounce a switch for both positive and negative transitions.

2. If the following sequences of signals are applied to the *SR* flip-flop, what are the final values of Q?

   a. $S = R = L$; $S = L$, $R = H$, $S = L$, $R = L$

   b. $S = R = L$; $S = H$, $R = L$; $S = L$, $R = L$

   c. $S = R = L$; $S = L$, $R = H$; $S = H$, $R = L$; $S = L$, $R = H$

3. An edge-triggered *JK* flip-flop with a positive transition clock input has the given sequence of input signals. Fill in the correct value of Q after each clock transition.

   $$J = 0, \quad K = 1, \quad CK{\uparrow}, \quad Q = ?$$
   $$J = 1, \quad K = 0, \quad CK{\downarrow}, \quad Q = ?$$
   $$J = 1, \quad K = 1, \quad CK{\uparrow}, \quad Q = ?$$
   $$J = 1, \quad K = 1, \quad CK{\downarrow}, \quad Q = ?$$
   $$J = 1, \quad K = 0, \quad CK{\uparrow}, \quad Q = ?$$
   $$J = 1, \quad K = 1, \quad CK{\downarrow}, \quad Q = ?$$

4. What values must be applied to the J and K inputs to cause the flip-flop to change from $Q = 0$ to $Q = 1$? From $Q = 1$ to $Q = 0$? From $Q = 0$ to $Q = 0$? From $Q = 1$ to $Q = 1$?

B. The Monostable Multivibrator

The monostable multivibrator or one-shot produces an output pulse of precise width, initiated by an input trigger signal. The width of the pulse is determined by a resistor value and a capacitor value; thus this width can be selected to be any desired value. We will first consider a specific implementation of the monostable using the popular 555 timer circuit.

The 555 One-Shot

A block diagram of the 555 is shown in Fig. 4.21. The three equal values of resistance establish the reference voltages of $V_{CC}/3$ and $2V_{CC}/3$. The trigger comparator compares the trigger input to the reference voltage $V_{CC}/3$. As long as the input voltage is greater than $V_{CC}/3$, the comparator output is not asserted. When the input signal drops below $V_{CC}/3$, the comparator output is asserted and drives the flip-flop to the $Q = 0$ state.

If the trigger input voltage exceeds $V_{CC}/3$ and the threshold input voltage is below $2V_{CC}/3$, neither comparator is asserted. When the threshold input voltage increases above $2V_{CC}/3$, this comparator output is asserted and

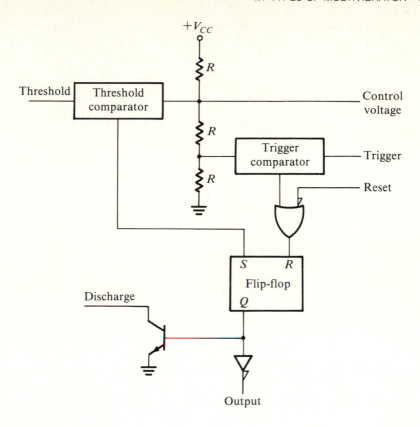

Figure 4.21 The 555 timer.

sets the flip-flop to the $Q = 1$ state. If both comparators are asserted simultaneously, requiring that the trigger voltage is less than $V_{CC}/3$ and the threshold voltage is greater than $2 V_{CC}/3$, the reset input overrides the set and forces the flip-flop to the $Q = 0$ state. This is called a reset-overrides-set flip-flop for obvious reasons.

The output buffer inverts the flip-flop output to cause a high-level output when the flip-flop is reset and a low output when the flip-flop is set. When the flip-flop is set, it provides base drive current to the discharge transistor allowing this transistor to saturate when the collector circuit is completed. We note that the flip-flop can also be reset by applying a low voltage to the reset input.

The circuit of Fig. 4.22 shows the connections required to convert the 555 timer into a one-shot. The 0.01 μF capacitor is connected to the control voltage input to hold this voltage constant even if a transient voltage appears on the power supply. The normal state of the flip-flop is the set state which saturates the discharge transistor. This places a short circuit across the capacitor holding the capacitor voltage to 0 V. Since the threshold voltage is tied to this point, the threshold comparator is not asserted. The trigger input is held at a high level until the output pulse is to be initiated.

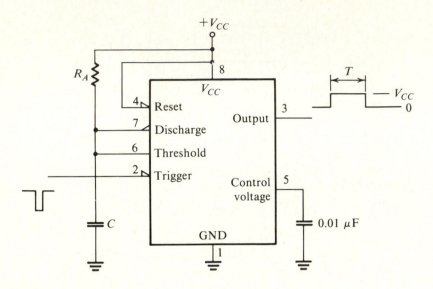

Figure 4.22 The 555 timer used as a one-shot.

When the trigger input drops below $V_{CC}/3$, the flip-flop changes state. The output is now at the high level and the discharge transistor is turned off, becoming an open circuit. The capacitor can now charge through R_A toward the voltage V_{CC}. As the capacitor charges, the threshold voltage reaches a value of $2V_{CC}/3$ causing the flip-flop to return to the set state. This ends the period, dropping the output to 0 V and again saturates the output transistor. The period T is the time taken for the capacitor to charge from 0 V to a voltage of $2V_{CC}/3$ with a target voltage of V_{CC}. This time can be calculated from the general equation for a charging capacitor.

$$v_c = v_i + (v_t - v_i) \left[1 - \exp \frac{-t}{t_c} \right] \tag{4.1}$$

Here v_i is the initial voltage, v_t is the target voltage, and t_c is the time constant. For the case under consideration $v_i = 0\,V$, $v_t = V_{CC}$, and $t_c = R_A C$. The period T is then found to be

$$T = R_A C \ln 3 = 1.1 R_A C \tag{4.2}$$

The trigger pulse width must be less than T for Eq. (4.2) to be valid. If the trigger input remains low after the period is over, the output remains high since the reset of the flip-flop overrides the set. When the input trigger pulse is longer than T, an RC differentiator should be used to ensure that the pulse reaching the trigger terminal is short enough. The resistor connects from V_{CC} to pin 2, and the capacitor is inserted between the input line and pin 2. The trigger input is then held positive until the leading edge of the trigger pulse arrives. This transition is coupled through the capacitor to initiate the period. Pin 2 then charges toward V_{CC} with a time constant largely determined by the RC product of the trigger circuit. This value should be smaller than that of the one-shot period.

The period T can be varied from a minimum of approximately 10 μs to a maximum of several hours by varying the elements of R_A and C. Values of $R_A = 10$ kΩ and $C = 0.001$ μF lead to $T = 11$ μs, whereas $R_A = 10$ MΩ and $C = 100$ μF result in $T = 1100$ s. The 555 timer is popular for lower frequency applications, but it cannot be used when smaller rise and fall times are required.

TTL One-Shots

All one-shot circuits can be classified as retriggerable or nonretriggerable. The nonretriggerable type can be triggered only once during the period T. Any subsequent trigger pulses are ignored until the period has ended. The period T is not affected by the input signals that occur after the period is initiated by a single pulse. The 555 is nonretriggerable. A retriggerable one-shot can be triggered several times during the period. Each time a trigger pulse occurs, the period is adjusted as though it had started at the time of the last input pulse. If the period is T, the output signal will last T seconds after the final trigger pulse. Figure 4.23 shows the difference between the retriggerable and nonretriggerable circuit.

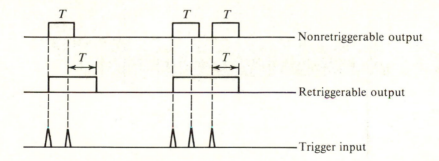

Figure 4.23 Outputs of retriggerable and nonretriggerable one-shots.

The 74121 of Fig. 4.24 is an example of a TTL, nonretriggerable one-shot whose period can be varied from 30 ns to 40 s. If the R_{INT} pin is connected to V_{CC} and the C_T and C_T/R_T pins are left open, period T is typically 30 ns. If the R_{INT} pin is left open and a capacitor is connected from the C_T to the C_T/R_T pin (positive to C_T/R_T) and a resistor R_T is connected from the C_T/R_T pin to V_{CC}, the one-shot will have a period given by

$$T = C_T R_T \ln 2 = 0.69 C_T R_T \tag{4.3}$$

The circuit is triggered by holding either A_1 or A_2 low and taking B from a low to high level. For triggering on a negative transition, A_1, A_2, and B are all high; then either A_1 or A_2 is taken to a low level to initiate the period. The input pulse width can be longer or shorter than period T with no effect on the period duration.

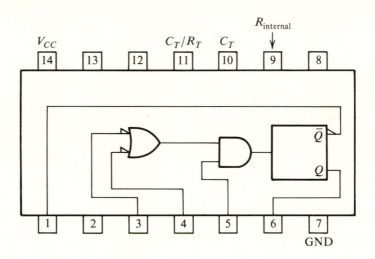

Figure 4.24 The 74121 one-shot.

The 74123 is an example of a retriggerable one-shot that is similar to the 74121 except that if a second trigger pulse occurs before the period ends, the output pulse is extended. The one-shot has a clear input that can, when required, override the timing mechanism to terminate the output period prematurely.

Drill Problems: Sec. 4.1B

1. Derive Eq. (4.2) from Eq. (4.1).

2. Select R_A and C for the 555 timer one-shot to result in a period of 6.4 ms. Are there other values of R_A and C that would give this same period? Explain.

C. The Astable Multivibrator

The astable multivibrator is used to generate a repetitive output with no input trigger signal. One method of creating a gated astable system is to use two one-shots and a gate as shown in Fig. 4.25. When the start gate signal is low, the circuit is inactive. Gate input B will be high at this time. When input A swings positive to open the start gate, the NAND gate output goes negative to initiate the period T_1. At the end of T_1, the negative transition occurring at the input of the second one-shot initiates the period T_2. During T_2 the NAND gate is closed and has a high-level output. At the end of T_2 the NAND gate output swings negative, initiating the period T_1. This sequence will be repeated until the start gate signal drops to the low level. Section 4.2 considers an application of this gated astable circuit.

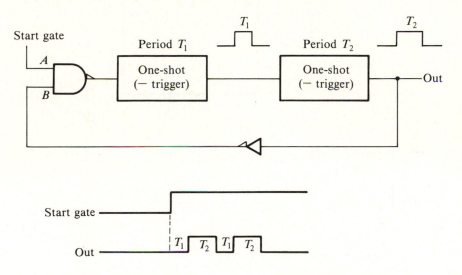

Figure 4.25 Generation of repetitive waveform with one-shots.

The 555 timer can also be used for an astable with the connections shown in Fig. 4.26. The capacitor voltage can swing between $V_{CC}/3$ and $2V_{CC}/3$. As the capacitor voltage rises toward $2V_{CC}/3$, it does so with a target voltage of V_{CC} and a time constant of $C(R_A + R_B)$. When it reaches $2V_{CC}/3$, the threshold comparator causes the flip-flop to change state (see Fig. 4.21). The discharge transistor saturates, and the capacitor voltage heads toward ground with a time constant of CR_B. When the voltage drops to $V_{CC}/3$, the trigger comparator changes the state of the flip-flop and shuts off the discharge transistor. The capacitor again charges toward V_{CC}. The output and capacitor voltage waveforms of the 555 are shown in Fig. 4.27.

The duration of the positive portion of the waveform is called t_1 and is calculated from Eq. (4.1) with $v_i = V_{CC}/3$, $v_t = V_{CC}$, and $t_c = (R_A + R_B)C$. This gives

$$v_c = \frac{2V_{CC}}{3} = \frac{V_{CC}}{3} + \left[V_{CC} - \frac{V_{CC}}{3}\right]\left[1 - \exp\frac{-t_1}{t_c}\right]$$

which can be solved for t_1 to yield

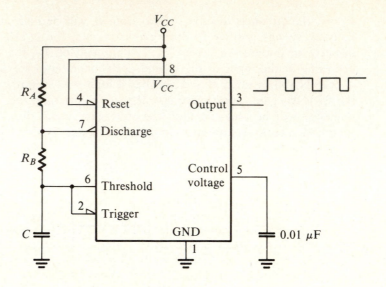

Figure 4.26 The 555 timer as an astable.

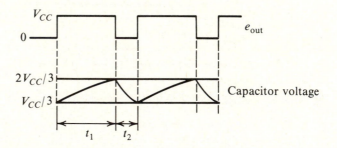

Figure 4.27 Timing waveforms.

$$t_1 = C(R_A + R_B) \ln 2 = 0.69C(R_A + R_B) \tag{4.4}$$

Likewise the portion of time t_2 can be found to be

$$t_2 = 0.69CR_B \tag{4.5}$$

The duty cycle of the astable is defined as the ratio of the duration of the positive portion of the period to the total period, or

$$\text{Duty cycle} = \frac{t_1}{t_1 + t_2} = \frac{R_A + R_B}{R_A + 2R_B}$$

If the value of R_B becomes much greater than R_A, the duty cycle will approach a minimum value of 50 percent.

Some manufacturers give an alternative definition of duty cycle as the ratio of the time the output transistor is on (low voltage) to the total period. This definition leads to

$$\text{ON duty cycle} = \frac{t_2}{t_1 + t_2} = \frac{R_B}{R_A + 2R_B}$$

Another configuration used to obtain a 50 percent or even smaller duty cycle, with a smaller spread of timing resistor values, is shown in Fig. 4.28. The value of t_1 is found by noting that the capacitor charges through R_A only, rather than $R_A + R_B$. This value is then

$$t_1 = 0.69CR_A \qquad (4.6)$$

During t_2 the capacitor discharges from $2V_{CC}/3$ to $V_{CC}/3$. The target voltage and discharge resistance can be calculated from the circuit of Fig. 4.29. (Note that the discharge transistor is saturated.)

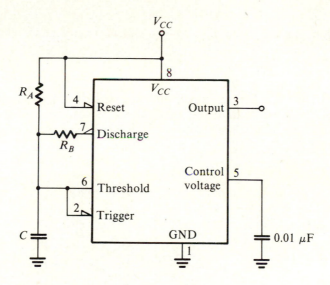

Figure 4.28 Astable capable of 50 percent duty cycle.

The Thevenin equivalent voltage, which is also the target voltage, is

$$v_1 = \frac{R_B V_{CC}}{R_A + R_B}$$

and the discharge resistance is

$$R_{th} = \frac{R_A R_B}{R_A + R_B}$$

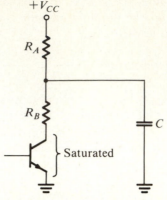

Figure 4.29 Equivalent circuit during t_2.

Using Eq. (4.1) with $v_i = 2V_{CC}/3$, $v_t = V_{CC}R_B/(R_A + R_B)$, $t_c = R_{th}C$, and solving for the time for the capacitor voltage to reach $V_{CC}/3$ results in

$$t_2 = R_{th}C \ln \frac{2R_A - R_B}{R_A - 2R_B} \tag{4.7}$$

The resistor R_B can be adjusted to result in a t_2 that equals t_1. In using Eq. (4.7), we must limit the value of R_B to lead to a target voltage of less than $V_{CC}/3$ or the circuit will not oscillate. This imposes an upper limit on R_B of $R_A/2$.

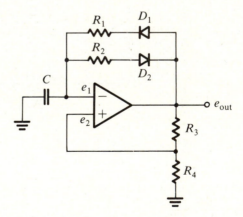

Figure 4.30 An astable circuit using an op amp. (From Ref. 5.)

An op amp can also form the basis of an astable multivibrator. Figure 4.30 shows one possible configuration. Let us assume that e_2 is greater than

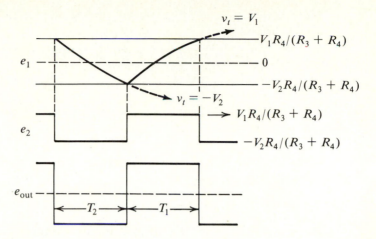

Figure 4.31 Waveforms for the astable of Fig. 4.30.

e_1. If this is the case, the output will be at the positive limit of the active region V_1. The capacitor will take on charge through D_1 and R_1, with a target voltage of V_1. When e_1 becomes equal to e_2, which is given by $V_1R_4/(R_3+R_4)$, the op amp output will switch to the negative limit of the active region, $-V_2$, and will drop e_2 to $-V_2R_4/(R_3+R_4)$; C will be discharged through D_2 and R_2, with a target voltage of $-V_2$. When e_1 drops to a value of $-V_2R_4/(R_3+R_4)$, the output returns to the positive state. The waveforms of Fig. 4.31 can be used to calculate the periods. The period T_1 is calculated by using the standard capacitor equation [Eq. (4.1)] with $v_1 = -V_2R_4/(R_3+R_4)$, $v_t = V_1$, $t_c = R_1C$, and by solving for the time required for e_1 to reach $V_1R_4/(R_3+R_4)$. This results in

$$T_1 = R_1C \ln\left[1 + \frac{R_4}{R_3} + \frac{V_2R_4}{V_1R_3}\right] \tag{4.8}$$

T_2 is found to be

$$T_2 = R_2C\ln\left[1 + \frac{R_4}{R_3} + \frac{V_2R_4}{V_1R_3}\right] \tag{4.9}$$

Typically V_2 and V_1 will be equal, reducing the preceding equations to

$$T_1 = R_1C \ln\left[1 + 2\frac{R_4}{R_3}\right] \tag{4.10}$$

and

$$T_2 = R_2C \ln\left[1 + 2\frac{R_4}{R_3}\right] \quad \text{for } V_2 = V_1 \tag{4.11}$$

Note that if V_1 and V_2 are equal and R_4 is selected to equal R_3, the total period becomes

$$T = T_1 + T_2 = C(R_1 + R_2) \ln 3 = 1.1 C (R_1' + R_2)$$

This expression is similar to Eq. (4.2), which gives the period of a one-shot with a 555 timer.

In addition to the circuits discussed previously, the astable oscillator circuits of Fig. 4.32 can also be used to generate repetitive output signals [3]. These circuits are based on either the TTL family or the CMOS logic family.

Drill Problems: Sec. 4.1C

1. Derive Eq. (4.4) from Eq. (4.1).

2. Select C, R_A, and R_B to give the 555 timer astable circuit of Fig. 4.26 values of $t_1 = 2.4$ ms and $t_2 = 1.0$ ms.

3. Repeat problem 2 for $t_1 = 2.4$ ms and $t_2 = 2.2$ ms.

4. Select C, R_A, and R_B of Fig. 4.28 to give $t_1 = 2.4$ ms and $t_2 = 2.2$ ms.

5. Repeat problem 4 for $t_1 = t_2 = 2.4$ ms.

4.2 MULTIVIBRATOR APPLICATIONS

Section overview: The astable multivibrator is often used to generate a rectangular output waveform that serves as a timing reference signal. This circuit is called a clock and is the first topic discussed in the section. We then proceed to the generation of timed gates using one-shot circuits. Counter and register circuits based on flip-flops are covered to demonstrate some important applications of this device.

A. Clocks

In Sec. 4.1 we discussed the use of the astable multivibrator to produce rectangular waveforms. This type of signal is often used in a digital system to establish a time reference for various operations carried out within the system. This repetitive waveform is called a clock signal since it is used to create a time reference. Some systems require very accurate clocks in which case a crystal-controlled astable circuit may be used. In computers, the master clock will generally consist of such a circuit. Several other digital systems do not require the precision of the master clock and can use the simpler astable circuits. The duty cycle of most clock signals used in digital systems is 50 percent although some microprocessor chips require a two-phase clock with a lower duty cycle.

We shall see that the clock transitions are used to initiate actions in various components of a digital system. Because a transition occurs only every half-period of the clock, it is sometimes useful to produce a second clock

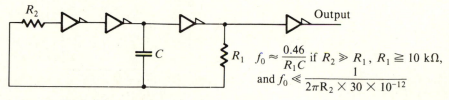

$R = 390 \ \Omega$

C

Output

$$f_0 \cong \frac{1}{3RC}$$

Typical inverters 7404, 74LS04 (TTL)

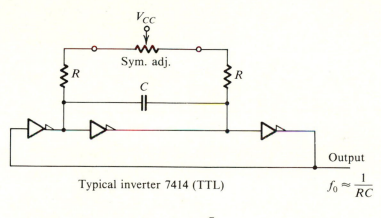

R_2

Output

C

R_1 $f_0 \approx \dfrac{0.46}{R_1C}$ if $R_2 \gg R_1$, $R_1 \geqq 10 \text{ k}\Omega$,

and $f_0 \ll \dfrac{1}{2\pi R_2 \times 30 \times 10^{-12}}$

Typical inverter 74CO4, 74C14 (CMOS)

V_{CC}

Sym. adj.

R

C

R

Output

Typical inverter 7414 (TTL)

$$f_0 \approx \frac{1}{RC}$$

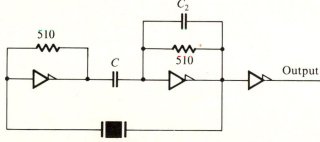

C_2

510

C

510

Output

Quartz crystal frequency $= f_0$
Typical inverter 7404, 74LS04 (TTL)

Figure 4.32 Some typical clock circuits.

signal that is delayed one-quarter of the clock period. The presence of both the clock signal and this delayed clock signal allows more choice of times when transitions occur.

The delayed clock signal DC which lags C by one-quarter cycle is produced by first inverting $2C$ and then applying this signal to the clock input of a T flip-flop. This scheme is shown in Fig. 4.33 along with pertinent timing diagrams.

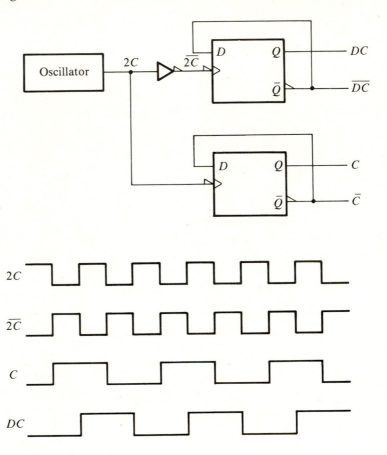

Figure 4.33 Clock system with delayed clock signal.

The signal DC now provides a clock transition at the one-quarter and three-quarter points of the period while C provides a transition at the beginning and at the one-half point of the period. For very high frequency clocks this method is inappropriate since the oscillator is required to produce an output frequency equal to twice that of the desired clock frequency.

An important observation here is that the frequency of a clock signal applied to the input of a toggle flip-flop produces an output having a frequency equal to one-half that of the input. The toggle flip-flop divides the input frequency by a factor of 2. A pair of cascaded flip-flops divides the frequency by a factor of 4. This concept can be extended to divide the frequency by any power of 2.

Drill Problem: Sec. 4.2A

Design a clock circuit that produces a signal delayed from a reference clock signal by one-eighth of the clock period.

B. Timed Gates

There is often a need in logic circuits to establish a fixed time between the occurrence of one event and another. We will see in Sec. 4.3 that a clock signal may be used in conjunction with other circuits to accomplish this if the time between events is a multiple of the clock period. In some instances though, there is no relation between the time period required and the clock period. In such a case the one-shot circuit can be used.

An example of this occurs in receiving a particular pattern of level transitions on an incoming data line. A word or series of bits may be transmitted sequentially over a pair of lines. Two common formats for transmission over short lines (for example, 100 ft lines) are shown in Fig. 4.34. These are the return to zero or RZ code and the nonreturn to zero or NRZ code. The RZ code represents the bit value of the code during the first half of the clock period and always returns to a zero value during the second half of the clock period. A 1 is indicated by a high level during the initial half of the clock period, while a 0 is indicated by a low level during this time. The NRZ code remains high for the entire clock period when a 1 is present and remains low for the clock period to represent a 0.

When a word is transmitted from one point to another, it is often accomplished over a single pair of wires (a signal line and a common line). A clock signal cannot be transmitted from the source to the receiver unless a third wire is used. It is more economical to use a two-wire system and require that the clock signal be generated by the receiver. To help the receiver establish the proper time reference, the word is usually preceded by a start bit. This bit signals the receiving system that a data word follows and allows the receiver to begin generating the clock signal.

One difficulty that can arise in this type of system is the ever-present problem of noise. Although we will not develop any theory relative to the origin of noise, we will state that noise is a very troublesome problem in practical systems. A transient signal may appear on the incoming line due to noise which the receiver interprets to be the start bit. The clock may then start and the receiver accepts an incorrect word that consists only of 0 bits.

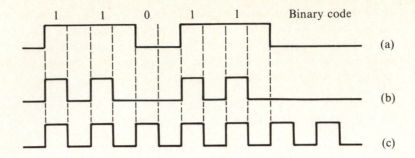

Figure 4.34 Serial transmission of a binary word: (a) NRZ code; (b) RZ code; (c) clock signal.

One method of decreasing the probability of initiating incorrect word reception due to noise is to generate a unique start bit that cannot easily be mistaken for noise. For example, the start bit could appear as shown in Fig. 4.35, assuming NRZ code follows. To detect the start bit, a one-shot of period $0.7T$ might be triggered on the positive transition of the start bit. At the end of this period, a second one-shot with period equal to $0.1T$ is triggered. If a negative transition occurs on the data line during this $0.1T$ window, the system assumes a valid word will follow. The clock is then started $0.25T$ after this negative transition occurs. A circuit to accomplish this start bit check is shown in Fig. 4.36.

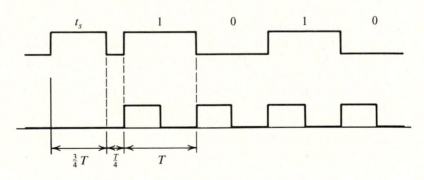

Figure 4.35 A start bit with duration equal to three-quarter clock period.

The second one-shot establishes the $0.1T$ window by enabling the AND gate during this time. If the data line makes a negative transition while the window is open, one-shot 3 is triggered. This circuit has a period equal to $0.25T$; thus the receiving clock can be started on the trailing edge of this one-shot output. The clock will then begin its first period near the start of the first data bit.

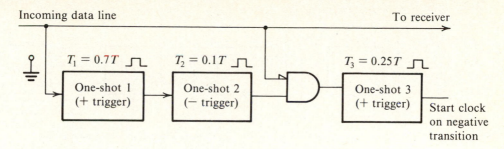

Figure 4.36 A start bit detector.

C. Binary Counters

A binary pulse counter can be constructed with a T flip-flop circuit as its basic element. One flip-flop is required for each column of the maximum binary number to be counted. For example, if the counter must count from 0 to decimal 1000, ten stages are required since nine stages can reach a maximum decimal number of $2^9 - 1 = 511$. Ten stages can reach binary 1111111111 which corresponds to decimal $2^{10} - 1 = 1023$.

Figure 4.37 shows a four-position binary counter that will proceed from 0 up to decimal $2^4 - 1 = 15$. The first negative transition of the counter input line will cause $Q1$ to assume a high level. All other stages are still in the 0 state. As $Q1$ makes this transition from 0 to 1, a positive transition is presented to CK2 with no corresponding change in state of FF2. The counter now reads 0001. A second negative transition causes $Q1$ to switch from 1 back to 0. This negative transition is presented to CK2, resulting in a change in state to $Q2 = 1$. The counter now reads 0010 corresponding to decimal 2. As additional negative input transitions occur, the count advances to consecutive binary numbers. Figure 4.37(b) shows a timing chart for the counter input and all outputs.

The 7493 is a direct implementation of the 4-bit counter of Fig. 4.37 with two minor differences. This circuit, shown in Fig. 4.38, incorporates a means of directly resetting the counter to contain all zeros. To do so requires assertion of inputs R_1 and R_2 simultaneously. As long as these inputs are both asserted, all flip-flops will remain in the 0 state regardless of the signal presented to the clocked inputs. Direct sets or resets always override clocked operation in flip-flop circuits.

We note that the output of FFA is not connected to the input of FFB in the 7493. If a 4-bit counter is desired, Q_A must be externally wired to input B and the counter input becomes input A. A 3-bit counter results if input B is taken as the counter input with Q_B, Q_C, and Q_D as the outputs. A 1-bit counter results if input A is used as the counter input and Q_A is used as the

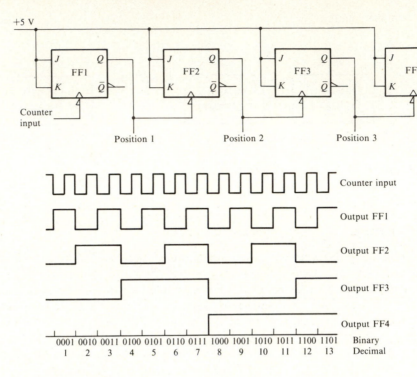

Figure 4.37 A binary counter and waveforms.

counter output. A 4-bit binary counter is also called a divide-by-16 counter because the output frequency equals the counter input frequency divided by 16. One-bit, 2-bit, and 3-bit counters are called divide-by-2, divide-by-4, and divide-by-8 counters, respectively.

The 7493 presents speed problems in very high frequency applications. If three 4-bit circuits are used to create a 12-bit counter and all stages are in the 1 state, the next negative input transition will cause all stages to revert to 0. Unfortunately, all stages do not change simultaneously. The first stage output changes to 0 only after a propagation delay time. This signal then causes the second stage to revert to 0 after a second propagation delay time. Each flip-flop stage adds to the delay so that the transition of the last stage is delayed from the input transition by 12 times the average propagation delay time per stage. It is possible to decrease this delay in critical applications, but at the expense of circuit simplicity.

With ICs, circuit complexity is considerably less important than in discrete elements. Thus, the IC circuitry required to implement the synchronous counter of Fig. 4.39 is on a single chip that includes considerably more circuits than shown here. The synchronous counter applies the input line to all stages through AND gates. The negative transitions of the input signal will

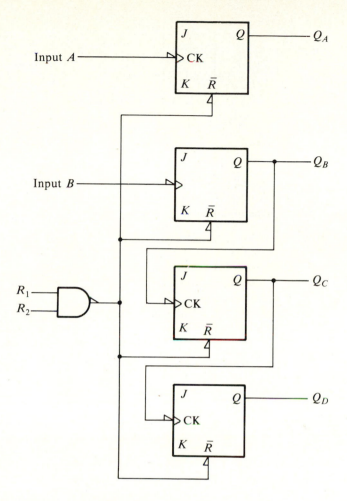

Figure 4.38 A 1-, 3-, or 4-bit counter.

not reach the T input of a given stage unless the appropriate conditions are present to open the gate. For example, after six negative transitions, the counter contains 0110. None of the AND gates are open; hence on the seventh transition, only Q1 changes to binary 1 giving a count of 0111. Now all gates are opened so that the eighth transition reaches all four T inputs, and the resulting count is 1000. All stages of the synchronous counter that must toggle see the input transition at the same time resulting in minimum delay. More will be said about synchronous design in Chap. 5.

There are occasions in digital systems that require a backward or down counter. A binary number is set into the counter that then counts toward zero as input pulses occur. Figure 4.40 shows a simple method of constructing a down counter. The timing chart assumes that the original count was

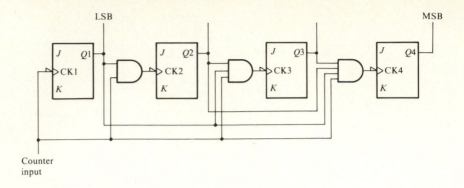

Figure 4.39 A synchronous counter.

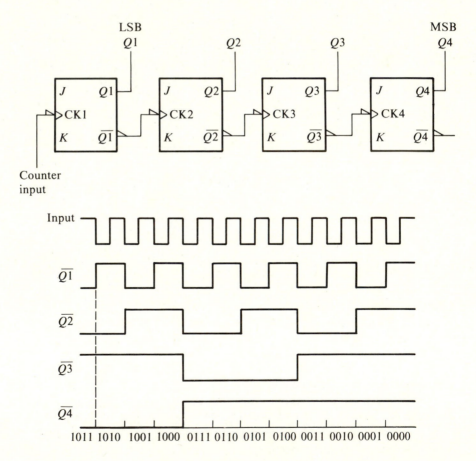

Figure 4.40 A down counter and timing diagram.

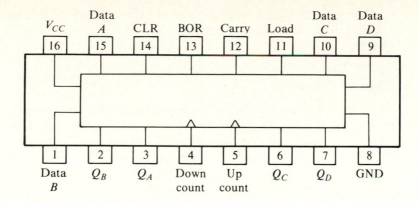

Figure 4.41 The 74193 synchronous up-down counter.

1011 or decimal 11. After 11 negative transitions of the input signal the counter contains a count of 0000.

An example of an IC up-down counter is the 74193. This is a synchronous, 4-bit, up-down counter with a clear input, and borrow and carry outputs. Data can be loaded into the counter in parallel if a preset count is desired. Figure 4.41 shows the block diagram for this circuit, which has a complexity equal to approximately 55 gates. This counter can be cascaded with other stages to increase the bit capacity by 4 per chip. The carry and borrow outputs connect directly to the up-count and down-count inputs, respectively, of the adjacent chip. The clear input resets all bits to 0 at any time this input is activated.

Drill Problems: Sec. 4.2C

1. Show how to connect the 7493 counter of Fig. 4.38 to construct a divide-by-8 counter. Show the output waveform referenced to a clock signal input. Show at least 10 clock periods of input.

2. Repeat problem 1 for a divide-by-16 counter. Show at least 18 clock periods of input.

D. Registers

A register is a collection of one or more flip-flops designed to store binary information. A single flip-flop register is often called a flag. Typically a register will store binary words. Two basic types of register are used in computers or other digital systems: parallel registers in which all positions are filled simultaneously and dynamic shift registers that can be filled or emptied serially from one end of the register. In several instances a shift register will combine parallel and serial operation. For example, the register may be filled serially from a receiving line and shifted into a memory in parallel mode.

Parallel Operation

The 8-bit register of Fig. 4.42 demonstrates the fundamental operation of parallel registers. The reset line is activated to clear the register before any data are entered. When the input data lines are set, a strobe pulse appearing on the input data strobe line will set the data into the register. After the data word has been entered, it can be shifted to another section of the computer. Activating the output data strobe causes the data to appear on the output lines during the strobe. Several sets of output gates can be used to shift the data into one of several different locations, depending on which output data strobe is activated.

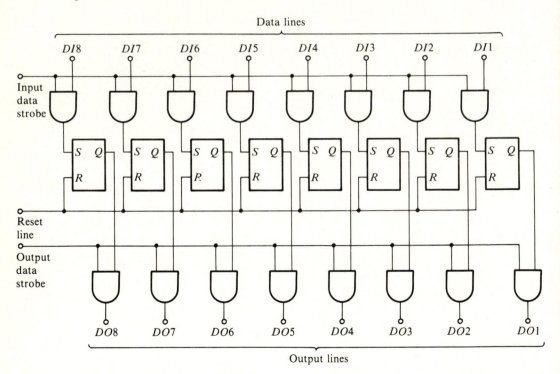

Figure 4.42 Parallel register.

Parallel registers are rather fast, since they can be filled or read in a very short time, but they require a large number of gates for control. The gated D flip-flop can be used to decrease the number of required gates (Prob. 4.28).

Serial Operation

A shift register is used to receive or transmit serial information. Often it is designed to also receive or transmit parallel information. Some applications require the register to receive in parallel and transmit serially or to receive serially and transmit in parallel. We will here be concerned primarily with the serial mode of operation.

Two simple serial shift registers are shown in Fig. 4.43; one is based on *D* flip-flops, the other on *JK* flip-flops. The shift pulses are applied simultaneously to all flip-flops of the register. When a negative transition is applied to the shift input, each flip-flop is filled with the information from the adjacent stage to the left. In this case, all information shifts one position to the right. The bit in FF1 disappears, and a 0 shifts into FF8 when connected as shown.

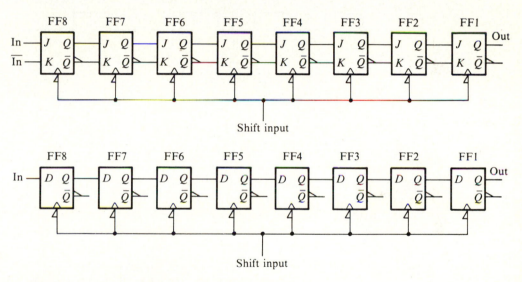

Figure 4.43 Shift registers.

An 8-bit word contained by the register can be transmitted onto a line simply by connecting Q of FF1 to the line and applying eight negative transitions to the shift input. Data can be received by connecting the D gate of FF8 to the input line and applying the proper number of negative transitions to the shift input. For the *JK* register $J8$ is connected to the incoming data line while $K8$ is connected to the inverted data signal. Of course, positive transition sensitive shift registers are also available.

There are certain timing problems associated with transmitting and receiving information which will now be considered. Suppose we wish to transmit the 7-bit ASCII code for the character "J" which is 1001010. We will proceed from the LSB bit to the MSB as we transmit, but we must precede the LSB with a start bit of 1 to signal the receiver that a digital word follows.

One method of transmitting this information onto a line uses nine flip-flops. The register is loaded with 100101010. The leading 0 is not part of the data word but is added to produce equal spacing of the bits. The timing chart for the transmission of this word is shown in Fig. 4.44. The first negative transition sets the start bit onto the line. Succeeding transitions shift the remaining 7 bits onto the line, while the last transition clears the line. If the start bit had been in the first flip-flop rather than the second, the line would have been high prior to the application of the first negative transition.

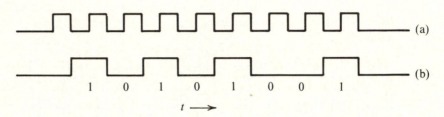

Figure 4.44 (a) Gated clock applied to shift input; (b) serial code appearing on Q of FF1.

We note that the code produced by the shift register is NRZ code. If RZ code is desired, the circuit of Fig. 4.45 can be used. The inverter must delay the inverted clock signal applied to the gate enough to allow the shift register to change before the gate is opened. This prevents possible slivers from occurring in the output RZ code at the points marked a. In order to use the shift register as a serial receiver, an input data line is connected to the J input of the leftmost flip-flop, and an inverter is connected to K. Note that D flip-flops could be used instead of JK flip-flops to construct this register. If an 8-bit word is to be received, exactly eight shift pulses must be applied to the shift input to fill the register. Furthermore, the proper time relationship must exist between the incoming data and the shift pulses. The gates $J8$ and $K8$ must be set prior to the application of a shift pulse. (The manufacturers of flip-flops and registers specify gate setup times.) Thus, shift pulses must occur at intervals of 1 bit time, but must also occur after the incoming data transitions. Based on the assumption that the register responds to positive shift input transitions, Fig. 4.46 demonstrates this point.

For NRZ code, the shift transition can occur at any point after each data bit has had enough time to set the gates, but it must occur before the data bit changes to the next value. A reasonable choice is to place the shift transition at the midpoint of the data bit as shown in Fig. 4.46(a).

For RZ code, the choice is somewhat restricted, since the 1 bit lasts only

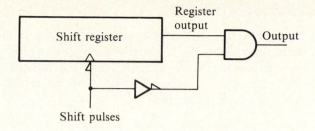

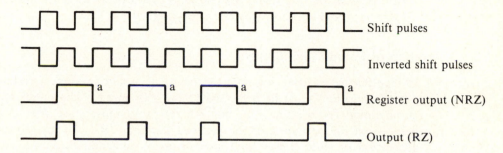

Figure 4.45 Conversion of NRZ to RZ code.

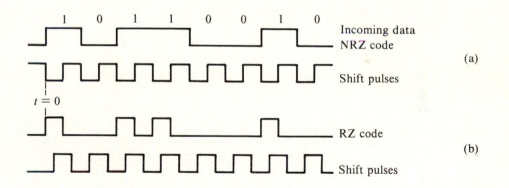

Figure 4.46 Relationship between shift pulses and incoming data: (a) NRZ code; (b) RZ code.

one-half of the clock period. The shift transition here must occur before one-half of the clock period has expired. A reasonable choice for RZ code is to place the shift transition one-fourth of a clock period after the bit transition, as shown in Fig. 4.46(b). After the correct number of shift pulses are applied, the incoming data reside in the shift register.

Drill Problems: Sec. 4.2D

1. An 8-bit serial shift register contains the word 10110011. The register shifts on positive clock transitions. If the rightmost flip-flop output connects to the line, draw a timing chart for this line when eight cycles of a clock are applied to the shift input of the register. How long does the first 1 bit exist on the line?

2. Repeat problem 1 if a ninth flip-flop, containing a 0, is added between the register and the line. Assume nine cycles of a clock are now applied. What is the purpose of this ninth flip-flop?

E. Generation of Gated Clock Signals

In transmitting or receiving data by means of a shift register, a certain number of transitions must be presented to the shift input. The pulses may start at any time for transmission, but for reception must start after a suitable delay from the beginning of the received start bit. We will first consider the generation of clock pulses for transmission of data and then proceed to the receiving register.

One of the simplest methods of generating the correct number of clock transitions uses a gated astable circuit made up of one-shot circuits as shown in Fig. 4.47. When the signal is to be generated, one-shot 1 is triggered. Both inputs of the NAND gate are now asserted which asserts the output of the NAND and triggers one-shot 2. At the end of the period of one-shot 2, one-shot 3 is triggered. The NAND gate is now deasserted until the end of the period of one-shot 3. As this period ends, the NAND gate output again goes low to retrigger one-shot 2. The two one-shots with period of $T/2$ continue to retrigger each other until one-shot 1 times out. Since the period of one-shot 1 is set to be $7\frac{3}{4}T$, where T is the desired clock period, eight negative transitions are produced as shown.

Another method of generating a clock gate is simply to gate the clock signal with a one-shot that is triggered at the correct time. Figure 4.48 shows this situation. The one-shot must be triggered at the beginning of a clock cycle. If the generate signal is not synchronized with the clock signal, the circuit of Fig. 4.49 can be used to produce a synchronized trigger signal. This signal can then be used to initiate the one-shot period of Fig. 4.48.

The one-shot is not considered to be as reliable as an edge-triggered flip-flop. A method of generating eight clock pulses without using a one-shot is shown in Fig. 4.50.

Again the trigger that opens the gate must be synced with the beginning of the clock period. The counter is at 0 count prior to this time; thus $\overline{Q}4 = 1$. The gate clock pulses are applied to the counter which responds to the negative transitions. The eighth negative transition switches the fourth flip-flop of the counter to the 1 state. This closes the gate since $\overline{Q}4 = 0$ after the eighth negative transition. The counter must be reset before it can

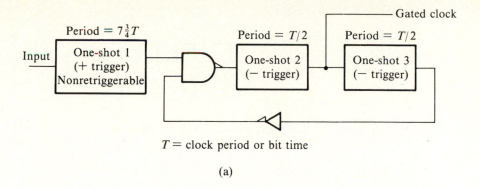

$T =$ clock period or bit time

(a)

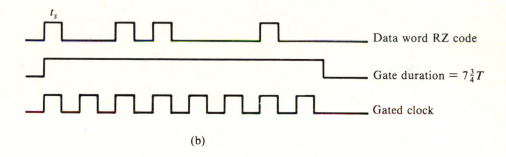

(b)

Figure 4.47 (a) Transmitting clock signal generation. (b) Timing chart for creation of eight negative transitions.

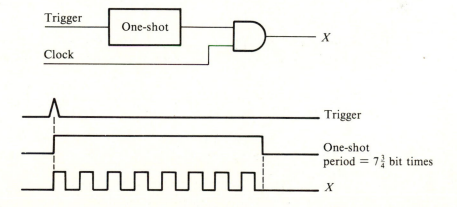

Figure 4.48 One-shot clock gating.

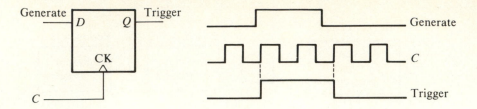

Figure 4.49 Trigger synchronization using a D flip-flop.

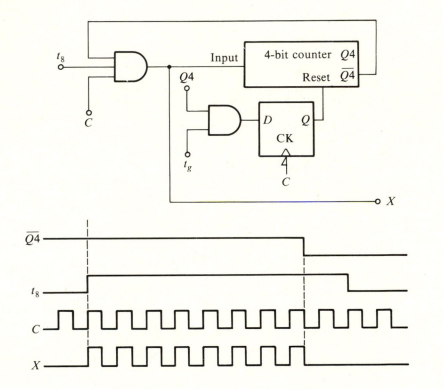

Figure 4.50 Using a counter to produce a gated clock signal.

be used again to gate the clock. The D-input of the flip-flop is gated to binary 1 when the trigger strobe falls, since $Q4 = 1$ at this point. The first negative clock transition after t_g has returned to a 0 value raises the reset line to reset the counter to 0. Since $Q4$ and the D-input now equal 0, the following negative clock transition returns the flip-flop to the 0 state, deactivating the reset line. The circuit is now ready to generate another series of clock pulses.

Figure 4.51 shows an alternate method that uses a counter to generate nine negative transitions of a subclock (the reset circuitry is not shown for the sake of simplicity). Assuming a zero count on the counter, the NAND gate output will be high. When t_g goes positive, the astable clock is gated on. After nine negative clock transitions the NAND gate input will consist of $Q1 = \overline{Q2} = \overline{Q3} = Q4 = 1$. The output of the NAND now becomes negative, closing the clock gate. The counter must be reset after t_g falls in order to be ready to generate the next series of pulses.

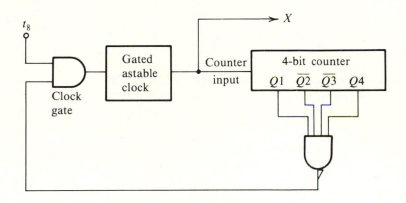

Figure 4.51 Alternative method of clock gating.

At this point we have examined several methods of generating a series of clock pulses that can be applied to the shift input of a shift register to transmit the register contents onto a line. Now we will consider the problem of generating the correct number of pulses for a receiving shift register. This problem is slightly more difficult than the transmitting problem. In transmitting, the generate signal can occur at any time to initiate the shifting process. In receiving, we must detect the start bit, then generate the shift pulses so that after each individual bit appears on the incoming data line, a shift pulse occurs. When receiving NRZ code, the logical point for the shifting transition to occur is at the middle of the bit time. Since it is impossible to generate a receiving clock with the exact frequency as the transmitting clock, it is necessary to synchronize the receiving clock with the start bit, and then place the shifting transitions at the midpoint of the bit times. If the receiving clock drifts from the correct frequency, an error of ± 0.5 bit times must accumulate to cause a bit error.

One method of starting the clock at the correct time uses the circuit of Fig. 4.47(a). The input to one-shot 1 is now the incoming data line. One-shot 1 must be nonretriggerable so that the period starts with the leading edge of the start bit and ends 7¾ bit times later for 8-bit reception. The incoming data line also is connected to the shift register. The output of one-shot 2 goes to the shift input of the register. If RZ code is being received, the period of one-shot 2 can be adjusted to $T/4$ and that of one-shot 3 to $3T/4$.

Another approach to generating the correct number of clock transitions at a receiver is shown in Fig. 4.52. This method replaces the one-shots with a counter. Prior to reception of a word the counter must contain a count of 1000, leading to $Q4 = 1$ and $\overline{Q}4 = 0$. The flip-flop must also be reset. The gated clock is now off. When the start bit occurs, the flip-flop is set and, after a slight delay, the counter is reset. This drives $\overline{Q}4$ to 1 which initiates the output of the gated clock. The assertion of $\overline{Q}4$ also resets the flip-flop which now removes the reset signal from the counter. The gate to the FF is closed to prevent the assertion of the S input during the time the word is being received. The output pulses of the gated clock are applied to the receiving register and are also counted by the counter. When a count of 8 is reached, $\overline{Q}4$ goes low, terminating the gated clock output. The system is now ready to receive another word. Two inverters are used to delay the signal applied to the counter reset. This delay is necessary to stretch the width of the reset pulse to an acceptable level. If the FFQ signal were connected directly to the counter reset, it would disappear immediately upon the counter's beginning to reset. The signal $\overline{Q}4$ would start positive and $Q4$ would start negative, which would reset the flip-flop. With the reset signal removed, the counter might not reset properly—which brings up an important point about logic design that deserves emphasis. A signal that drives a circuit reset should not be removed when the circuit is reset, because oscillations or improper reset may occur. Using the delaying inverters allows the reset pulse width to stretch to a value slightly greater than the total delay, ensuring proper reset.

There is a more reliable method of designing the transmitting and receiving clock circuitry discussed in this section. This method is state machine design which will be covered in the following three chapters. We have not attempted to propose design principles here; we have simply proposed systems to produce the gated clock signals. In Chaps. 5 and 6 we will develop principles that allow the orderly design of control circuits that can be used for many different purposes.

Drill Problems: Sec. 4.2E

1. Redesign the system of Fig. 4.47(a) to receive RZ code at a 1 kHz bit rate. Each word is a 12-bit word including the start bit.

2. Redesign the system of Fig. 4.52 to receive a 12-bit word that includes the start bit.

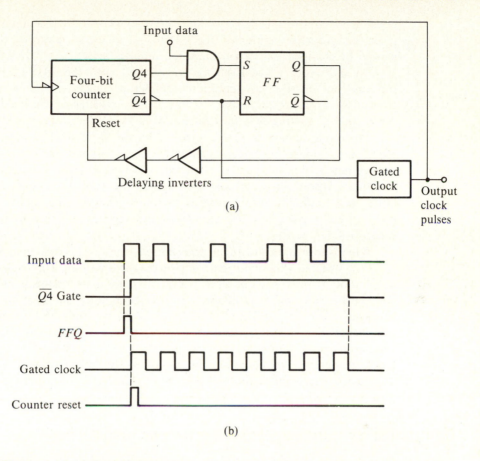

Figure 4.52 (a) Receiving clock signal generator using counter. (b) Timing chart.

F. Integrated Circuit Registers

IC shift registers are available from several manufacturers. Several hundred bit registers can be fabricated on a single chip for serial-in serial-out registers. Obviously there cannot be lines available for direct sets to each stage. There are various types of 8-bit registers available in monolithic circuits. The 74164 is an 8-bit, parallel-out, serial shift register; the 74165 is a parallel-load shift register. These two units can be used in a digital communication system, with the 74164 as the sending register and the 74165 as the receiving register. The sending register might be filled in parallel, then transmit serially to the receiver. When this register is filled, a parallel transfer can store its contents into the appropriate unit.

The 74194 is a 4-bit bidirectional universal shift register that can be used to construct larger registers. It features parallel input, parallel output, or serial input and output (right or left shift). Figure 4.53 includes the pin diagram and function table for the 74194. The mode controls, $S0$ and $S1$, should be changed only when the clock input is high.

An I/O register can easily be constructed with the 74194. Figure 4.54 indicates one possible configuration for a serial register capable of serial transmission or reception over a line. This register is designed for 8-bit, NRZ code consisting of a start bit followed by a 7-bit data word.

The mode of operation is controlled by the R/T latch. To receive data, the S input must be driven high while the R input remains low. In the receive mode, the output gate is closed, allowing nothing to shift to the output line. Upon arrival of an input word, the start bit opens the one-shot clock gate, and data are presented to the register by the input gate. After eight positive transitions of the clock, the register is full and the clock is turned off.

When more data are to be transmitted, the R/T latch must be reset to the $\overline{Q} = 1$ state. The input gate is closed, and one-shot 1 is triggered. After a delay of ½-bit time, one-shot 2 and the clock gate one-shot are triggered. This ½-bit time delay is necessary to create a start bit that lasts a full bit time, since the first shift pulse of the clock occurs only ½-bit time after the clock gate is initiated. Information is serially shifted onto the line through the output gate, which has been enabled by the R/T latch.

It is possible to eliminate the one-shots by using other logic elements such as counters and additional flip-flops. Although this conversion is by no means a trivial task, you may enjoy it as a challenging exercise (Prob. 4.33).

4.3 RELIABILITY CONSIDERATIONS OF THE ONE-SHOT

Section overview: This section considers some reliability problems with the one-shot circuit.

Some digital systems must be completely reliable in order to be useful, while others may allow a small percentage of errors to occur and still not detract from the effectiveness of the system. The control unit of a computer must be highly accurate while a multiplexed LED display can occasionally ignore a current pulse to a display with no serious effect. Most control systems must be highly accurate even in the presence of extraneous noise signals that are often present in electronic circuits.

The one-shot circuit is generally avoided in high performance systems. This device does not function well in a noisy environment and also introduces timing problems in systems depending on precise time periods. The edge-triggered flip-flop can be considerably more reliable than the one-shot and is used in the popular state machine to be introduced in Chap. 5. The state machine is used almost exclusively in applications requiring high reliability.

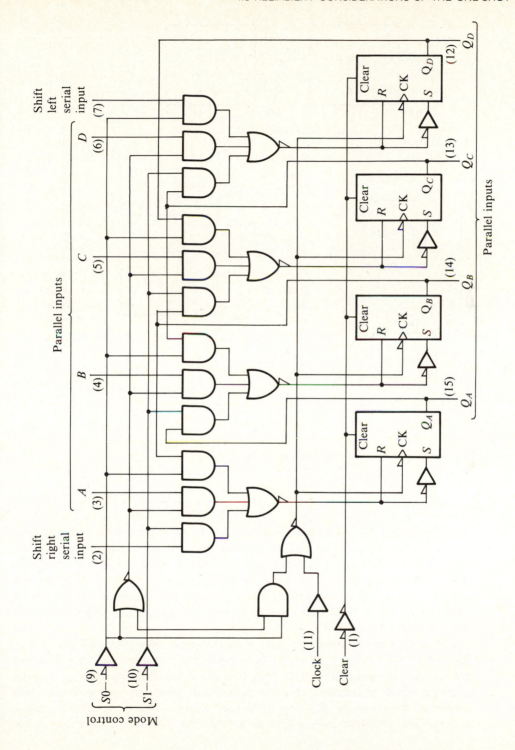

Figure 4.53 A 4-bit bidirectional universal shift register.

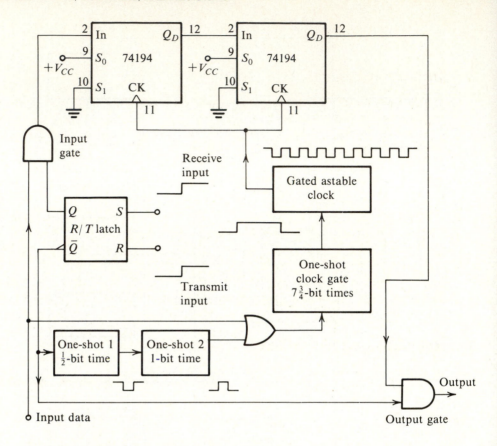

Figure 4.54 Serial I/O register and input code.

A. Noisy Data Lines

The generation of gated clock signals that can be applied to a receiving shift register was considered in the previous section. If the incoming data line is applied to the trigger input of a one-shot, the one-shot can be triggered by a noise pulse on the line. The clock generation circuit treats this noise pulse as a start bit and shifts an erroneous word into the receiving register. Any noise pulse with a great enough amplitude to trigger the one-shot leads to the reception of a false word.

It is possible to construct a receiving register system, based on clocked flip-flops, that rejects a relatively high percentage of noise pulses. This system accepts incoming data as a start bit only if the flip-flop input is asserted near the clock transition. A noise pulse must exist just prior to the clock transition to cause the flip-flop to change states. This requirement reduces the number of noise pulses interpreted by the system as start bits. The amount of reduction depends on the average noise pulse duration, clock frequency, and flip-flop setup times.

B. Synchronization of Incoming Bits

A second problem with a one-shot circuit is that asynchronous input data cause an asynchronous output. If there is a clock circuit in the receiver, the incoming data will have no fixed time relationship with the clock transitions. Hence, the one-shot output period is initiated with no fixed relationship to the clock signal. We learned earlier that an asynchronous input can be synchronized with a clocked flip-flop when synchronization is important.

The third problem is that the period of the one-shot is determined by imprecise elements that vary with temperature. Even if the trigger input is synchronized to the clock signal, the trailing edge of the one-shot output will not be synchronized. If this is an important consideration, a flip-flop circuit must again be used.

Although the last section suggests several applications of the one-shot, care must be taken to determine if the poor reliability can be tolerated in each specific application under consideration.

SUMMARY OF SIGNIFICANT POINTS IN CHAP. 4

1. There are three types of multivibrator: the monostable, the astable, and the bistable.

2. The monostable and astable are used for timing purposes.

3. The bistable flip-flop is the most important logic element in use today. This circuit is used in counter and register systems and forms the basis of the state machine to be considered in the remaining chapters.

4. Serial transmission and reception of binary words can be accomplished with shift registers and properly timed gates.

REFERENCES AND SUGGESTED READING

1. D.J. Comer, *Electronic Design with Integrated Circuits*. Reading, Mass.: Addison-Wesley, 1981, chap. 3.
2. Engineering Staff, *The TTL Data Book for Design Engineers*. Dallas: Texas Instruments, 1983.
3. W.I. Fletcher, *An Engineering Approach to Digital Design*. Englewood Cliffs, N.J.: Prentice-Hall, 1980, chap. 5.
4. R. Sandige, *Fundamentals of Digital Analysis*. New York: McGraw-Hill, 1978, chap. 11.
5. D.F. Stout and M. Kaufman, *Handbook of Operational Amplifier Circuit Design*. New York: McGraw-Hill, chap. 20.
6. Signetics Staff, *Signetics Analog Data Manual*. Sunnyvale, Calif.: Signetics, 1982.

CHAP. 4 PROBLEMS

Sec. 4.1A

*4.1 Construct the characteristic table for the circuit shown in Fig. P4.1 with A and B as inputs and X as output.

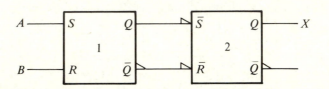

Figure P4.1

4.2 Repeat Prob. 4.1 after reversing the connections between the flip-flops.

*4.3 Explain why the resistors are present in the switch debouncer circuit of Fig. 4.6. Would the circuit function if these resistors were deleted?

4.4 Use the 74279 chip (flip-flop 1) to debounce a switch. Show all pin connections used.

In Probs. 4.5 to 4.15, plot the output waveform referenced to the clock signal assuming the initial contents of all flip-flops is $Q = 0$.

4.5

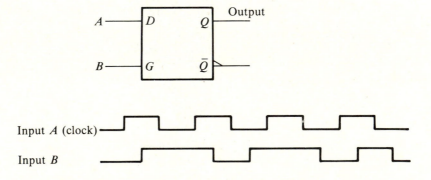

Figure P4.5

*4.6

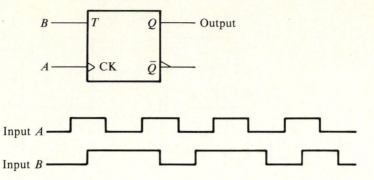

Figure P4.6

4.7

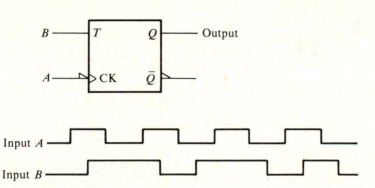

Figure P4.7

4.8

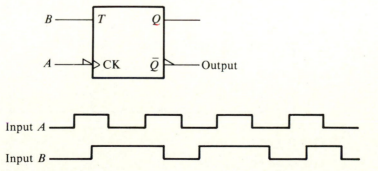

Figure P4.8

*4.9

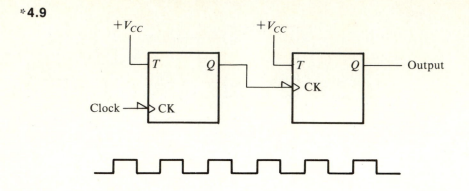

Figure P4.9

4.10

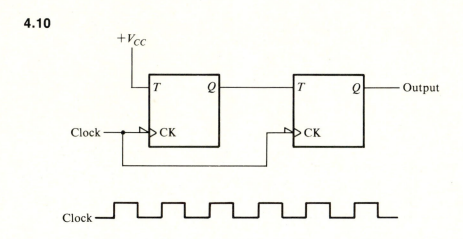

Figure P4.10

4.11

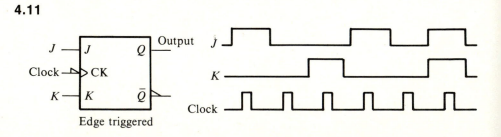

Figure P4.11

* 4.12

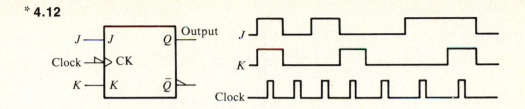

Figure P4.12

4.13

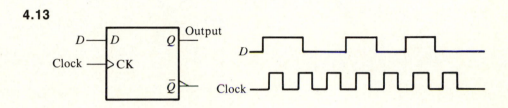

Figure P4.13

4.14

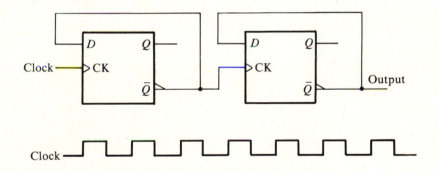

Figure P4.14

4.15

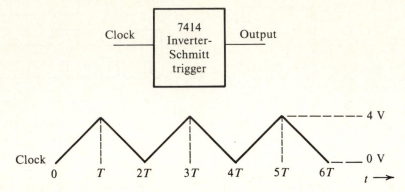

Figure P4.15

Sec. 4.1B

* **4.16** Select R_A and C in Fig. 4.22 to give the one-shot a 1 ms period.

4.17 Select R_A and C in Fig. 4.22 to give the one-shot a 100 ms period.

* **4.18** If a 2 ms trigger pulse must initiate a 1 ms period for the one-shot of Fig. 4.22, select proper values for a differentiator trigger circuit.

Sec. 4.1C

4.19 Use 74121 chips to implement the gated astable of Fig. 4.25. Both T_1 and T_2 are to be 1 ms. Show all pin connections.

4.20 Repeat Prob. 4.19 for $T_1 = 100$ ms and $T_2 = 50$ ms.

4.21 Select R_A, R_B, and C for the astable of Fig. 4.26 to result in $t_1 = 2$ ms and $t_2 = 1$ ms.

* **4.22** Repeat Prob. 4.21 for $t_1 = 50$ ms and $t_2 = 40$ ms.

4.23 Calculate t_1 and t_2 for the circuit shown in Fig. P4.23.

4.24 Select R_A, R_B, and C in the astable of Fig. 4.28 to result in $t_1 = t_2 = 100$ ms.

Sec. 4.2A

* **4.25** Sketch the waveforms appearing at points A, B, and C of Fig. P4.25.

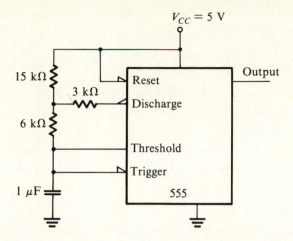

Figure P4.23

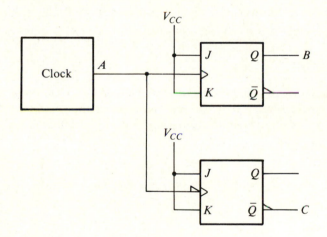

Figure P4.25

Sec. 4.2C

4.26 Show how to connect the 7490 counter to construct a divide-by-5 counter. Show the output waveform referenced to a clock signal input. Show at least eight clock periods of input.

4.27 Repeat Prob. 4.26 for a divide-by-10 counter. Show at least 15 clock periods of input.

Sec. 4.2D

4.28 Show how to eliminate the input data gates of Fig. 4.42 by replacing the *SR* flip-flops with gated *D* devices. Is a reset pulse necessary in this case?

*__4.29__ Develop a simple circuit to convert RZ code to NRZ code. Assume the presence of a clock signal synced with the RZ code.

Sec. 4.2E

4.30 Design a system that produces 14 negative clock transitions on a line after which the line remains at the 0 level. Assume that a start strobe is available that is synced with the positive half-cycle of the clock.

4.31 Repeat Prob. 4.30 assuming 17 transitions are desired and the start strobe is asynchronous.

4.32 If we know the clock period is 1 ms in Prob. 4.30, design the system using 74121 one-shot chips.

Sec. 4.2F

4.33 Redesign the I/O register of Fig. 4.54 to eliminate all one-shots from the design. Show timing chart when the character 11011010 is transmitted or received. State any assumptions made.

CHAPTER 5

Design of Simple State Machines

5.1 THE NEED FOR STATE MACHINES

Before we begin developing state machine theory, we should establish that this is an important area of digital system design. We will do this by emphasizing three advantages of the state machine approach.

The first advantage is that many electronic systems require the type of sequential operation exhibited by state machines. Therefore, state machine design can be applied to the solution of a wide variety of practical circuit problems.

The second advantage is that state machine design methods lead to minimal design. In combinational circuits we found that the Karnaugh map was useful in minimizing the number of gates required to implement a logic function. Although the importance of this tool has perhaps diminished as a result of the availability of MUXs, PLAs, ROMs, and decoders, it remains a significant method in combinational design. The state machine design procedure is to sequential circuits what the K-map is to combinational circuits. This design results in the minimum number of required flip-flops and can minimize other circuitry in the system as well.

The last advantage of the design method is that it is a well-developed, orderly procedure that anticipates and solves commonly occurring problems of sequential circuits. Other design procedures often result in the appearance of very narrow unwanted pulses or glitches on output lines or occasional oscillation problems. State machine methods eliminate these problems and reduce the time taken to debug the implemented hardware.

The preceding three advantages of state machine design over other types of design make it the predominate method used in sequential logic system design.

5.2 THE STATE MACHINE

Section overview: The state machine is defined and classified in this section. Three applications of the state machine as a system controller are considered.

A. State Machine Definition

A state machine, also called a sequential machine, is a system that can be described in terms of a set of states that the system may enter [1]. Once in a particular state, the system must be capable of remaining in that state for some finite period of time even if the system inputs change. This requirement dictates memory capability for the state machine. Furthermore, the state machine must have a set of inputs and a set of outputs. As the system progresses from one state to another (from present state to next state), the next state reached depends on the inputs and the present state. The outputs also depend on inputs and the present state.

Before leaving this definition of state machines, let us consider the practical meaning of the existence of states. A state may be a set of values measured at various points in a circuit. A simple flip-flop has two states in which it can exist: the output Q can equal either 1 or 0. A set of n flip-flops can produce 2^n possible unique output codes and thus could have 2^n possible unique states of existence. We note here that n lines driven by combinational variables can also take on 2^n different unique codes, but these are not states since they cannot exist independently of the input variables. These n lines change immediately as the input variables change and cannot remain at any code different from the input values. A set of n flip-flops can exist in a particular state different from the set of input values, and this is a requirement of the state machine.

B. Classification of State Machines

It was earlier stated that a sequential machine or state machine must possess memory capability. A large majority of practical state machines use clocked flip-flops as the storage elements. The code that defines each state then corresponds directly to the code contained by the flip-flops.

The general model of the sequential machine is shown in Fig. 5.1. This model is also called the Mealy machine after the man who first proposed the model [1,2]. The input forming logic (IFL) and the output forming logic (OFL) sections are made up of combinational logic circuits. The memory section contains the state of the system. A path is provided from memory output to the IFL. Both input signals and present state signals drive the IFL to determine the next state of the system. The outputs are determined by the present state and the system inputs. A slight variation of the Mealy machine is the Moore machine [1,2] which uses only the memory to drive the OFL. In this case, the output is a function only of the state of the system.

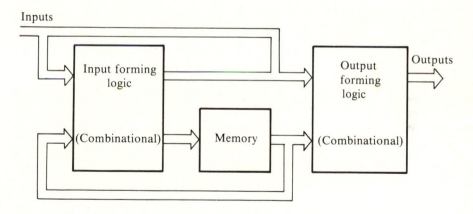

Figure 5.1 General model of a sequential machine.

Another important characteristic of the state machine depends on whether the system is clock-driven or not. In many digital systems a timing reference signal is required. Some sort of astable multivibrator is generally used to produce a continuous clock signal. We refer to a variable as clock-driven if that variable changes value only at the time of a clock transition. It need not change at every clock transition to be clock-driven, but when it does change it must do so as the clock is changing values. When a variable is clock-driven, it is considered a synchronous variable.

If a variable can change at a time not related to transitions of the reference clock, it is called an asynchronous variable. This variable might change as a human operator closes a switch, as an analog voltage reaches a trigger level, or as a vehicle passes a checkpoint. There is no fixed relationship between an asynchronous variable change and the reference clock change. We should also note that a variable might be clock-driven by the clock in one system, but still be considered asynchronous with respect to the reference clock of a second system.

State machines are normally classified as synchronous or asynchronous; however, we will expand the number of categories to three. If a machine changes state in response to the clock and all inputs are synchronous, we will classify that circuit as a synchronous system. If the state changes occur in response to the clock, but one or more inputs are not clock-driven, the machine will be called a mainly synchronous system. If the state changes are input-driven rather than clock-driven, the system is an asynchronous one.

Mainly synchronous systems are encountered more than any other type in the digital field. Synchronous systems are the easiest to design and asynchronous systems are the most difficult. We will consider synchronous systems first and then proceed to mainly synchronous machines in this chapter. Appendix 1 considers asynchronous design.

C. The State Machine as a Sequential Controller

Before we proceed to the fundamental ideas of state machine design, we wish to provide some motivation for this study by considering a few applications of the state machine.

In the control of certain devices or processes it is necessary to generate unique sets of controlling signals during specified time periods. Control of machinery may require time slots ranging from milliseconds to seconds, whereas control of a computer or electronic device may use time slots as short as nanoseconds.

A given process may involve controlling the speed or position of electric motors. For example, a computer-controlled vehicle may use electric motors in the guidance system. The access mechanism in a magnetic disk drive might be powered by an electric motor of some sort. Another process could involve timing of events. The control of stop signals at an intersection to maximize traffic flow on both streets is a typical application of a sequential controller. Here timing is a variable depending on the density of vehicles at various points near the intersection. Control signals must be developed to assert the proper lights at appropriate times.

The process being controlled is referred to, quite reasonably, as the controlled process. The device generating the controlling signals is called a sequential controller or simply a controller. Generally a controller will produce output values depending on certain events or inputs that have occurred earlier or are now present. In these applications, the input signals are generated by commands or by sensors that monitor the process. These inputs help determine the sequence of states through which the system proceeds. The time that the system resides in each state provides the time slots required for driving the controlled system. A different set of outputs can be produced at each state.

A significant requirement of a controlled process is that it must be repeated over and over again with only slight variations. The process is then cyclic in nature. So also is a state machine. Although it may run through a different sequence of states before returning to the starting state, the system has a finite number of such paths and repetition is inevitable.

We will demonstrate these ideas by considering three practical problems in the digital field. The first problem can occur in the digital communication area. Suppose some sort of measurement data is being gathered at a location away from a central computer. This information is transmitted to the central computer to be processed and stored. There may be several sets of variable length data transmitted. For example, one set of data may correspond to pressure measurements, another set may correspond to temperature values, another may correspond to radiation measurements. The data may only be meaningful when the measured value exceeds some threshold; thus each set of data is a variable length, depending on the process being monitored.

In order to separate the sets of data, a frame separator word can be transmitted. This particular word will be unique to avoid confusion with any transmitted data word. It may be transmitted first to signal the computer that data follows. The first set of data is then transmitted followed by the frame separator word. When the receiver identifies this word, the computer recognizes that the next data set corresponds to another measurement. The frame separator word precedes each data set and terminates the final one. Knowing the order of transmission of data sets allows the computer to then process each set appropriately.

The identification of the frame separator word can be accomplished in various ways. One method would be to continually run the incoming words through a shift register and check the register contents at each word boundary for the frame separator word. Conceptually, this is a very simple method. If the word length is 16 bits, a 16-bit shift register is required by this method. A 4-bit counter that keeps track of the word boundaries is also required along with a 16-input decoding circuit.

Another method that can be used is the state machine approach. In this instance, the design procedure might be more time consuming, but the resulting circuit will require less components. The basic goal here would be to create 16 time slots in which the value of each bit of each incoming word is checked. As bits are checked, if they correspond to the bits of the frame separator word, a particular sequence of states is entered. If all 16 states of this sequence are entered, an output is generated that signifies the occurrence of the frame separator word. At any point in the sequence, the detection of a bit that does not correspond to that of the frame separator word sends the system into a sequence of states that will not generate an output. This sequence is designed only for timing purposes. Although the occurrence of an incorrect bit immediately tells the system that the word being checked is not the frame separator word, the system must not start a new check until the beginning of a new word boundary. The use of states for timing purposes eliminates the need for counting circuits.

This particular application would require only five flip-flops and some gating circuits to implement with a state machine. This can be compared to 20 flip-flops plus gating in the shift register scheme. Such a comparison clearly demonstrates the advantage of component minimization in state machine design.

A second problem that can be considered is that of performing serial addition with a single full adder without using a shift register. The organization of a digital system might be primarily a parallel register structure. In order to conserve components, a single full adder rather than a parallel adder may be used. Sixteen-bit words require a 16-bit adder for parallel addition. Gating circuitry to store the sum must also be used. A single-bit full adder and a state machine can accomplish the same result with fewer components. Obviously a serial addition will require more time than a parallel addition.

Let us consider the sequence of operations needed to accomplish a 4-bit addition using a single-bit full-adder. The following steps should accomplish this procedure. Here we assume that one word is stored as bits $A3$ $A2$ $A1$ $A0$ while the other is stored as $B3$ $B2$ $B1$ $B0$. The full-adder has inputs $I1$ and $I2$ for the two operands and CIN, the carry input. The outputs are SUM and COUT.

1. Present $A0$ and $B0$ to inputs $I1$ and $I2$. CIN must be 0.

2. Store SUM as $S0$.

3. Present $A1$ and $B1$ to inputs $I1$ and $I2$. If COUT in step 1 was 0, set CIN = 0; if COUT was 1, set CIN = 1.

4. Store SUM as $S1$.

5. Present $A2$ and $B2$ to inputs $I1$ and $I2$. If COUT in step 3 was 0, set CIN = 0; if COUT was 1, set CIN = 1.

6. Store SUM as $S2$.

7. Present $A3$ and $B3$ to inputs $I1$ and $I2$. If COUT in step 5 was 0, set CIN = 0; if COUT was 1, set CIN = 1.

8. Store SUM as $S3$.

9. Store COUT as $S4$.

The final result is $S4$ $S3$ $S2$ $S1$ $S0$.

To implement this set of procedures with a state machine requires that we choose states to accomplish the various steps listed. We would require a state or time slot in which to present $A0$, $B0$ and CIN = 0 to the adder. Still another would present $A1$, $B1$, and CIN to the adder. As the process begins, the first state would create and store $S0$ and the carry COUT. Based on the value of COUT in this state, the next state would be one that causes CIN = 0 or one that causes CIN = 1. As the system proceeds through the sequence of states, the path taken is determined by the value of COUT. Thus, we see the capability of the state machine to create time slots in which outputs are generated. These outputs depend on previous inputs that have occurred in the past and also on present inputs.

A final example of an application that can be satisfied using a state machine is that of a traffic light controller. At a busy intersection the movement of vehicles may be monitored by pressure-sensitive sensors buried in the roadways. The outputs of these sensors can be used to drive counters to determine the traffic density in each lane for some fixed time interval. Figure 5.2 shows the layout of the intersection and turn lanes.

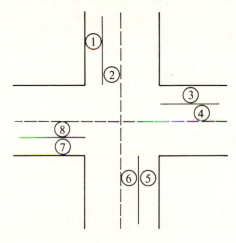

Figure 5.2 An intersection of streets.

The traffic lights are to operate in one of several modes. When there are no vehicles in the turn lanes (even-numbered lanes) and no significant traffic density imbalance, the odd-numbered lanes should see a 50 percent duty cycle for the lights. This means the lights controlling lanes 1 and 5 will be green 50 percent of the time as also will those lights controlling lanes 3 and 7. A second mode occurs if the total traffic density in lanes 1 and 5 differs from the density of lanes 3 and 7 by more than 50 percent. When this occurs, the lights controlling the higher density lanes should remain green for 60 percent of the time while the less dense lane lights are green for only 40 percent of the time.

If vehicles are detected in a single turn lane, the turn light for this lane should be green when the corresponding through lane light turns green. That is, if cars are detected in lane 6, but in no other even-numbered lane, this light will turn green when the lane 5 light turns green. Of course, the lane 1 light must remain red at this time. This condition will prevail for 5 percent of the total period, and then the turn signal will change to red while the lane 1 light turns green.

When cars are in facing turn lanes, either 2 and 6 or 4 and 8, the turn lights of one of these pairs change to green while all other lights are red. After a time duration of 5 percent of the total period, this pair of lights changes to red while the corresponding through lane lights change to green.

It is obvious that a state machine could serve as the controller for this system. Certain time slots must be created, the length of which will vary according to input information from the counters or sensors. The sequence of output signals required to drive the lights will also change in response to traffic patterns. The state machine can generate the correct sequence of outputs by allowing the inputs to influence the sequence of states assumed by the system.

From these three examples we can begin to see that the state machine is a very versatile and important device for digital control systems. Although other methods can be used to create controllers, the state machine is generally accepted as the most popular controller with the best design procedures available.

5.3 BASIC CONCEPTS IN STATE MACHINE ANALYSIS

Section overview: This section introduces the excitation table for the flip-flop. With this table, the signals necessary to produce desired flip-flop outputs can be found. The next state table is considered as an aid in analyzing a state machine. Next, state maps are also discussed. The state diagram is developed to graphically describe the possible state movement of a system.

A. Excitation Tables for Flip-Flops

Chapter 4 introduced the characteristic table for various flip-flops. These tables indicate the outputs resulting from each input combination and allow us to analyze circuits containing flip-flops. In designing circuits, we are given the output signals required and must find the inputs necessary to create these outputs. For flip-flops the excitation tables yield this information. These tables list the input conditions necessary to cause all possible output transitions of the flip-flop. Figure 5.3 shows the excitation tables for the D and JK flip-flops. The variable Q_n represents the Q output prior to the assertion of the CK input and Q_{n+1} is the output after CK has been asserted. The conditions on D or J and K are assumed to be stable prior to the assertion of CK.

Q_n Q_{n+1}	D
$0 \to 0$	0
$0 \to 1$	1
$1 \to 0$	0
$1 \to 1$	1

Q_n Q_{n+1}	J	K
$0 \to 0$	0	Θ
$0 \to 1$	1	Θ
$1 \to 0$	Θ	1
$1 \to 1$	Θ	0

Figure 5.3 Excitation tables.

The excitation table for the D flip-flop is rather straightforward. Regardless of the initial state Q_n, the state after the clock transition will equal the value of D. The JK table is somewhat more complex. If the initial state of the flip-flop is $Q_n = 0$, the value of $Q_{n+1} = 0$ can be obtained with two different sets of input conditions. If $J = 0$ and $K = 0$, the flip-flop will not change states, and thus the zero value for Q remains after the clock transition. A second set of inputs that results in this output condition is $J = 0$ and $K = 1$. This always forces the flip-flop to the $Q_{n+1} = 0$ state. Row 1 of the excitation table combines these input conditions to yield $J = 0$ and $K = \Theta$ ("don't care") for the 0 to 0 output.

The 0 to 1 transition can be caused by setting the flip-flop with $J = 1$ and $K = 0$ or by toggling with $J = 1$ and $K = 1$. These conditions can be specified by $J = 1$ and $K = \Theta$. The same type of reasoning leads to the conditions $J = \Theta$ and $K = 1$ for the 1 to 0 transition and $J = \Theta$ with $K = 0$ for the 1 to 1 output.

Drill Problems: Sec. 5.3A

1. Assuming the state changing clock transitions define the time slots, and $Q(t_0) = 1$, specify $D(t_0)$, $D(t_1)$, $D(t_2)$, and $D(t_3)$ to result in $Q(t_1) = 1$, $Q(t_2) = 1$, $Q(t_3) = 0$, and $Q(t_4) = 1$.

2. If the flip-flop of problem 1 is replaced by a JK device, specify the values of J and K required at t_0, t_1, t_2, and t_3.

3. Develop an excitation table for a T flip-flop.

B. The Flip-Flop as a State Machine

One of the most basic state machines is a flip-flop circuit. A D or JK flip-flop has a set of inputs (D or J and K), a set of outputs (Q), and a set of states ($Q = 0$ and $Q = 1$). Figure 5.4 shows one version of the clocked D flip-flop drawn to conform to the general model of a state machine given in Fig. 5.1.

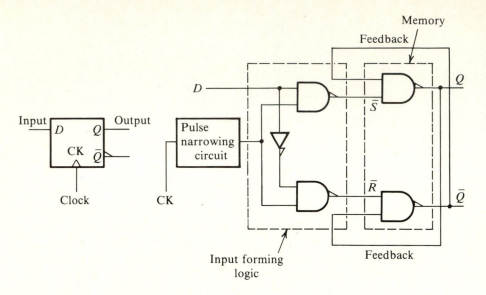

Figure 5.4 The D flip-flop as a state machine.

The $\overline{SR}$ flip-flop is the memory element and the two NAND gates and inverter comprise the input forming logic, or IFL. The feedback from memory to IFL is derived from the Q and $\overline{Q}$ outputs. There is no output forming logic, or OFL, for this state machine. Since the output Q is a direct function of the memory only, this is a Moore machine. The clock signal that drives the state changes is not considered an input signal although it is a necessary signal in the operation of synchronous and mainly synchronous state machines. It is simply understood that these types of systems will be driven by a reference clock signal.

While it is true that a single flip-flop is a state machine, its two possible states are too few to allow this device to be a very practical system. Two or more devices can be used to create a reasonable number of states to result in quite useful state machines.

C. Analyzing State Machines

Although our ultimate objective is state machine design, we can learn much about the operation of these systems through analysis of some rather simple versions. We can also see the value of certain design aids such as the next state table and next state map. We will consider the synchronous system of Fig. 5.5 consisting of two D flip-flops to introduce some basic ideas relating to state machine analysis.

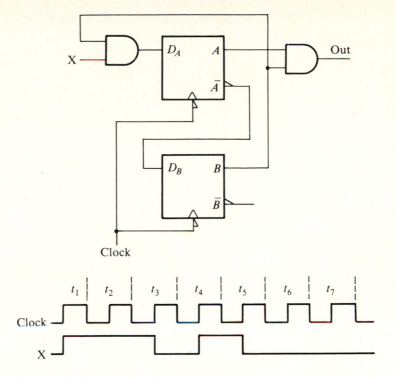

Figure 5.5 A state machine using D flip-flops.

The IFL consists of the AND gate that drives flip-flop A with the signal $D_A = XB$ and the wire that drives flip-flop B with $D_B = \overline{A}$. The OFL is the AND gate that results in Out $= AB$. We recognize in this circuit the basic characteristic of the state machine, that is, the next states are functions of the present state and the input X since X, A, and B determine the values of D_A and D_B.

It is worth noting that the input changes on the opposite clock transition to that which drives the state changes. The state changes take place on the negative clock transition while the value of X is updated on the positive transition. This allows the signals driving the flip-flop inputs to be stable prior to the state change.

In order to find the states through which the system moves, the next state table of Fig. 5.6 is constructed. This table first lists all possible present states and input values. A consideration of the input equations, $D_A = XB$ and $D_B = \overline{A}$, allows the flip-flop inputs to be listed for each present state and input combination. The next state for each combination can be found from the D flip-flop excitation table of Fig. 5.3. A negative clock transition causes the next state to take on the values presented to the D inputs prior to the transition. The output is determined from the equation Out = AB using present state values.

Row	Present State A	B	Input X	Flip-flop Inputs D_A	D_B	Next state A	B	Output Out
1	0	0	0	0	1	0	1	0
2	0	0	1	0	1	0	1	0
3	0	1	0	0	1	0	1	0
4	0	1	1	1	1	1	1	0
5	1	0	0	0	0	0	0	0
6	1	0	1	0	0	0	0	0
7	1	1	0	0	0	0	0	1
8	1	1	1	1	0	1	0	1

Figure 5.6 Next state table for the state machine of Fig. 5.5.

Once the next state table has been constructed, we can use it to trace the state movement of the system. Assuming that the initial state during t_1 of the waveform in Fig. 5.5 is $A = 0$ and $B = 0$, and noting that $X = 1$ prior to the first negative clock transition, we see that row 2 of the table describes these present state conditions. This row tells us that D_A will equal 0 and D_B will equal 1 for these conditions and the output will equal 0. The next state reached after a negative clock transition will be $A = 0$ and $B = 1$ or 01. After the negative clock transition, this state becomes the new present state. This state will persist during t_2. From the input waveform we see that $X = 1$ throughout t_2. Row 4 of the next state table applies to this case. This row indicates an output of Out = 0 and a next state of 11. After the negative clock transition, the present state will become 11 and will remain at this value during t_3. We note that X drops to 0 prior to the next negative clock transition, thus row 7 applies to this time period. The output Out = 1 and the next state will be 00. During t_4, the present state is 00 and X returns to 1 before the beginning of t_5. Row 2 is now used to predict an output of 0 and a next state of 01. This state of 01 lasts through t_5 with X dropping to 0 prior to t_6. Row 3 describes these conditions and dictates an output of 0 and a next state of 01. Since X remains low, the state 01 remains through t_6 and t_7. The timing chart of Fig. 5.7 reflects the information of the preceding discussion.

Since the input data X is updated at the midpoint of the state time, there may be some question as to which value of X to use in predicting the next state. The correct value is that which is in effect just prior to the state-changing clock transition. This value remains constant over the last half of the state time allowing any D inputs affected to be stable prior to state change.

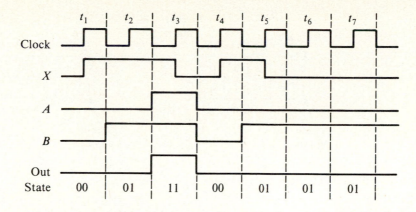

Figure 5.7 Timing chart for the circuit of Fig. 5.5.

Present state and next state maps can also be useful in state machine analysis and design. Before leaving this example, we will introduce these design aids. A present state map is used to define the state names. For the two flip-flop system of Fig. 5.5, there are four possible states. The present state map shows these possible states as functions of the flip-flop outputs A and B as in Fig. 5.8. This map defines state a as $A = 0$ and $B = 0$ or 00, state b as 01, state c as 10, and state d as 11. Although these are arbitrarily assigned in this case, we will later consider some factors that determine state assignment. The asterisk in state d identifies this as an output state.

	A 0	1
B 0	a	c
1	b	$d*$

Figure 5.8 Present state map.

The next state table plots the next state of each flip-flop as a function of present state and input. Figure 5.9 shows the next state maps for both flip-flops. This information can be plotted from the next state map or it can be generated from the equations for the flip-flop inputs. For a D flip-flop, the next state will correspond to the present state value appearing on the D input. We can equate the next state of A to the present value of D_A to generate the first map of Fig. 5.9, that is, $D_A = XB$. The second map is derived from $D_B = \overline{A}$.

During time slot t_1, the state is 00 and $X = 1$. These conditions define a specific location on the next state maps. When a state-changing clock

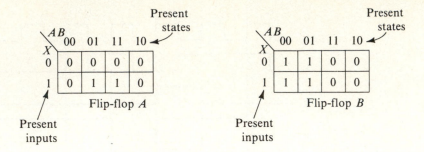

Figure 5.9 Next state maps.

transition occurs, the maps tell us that flip-flop A will remain at $A = 0$ while flip-flop B will move to $B = 1$. The state movement can be followed by means of the next state maps through each time slot. The next state map is very valuable in state machine design as will be demonstrated in a later section.

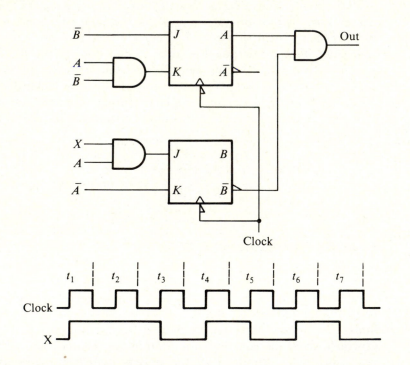

Figure 5.10 A *JK* state machine.

The D flip-flop is easier to analyze than the JK, but the latter generally requires less circuitry to implement a state machine than does the D flip-flop. Since this device is rather popular for state machines, we will analyze the JK

circuit of Fig. 5.10. The next state table is shown in Fig. 5.11. The expressions for flip-flop inputs are found by examining the IFL of the circuit. This gives $J_A = \bar{B}$, $K_A = A\bar{B}$, $J_B = AX$, and $K_B = \bar{A}$. The equation for output is seen to be Out $= A\bar{B}$. The excitation table for the JK flip-flop of Fig. 5.3 is used to determine the next state from a knowledge of the J and K inputs.

Row	Present State A	B	Input X	Flip-flop Inputs J_A	K_A	J_B	K_B	Next State A	B	Output Out
1	0	0	0	1	0	0	1	1	0	0
2	0	0	1	1	0	0	1	1	0	0
3	0	1	0	0	0	0	1	0	0	0
4	0	1	1	0	0	0	1	0	0	0
5	1	0	0	1	1	0	0	0	0	1
6	1	0	1	1	1	1	0	0	1	1
7	1	1	0	0	0	0	0	1	1	0
8	1	1	1	0	0	1	0	1	1	0

Figure 5.11 Next state table for circuit of Fig. 5.10.

The present and next state maps are shown in Fig. 5.12. We will apply the next state maps to analyze this state machine although the next state table could be used also.

Assuming an initial state during t_1 of 00 and $X = 1$, we enter the next state maps for these conditions. The maps indicate a next state of $A = 1$ and $B = 0$ which will exist during t_2; hence we move to the map locations specified by a state of 10 with $X = 1$. This indicates a next state of 01 which will exist during t_3. Prior to the end of t_3, X drops to 0. The maps indicate a state during t_4 of 00. Following this procedure through t_5, t_6, and t_7 leads to the waveforms of Fig. 5.13.

From these two examples we see that the next state table and next state maps are quite useful in analyzing state machines. The next section will discuss the state diagram which allows us to graphically summarize the state movement and output signals of the state machine.

Drill Problems: Sec. 5.3C

1. The state machine of Fig. 5.5 is modified by making $D_A = \bar{A}B$ and $D_B = X\bar{B}$. Develop the next state table for this system.

2. Develop the next state maps for the system of problem 1.

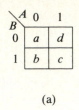

(a)

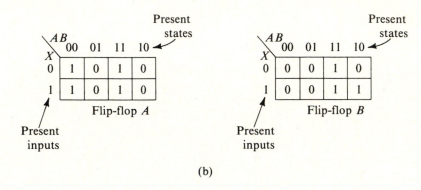

(b)

Figure 5.12 (a) Present state map. (b) Next state map.

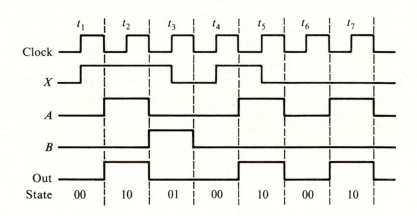

Figure 5.13 Waveforms for the circuit of Fig. 5.10.

D. State Diagrams

A state diagram allows the behavior of a state machine to be analyzed very rapidly. In addition, the state diagram can be quite useful in designing a machine from a set of specifications. We will, therefore, spend some time discussing the basic ideas underlying this important design aid. Figure 5.14 shows the symbols used in the state diagram.

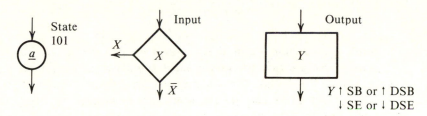

Figure 5.14 State diagram symbols.

The circular state symbol contains the name of the state with the state code adjacent to the symbol. The arrows indicate the entry path and exit path. Although there may be multiple entry paths, there will always be a single exit path. An input variable is represented by the diamond-shaped symbol containing the variable name. This symbol will always have a single entry path, but will have two exit paths determined by the value of the variable. When an output is to be generated, a rectangular block is used containing the name of the output variable. The upward arrow indicates that the output is asserted, while the downward arrow indicates deassertion. If the output is asserted at the beginning of the state, the notation $Y{\uparrow}$SB is used. If Y is asserted at mid-state, the notation $Y{\uparrow}$DSB is used indicating assertion is delayed from the beginning of state time. The notation $Y{\downarrow}$SE and $Y{\downarrow}$DSE means Y is deasserted at the end of state time and on the next clock transition after the state time end, respectively.

In order to illustrate its use, we will construct the state diagram for the circuit of Fig. 5.5. Using the next state table or next state maps and assuming the circuit can be reset to the a state, the state diagram of Fig. 5.15 results. From state a (00) the diagram indicates that the circuit will move to state b regardless of the value of X. After one negative clock transition the state of the system is b (01). If X is zero during the time the circuit is in state b, the negative clock transition would not change the state. Instead, state b would continue until X goes to 1. When this input condition occurs, the following negative clock transition will move the system to state d (11). From state d, the circuit will either move to state a, if $X = 0$, or to state c (10), if $X = 1$. When state d is entered, the output is asserted for one clock period. At the end of state c the circuit returns to state a, regardless of the input condition. This same information is presented by the next state table and maps, but it is represented more concisely by the state diagram.

Returning to the X input waveform of Fig. 5.5, we note that the system will move from state a to state b as a result of the first negative clock transition. Since $X = 1$ during t_2, the system then moves to state d at the beginning of t_3. The input X drops to 0 prior to the next negative clock

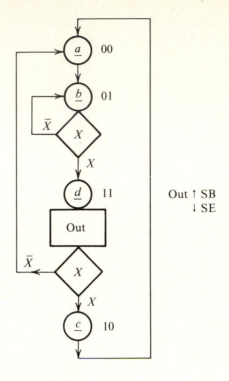

Figure 5.15 State diagram for circuit of Fig. 5.5.

transition moving the system back to state a during t_4. The system then moves to state b and remains there since $X = 0$ from that point on.

Not only is the state diagram useful in system analysis, it is indispensable in designing a controlling state machine from a set of specifications. A state diagram can consist of many states involving several inputs and outputs. In some cases, an input diamond must be repeated to provide the several possible output paths that may result in a state diagram. Figure 5.16 demonstrates such an example.

The four possible combinations of X and Y lead to four different next states following state a; consequently the duplicate input symbol for Y is required. If possible, all states that can be entered at a given time should be placed on the same horizontal level. States b, c, d, and e are on the same level since one of these states will be entered when state a is exited. We note that input X also controls entry into state h later in the diagram. This situation is not uncommon in state machines. The symbol W↓SE appears indicating that W is deasserted at the end of state h. Output P follows input X and will become asserted when in state f only if $X = 0$.

The symbol representing input S contains a shaded portion at the bottom

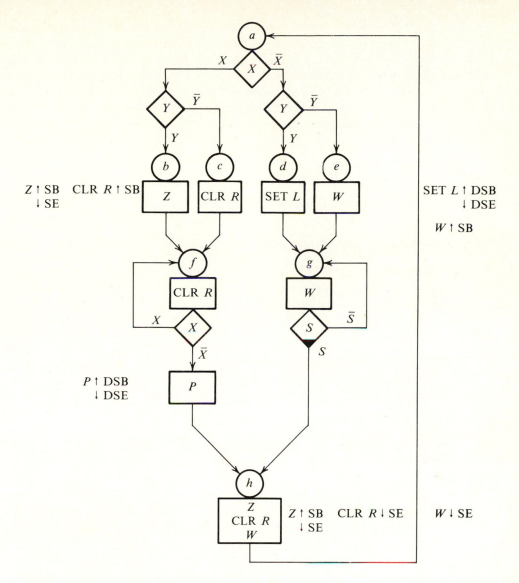

Figure 5.16 State diagram.

of the diamond. This indicates that input S is not a synchronous variable. S is therefore an asynchronous input that will change at some time that is not related to a clock transition. All inputs without this shaded region represent synchronous inputs. If state f is reached, the system will remain in this state as long as X is asserted. When X is deasserted, the system can move to state h.

Although the synchronous inputs change as the clock makes a transition, it is important to recognize that state changes should take place on the alternate clock transition. If this is true, the input conditions have one-half of a clock period in which to settle prior to a change in state.

Generally the state diagram will be obtainable from the specifications of the design problem. The circuit is then developed from the state diagram to satisfy the specifications using the techniques discussed in the following sections.

Drill Problem: Sec. 5.3D

Draw the state diagram for the system described in drill problem 5.3C-1.

5.4 SYNCHRONOUS STATE MACHINE DESIGN

Section overview: This is one of the most significant sections of the entire text. The basic principles of state machine design are covered in this material. Input forming logic design, output forming logic design, and criteria for state assignment are considered. Much of the succeeding material in this chapter and in Chap. 6 is based on concepts introduced in this section.

This section will discuss the basic design principles of state machines. The initial material will be rather general, but we will later discuss some very specific problems that arise due to nonideal effects of switching circuits. Each idea will be demonstrated by example to clarify its application. The first system considered is the sequential or synchronous counter. This system has no external inputs, other than direct sets or resets, and is therefore simple enough to serve as a good introductory circuit.

A. Sequential Counters

The sequential counter is used to generate some fixed sequence of states in a cyclic manner. For example, the sequence of states desired might be 000, 011, 101, 111, 100, 000, 011, 101, 111, 100, 000, ... Sequential counters are state machines with no inputs other than the preset inputs that initialize the system. The general block diagram for this system is shown in Fig. 5.17. This particular type of state machine has a next state that is determined by the present state only. Quite often, the output will be taken directly from the flip-flops eliminating the need for the OFL. The state diagram for the sequence previously listed is shown in Fig. 5.18. The next state table and appropriate maps are shown in Fig. 5.19. Gating information for both D flip-flops and JK flip-flops is also given. The excitation tables for the D and JK flip-flops are used to generate this table. When we are in state a, we see that flip-flop B must change from 0 to 1 on the clock transition. This requires that $J_B = 1$ and $K_B = \Theta$. The three next state maps for the flip-flops have been combined into a single map. The states marked with an asterisk were not defined from the specs in terms of the next state to follow. We would be tempted to call these "don't care" states since they would never be entered if the counter performed in the cyclic manner indicated by the state diagram. However, the counter might be functioning in the prescribed manner when an extraneous noise pulse occurs to send the counter to one of these undefined states. It would be possible for the counter to hang up in some repetitive sequence other than the desired one if these states are not designed to return the system to the prescribed sequence. For example, suppose that the next state after 001 is 010. Further, assume that the next state after 010 is 110 and the next state after 110 is 001. If this were true, the state diagram of Fig. 5.20 would apply. A noise pulse could send the counter to states f, g, or h, and the system would never return to the desired sequence unless the preset signal is applied or another random noise pulse accidently returns the state to a, b, c, d, or e.

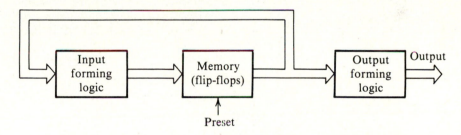

Figure 5.17 Block diagram of the sequential counter.

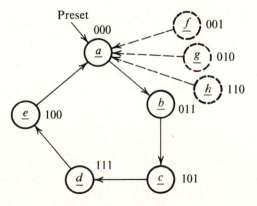

Figure 5.18 State diagram for sequential counter.

State	Present State			Next State			Flip-Flop Inputs								
	A	B	C	A	B	C	D_A	D_B	D_C	J_A	K_A	J_B	K_B	J_C	K_C
a	0	0	0	0	1	1	0	1	1	0	θ	1	θ	1	θ
f	0	0	1*	0	0	0*	0	0	0	0	θ	0	θ	θ	1
g	0	1	0*	0	0	0*	0	0	0	0	θ	θ	1	0	θ
b	0	1	1	1	0	1	1	0	1	1	θ	θ	1	θ	0
e	1	0	0	0	0	0	0	0	0	θ	1	0	θ	0	θ
c	1	0	1	1	1	1	1	1	1	θ	0	1	θ	θ	0
h	1	1	0*	0	0	0*	0	0	0	θ	1	θ	1	0	θ
d	1	1	1	1	0	0	1	0	0	θ	0	θ	1	θ	1

C \ AB	00	01	11	10
0	011	000	000	000
1	000	101	100	111

Next state ABC

Figure 5.19 Next state table and maps for the sequence counter of Fig. 5.18.

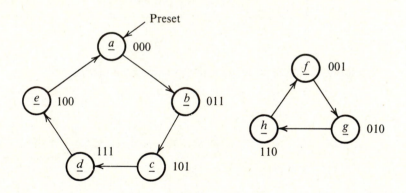

Figure 5.20 State diagram for a poorly designed sequence counter.

In order to avoid this problem, all unused states can be designed to have a next state of 000 or some other state contained in the desired sequence. If one of these states is inadvertently entered, the system will recover to the desired sequence when the next clock pulse is applied. This type of system is also referred to as a "self-starting" circuit. When the dc power is first turned on the circuit will come up in an arbitrary state. Regardless of which state comes up, the desired sequence will shortly be entered.

For a D flip-flop design we must produce the logic functions to drive D_A, D_B, and D_C. We begin the design process by plotting the maps of Fig. 5.21. The inputs to the IFL are the flip-flop outputs A, B, and C. These variables are used to plot the next state maps of Fig. 5.21 which are also the maps for D_A, D_B, and D_C. This counter can be constructed as shown in Fig. 5.22. The preset input can be taken low to start the sequence at state 000. Deassertion of this input will allow the counter to cycle through the prescribed sequence. The number of gates used for the IFL can be reduced further by using EXOR type circuits. We will leave these considerations to the student.

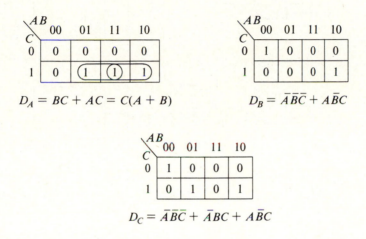

$$D_A = BC + AC = C(A + B)$$

$$D_B = \bar{A}\bar{B}\bar{C} + A\bar{B}C$$

$$D_C = \bar{A}\bar{B}\bar{C} + \bar{A}BC + A\bar{B}C$$

Figure 5.21 D input maps for sequence counter.

The JK realization is more difficult to find, requiring six maps as shown in Fig. 5.23 rather than three. These maps are based on the next state or D maps of Fig. 5.21. The JK excitation table aids in this construction. Although more maps are required in JK flip-flop design, fewer gates and/or inputs are sometimes needed for the input forming logic as demonstrated by the counter of Fig. 5.24. Again if EXOR or NEXOR circuits are used, the IFL can be reduced still further.

There is a shortcut method that can be used to construct the JK maps from the next state or D maps. Returning to the excitation tables of Fig. 5.3, we note that for a present state of $Q_n = 0$, K is always equal to Θ while $J = D$ for both values of next state. For a present state of $Q_n = 1$, J always equals Θ, while $K = \bar{D}$ for both values of next state. From these relations we can develop an algorithm to convert the next state or D map to J and K maps. This algorithm follows.

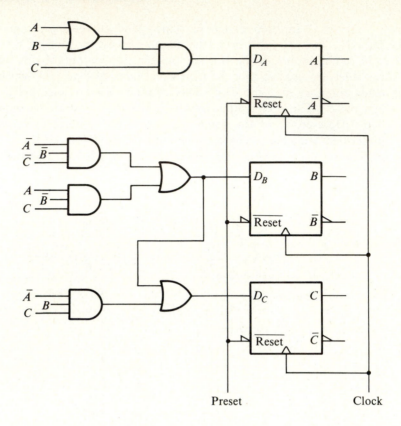

Figure 5.22 Sequence counter with *D* flip-flops.

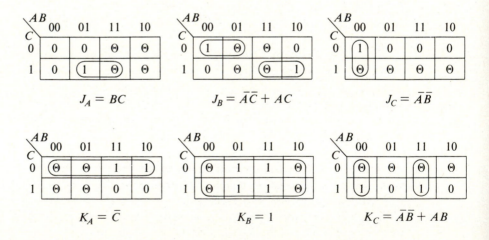

Figure 5.23 *JK* inputs for sequence counter.

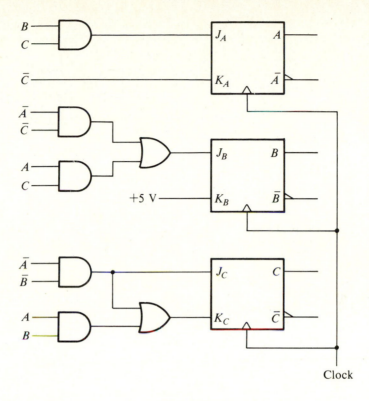

Figure 5.24 Sequence counter with *JK* flip-flops.

1. Place a Θ in all locations of the K-map corresponding to a present state of 0. Place $K = \bar{D}$ values in the remaining locations.

2. Place a Θ in all locations of the J-map corresponding to a present state of 1. Place $J = D$ values in the remaining locations.

Since the next state maps will always be known prior to the development of the flip-flop maps, this method can decrease the time required to plot the J- and K-maps.

In Chap. 4, the ripple counter was discussed. Although this counter is useful in lower frequency counting applications, the ripple effect limits its use in higher frequency work. The sequential counter solves this problem and can be used at higher operating speeds. In this situation, the states are chosen to be consecutive starting at the zero state. The resulting counter is called a synchronous counter, and the standard 74163 is shown in Fig. 5.25. This counter can be preset to any desired initial state and includes provision for cascading chips to create larger counters.

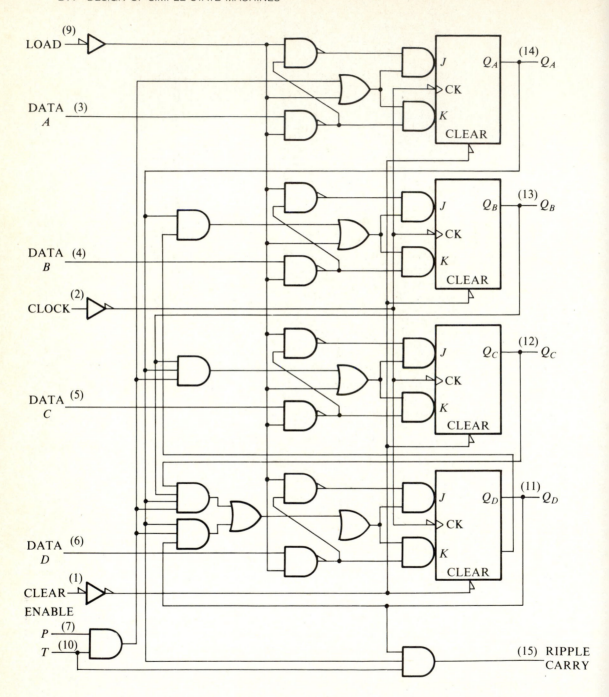

Figure 5.25 The synchronous binary counter.

The sequential counter is a synchronous system, but it is a special case that involves no external inputs other than the preset input. Only the state flip-flop outputs are used to drive the IFL. We will now consider the design of systems having external inputs that influence the next states of the circuit.

Drill Problems: Sec. 5.4A

1. Design a synchronous counter using D flip-flops to count in the sequence 01, 11, 10, 01, 11, 10, ...

2. Repeat problem 1 using JK flip-flops.

B. State Changes

The flip-flops of the state machine change to a new value as a result of a clock transition. At this transition, the information represented by the flip-flop inputs is set into the flip-flops. There is a requirement called data-to-clock setup time that tells us that the data applied to the D or JK inputs must be stable prior to the occurrence of the state-changing clock transition. A typical minimum setup time for a conventional TTL flip-flop is 10 ns. For low power Schottky this time is 20 ns, while a value of 3 ns might be typical of Schottky TTL. Figure 5.26 demonstrates the problem that can arise if we do not observe this specification.

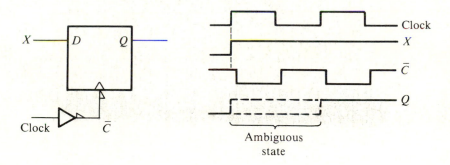

Figure 5.26 A D flip-flop with synchronous input.

If X changes on the positive clock transition, the D input will change slightly before this transition reaches the CK input. The signal X may reach D soon enough to satisfy setup time requirements for the flip-flop, depending on the propagation delay time of the inverter. If so, the Q output will switch to 1 at this time. If the propagation delay of the inverter is not long enough to allow an appropriate setup time, the flip-flop will not change state until the next positive clock transition.

The ambiguous situation depicted in Fig. 5.26 must be avoided in order to achieve reliable state machine operation. Although not the only solution, one method that avoids this problem is stated as the following design principle:

> When inputs are synchronous, state changes should occur on the alternate clock transition to input data changes.

This allows a half-period of the clock for data lines to become stable before the state change takes place.

It is possible to solve this problem in other ways. For example, a series of delaying inverters on the clock input can guarantee the necessary delay to allow data to stabilize before state changes. This method can be used in situations wherein timing is very critical, but we will always observe the principle that data and state changes take place on alternate transitions of the clock signal. If input data changes on the positive transition, we will design the system to change state on the negative transition.

Before we leave the topic of state changes it is important to realize that the state machine is more reliable than many other digital systems as a result of the fact that output changes occur only during clock transitions. Input data changes have a half-period to settle, and thus any hazard produced by inputs are ignored by the state flip-flops. When a state change takes place, the IFL may generate hazards since hazard covers are not used in this section of the system. These hazards will appear after the clock transition that drives the state change take place and will not affect the operation of the state machine. This discussion assumes that the flip-flops used are edge-triggered devices rather than "ones-catching" master-slave circuits.

Even in the presence of noise the state machine operates more reliably than other systems. Not only must a noise pulse be large enough to cause problems of level interpretation, it must occur slightly before the state changing clock transition to cause an erroneous state to be reached. The list of advantages of the state machines over other types of digital systems must include that of highly reliable operation.

C. Number of State Flip-Flops

Another matter that must be considered prior to IFL design is that of the number of flip-flops required for a state machine. This is easily found from the state diagram which shows the number of states needed for the system. Given n flip-flops, the number of unique codes generated is 2^n states.

In order to choose n, the number of states of the system is first determined and the smallest value of n that satisfies the relation

Number of states less than or equal to 2^n

is selected. For example, if 12 states are needed, four flip-flops are required since $2^4 = 16$. If we used three flip-flops, we see that only $2^3 = 8$ is the maximum number of states possible.

D. Input Forming Logic

The IFL consists only of combinational logic. This section is driven by system inputs and by the outputs of the state flip-flops. With these input signals the IFL produces a set of variables that will drive the flip-flops to the proper next state when the clock transition takes place. If D flip-flops are used in the state machine, one input per flip-flop is required. JK devices require two inputs per flip-flop. Regardless of the type of flip-flop used, the IFL must generate combinational logic functions and can therefore be constructed from gates, MUXs, decoders, PLAs, or ROMs. After discussing the use of some of these building blocks we will consider the factors involved in choosing a particular implementation.

IFL Using Gates

The traditional method of IFL realization uses gates. This method is still significant in applications using standard IC chips since component cost can be minimized with gates. When IFL is constructed from gates, this circuitry can be minimized by observing the following two principles:

> Principle 1. States having the same next state for a given input condition should have logically adjacent state assignments.

> Principle 2. States that are next states of a single state should have logically adjacent assignments.

Two examples will demonstrate the use of these principles before we consider why they lead to a minimal realization of IFL.

A problem that occasionally occurs in digital communications is that of identifying certain bit patterns within a serial data stream. We will simplify this problem to the point of impracticality to introduce the basic ideas of IFL design.

■ *Example 5.1* A serial data line X is allowed to change on the falling edge of the clock signal. A state machine is to be designed to detect every next value of X. If $X = 1$, output Y is to be asserted. If $X = 0$, output Z is to be asserted. The output value should be asserted for one state time, and then the next value of X should be checked.

□ *Solution:* The state diagram reflecting the specified behavior is shown in Fig. 5.27. Since data changes take place on the negative clock transition, state changes must occur on the positive transition. Depending on the value of X at the first positive clock transition, the system moves to a specific state. If $X = 1$, the system moves to state b; if $X = 0$, it moves to state c. The output in either of these states exists for one clock time after which the system returns to state a to check the alternate data bit.

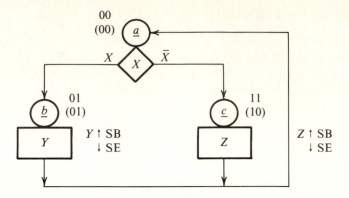

Figure 5.27 State diagram for Example 5.1.

A timing diagram is useful to indicate the relationship between data input, clock, state times, and outputs. This chart is shown in Fig. 5.28.

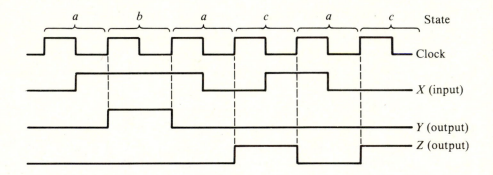

Figure 5.28 Timing diagram for Example 5.1.

Three states are required to satisfy this problem, and thus two state flip-flops are used. We will label these flip-flops A and B, where A represents the MSB of the state code.

Prior to designing the IFL, the state assignment must be made. This selection is made in accordance with the two principles stated at the beginning of the section. States b and c both go to state a regardless of the value of X. Therefore these states should be adjacent according to Principle 1. Principle 2 also dictates the adjacency of these two states since both follow state a.

In order to demonstrate the results of a correct state assignment, we will assign the states two different ways as shown in Fig. 5.29. The first map selects b and c as adjacent states while the second map chooses these states to be nonadjacent.

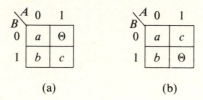

(a)   (b)

Figure 5.29 Present state maps for Example 5.1: (a) correctly chosen states; (b) for incorrectly chosen states.

The next step of the procedure is to develop the next state maps (or tables) for flip-flops A and B. This is simplified by adding the assigned codes to the state diagram as shown in Fig. 5.27. The improperly chosen states are included in parentheses. Figure 5.30 includes the next state maps for both cases of state assignment.

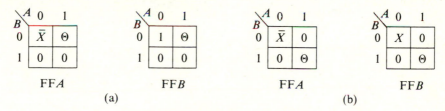

(a)   (b)

Figure 5.30 Next state maps for Example 5.1: (a) for correctly chosen states; (b) for incorrectly chosen states.

The next state map for properly chosen states is created by considering the conditions prevailing at each state. At state a if $X = 0$, the next state code will be 11; if $X = 1$, this code will be 01. Flip-flop B will change to 1 regardless of the value of X. This fact is reflected by placing a 1 in the map for flip-flop B at the location corresponding to state a or 00. Flip-flop A will change to 0 if $X = 1$ and 1 if $X = 0$. This information is noted by placing an $\overline{X}$ in the map for flip-flop A at location 00. The other map locations are filled in a similar manner.

In location 00 of the incorrectly chosen state maps we see an $\overline{X}$ for flip-flop A and an X for flip-flop B. These values result because flip-flop A must be 0 if $X = 1$ and must change to 1 if $X = 0$. On the other hand, flip-flop B must change to 1 if $X = 1$ and remain at 0 for $X = 0$. The unused state leads to "don't care" conditions in the maps. Rather than use next state maps, we could use next state tables instead. The next state table for correctly chosen states is shown in Fig. 5.31.

Present State		Input	Next State		Outputs	
A	B	X	A	B	Y	Z
0	0	0	1	1	0	0
0	0	1	0	1	0	0
0	1	0	0	0	1	0
0	1	1	0	0	1	0
1	0	0	Θ	Θ	Θ	Θ
1	0	1	Θ	Θ	Θ	Θ
1	1	0	0	0	0	1
1	1	1	0	0	0	1

Figure 5.31 Next state table for Example 5.1 with correctly chosen states.

After developing the next state maps or table, we must decide which type of flip-flop to select for the state machine. The use of D flip-flops leads to a simpler design procedure since only one input per flip-flop is needed. JK flip-flops require a longer procedure but may result in less complex gating circuitry for the IFL. We will demonstrate the use of both types of flip-flops designing the system first with JKs and then with Ds.

We must consider the JK excitation table of Fig. 5.3 along with the next state maps of Fig. 5.30(a) to produce the gate input maps for the system. Figure 5.32 shows these maps. From the next state map for flip-flop A, we see that when the system is in state a, flip-flop A changes from 0 to 1 if $\bar{X} = 1$. Relating these conditions to the JK excitation conditions leads to $J = 1$, and $K = \Theta$ if $\bar{X} = 1$ and $J = 0$, $K = \Theta$ if $\bar{X} = 0$. This can be expressed as $J_A = \bar{X}$ and $K_A = \Theta$ in the map locations corresponding to state a. This same information could be generated more easily by using the algorithm of Sec. 5.4A for converting from next state map to J and K maps.

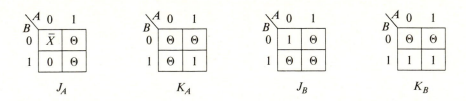

Figure 5.32 Gating maps for JK flip-flops.

When in state a, the next state of flip-flop B is always 1. The excitation table indicates that a transition from 0 to 1 requires $J = 1$ and $K = \Theta$ which is entered in the maps for J_B and K_B.

Moving to the proper next state from state b requires that flip-flop A remain at 0 and flip-flop B change from 1 to 0. The excitation table tells us that flip-flop A must have $J_A = 0$ and $K_A = \Theta$, while flip-flop B must have $J_B = \Theta$ and $K_B = 1$. To move from state c to state a requires that both flip-flops change from 1 to 0 resulting in $J_A = J_B = \Theta$ and $K_A = K_B = 1$. Of course, the unused state will contain "don't care" conditions for all flip-flop inputs.

The maps for the flip-flop inputs can now be reduced to give $J_A = \overline{X}B$, $K_A = 1$, $J_B = 1$, and $K_B = 1$. The IFL with JK flip-flops is shown in Fig. 5.33.

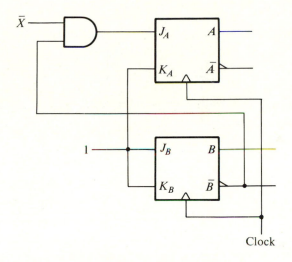

Figure 5.33 IFL using *JK* flip-flops for Example 5.1.

The D design is more straightforward than the JK design. Since the data on the input of a D flip-flop becomes the next state of the flip-flop after the clock transition occurs, the next state maps of the state machine flip-flops are also the maps for the D inputs. Figure 5.34 indicates these maps for flip-flops A and B. Maps for both correctly chosen and incorrectly chosen states are included. The D input maps for the correctly chosen states lead to the equations $D_A = \overline{X}B$ and $D_B = \overline{B}$. For the incorrectly-chosen states we get $D_A = \overline{X}A\overline{B}$ and $D_B = X\overline{A}\overline{B}$. The resulting IFL for the two cases is shown in Fig. 5.35. We note that the system of Fig. 5.35(a) uses only a single gate with two inputs, whereas the other system requires three gates and six inputs for the IFL.

Some designers prefer to work with a next state table instead of or in addition to next state maps. The flip-flop input information can be presented with a next state table as indicated in Fig. 5.36. Again we will emphasize that the D input information simply duplicates the next state information. We will defer the design of the OFL to a later section. ■

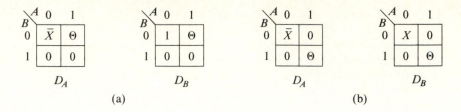

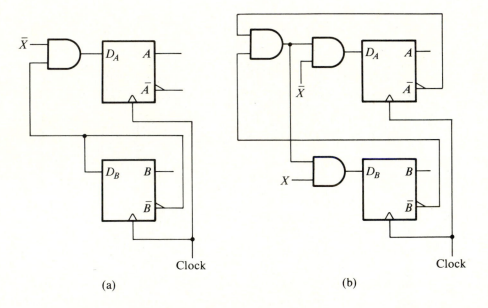

Figure 5.34 D input maps for Example 5.1: (a) for correctly chosen states; (b) for incorrectly chosen states.

Figure 5.35 Two designs for Example 5.1: (a) based on correctly chosen states; (b) based on incorrectly chosen states.

A second example will demonstrate a larger state machine and again compare designs using D and JK flip-flops.

■ *Example 5.2* A serial data line X is allowed to change on the falling edge of the clock signal. Design the IFL for two state machines, one using D and the other JK flip-flops, to detect the sequence 10 or 01 on the data line. If 10 occurs, output Y should be asserted for one state time before the next check is started. If 01 occurs, output W should be asserted for one state time. If 00 or 11 appear as the two sequential bits on X, no output

Present State A B	Input X	Next State A B	Flip-Flop Inputs D_A D_B J_A K_A J_B K_B
0 0	0	1 1	1 1 1 $\ominus$ 1 $\ominus$
0 0	1	0 1	0 1 0 $\ominus$ 1 $\ominus$
0 1	0	0 0	0 0 0 $\ominus$ $\ominus$ 1
0 1	1	0 0	0 0 0 $\ominus$ $\ominus$ 1
1 0	0	$\ominus$ $\ominus$	$\ominus$ $\ominus$ $\ominus$ $\ominus$ $\ominus$ $\ominus$
1 0	1	$\ominus$ $\ominus$	$\ominus$ $\ominus$ $\ominus$ $\ominus$ $\ominus$ $\ominus$
1 1	0	0 0	0 0 $\ominus$ 1 $\ominus$ 1
1 1	1	0 0	0 0 $\ominus$ 1 $\ominus$ 1

Figure 5.36 Next state and flip-flop input table for Example 5.1 with correctly chosen states.

should be generated and the next check should begin after a delay of one state time.

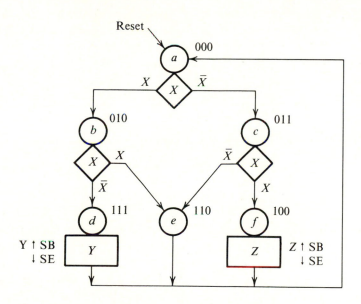

Figure 5.37 State diagram for Example 5.2.

□ **Solution:** The state diagram for this system is constructed as shown in Fig. 5.37. With six states, three flip-flops are required. This yields eight possible state codes. In order to make a proper state assignment, we note that states d, e, and f all proceed to state a for any value of X. Thus,

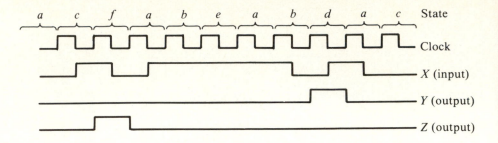

Figure 5.38 Timing chart for Example 5.2.

these states should be logically adjacent to satisfy Principle 1. States b and c both have e as a next state, but for different input conditions. States b and c are next states of a and should be logically adjacent as a result of Principle 2. States d and e are next states of b, and f and e are next states of c. These two pairs should also be logically adjacent. Figure 5.38 shows the timing diagram. An appropriate state assignment is given in Fig. 5.39.

C ╲ AB	00	01	11	10
0	a	b	e	f
1	Θ	c	d	Θ

Figure 5.39 Present state map for Example 5.2.

The next state, D, and JK input maps are then developed as shown in Fig. 5.40. The equations for the D design are $D_A = \bar{A}B$, $D_B = \bar{X}\bar{A} + \bar{A}\bar{C}$, and $D_C = \bar{X}A\bar{C}$. The equations for the JK design are $J_A = B$, $K_A = 1$, $J_B = \bar{A}$, $K_B = A + XC$, $J_C = \bar{X}A$ and $K_C = 1$. Both implementations are shown in Fig. 5.41. The D implementation uses five gates with 10 inputs while the JK implementation uses three gates with six inputs. ∎

Before we leave the subject of IFL design using gates we will consider why Principles 1 and 2 lead to a minimized system. Suppose we have a portion of a state diagram that appears as shown in Fig. 5.42. Let us further assume that this system has three D-type state flip-flops, A, B, and C. Figure 5.43 shows two of the many possible assignments for states c through g.

The next state maps for these assignments, which are the same as the D input maps, are included in Fig. 5.44. From each of the states c, d, e, and f the system moves to state g if $X = 1$. Thus, all flip-flops change to the 1 state if $X = 1$ from these four states. Since we do not know where the system moves from these states if $X = 0$, we cannot determine if an X or 1

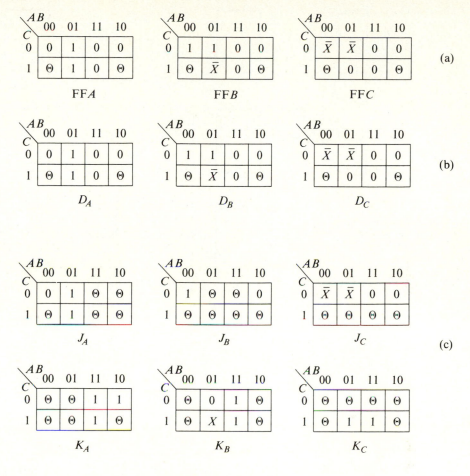

Figure 5.40 (a) Next state maps. (b) *D* input maps. (c) *JK* input maps.

belongs in the map locations corresponding to these states. We do see from Fig. 5.44(a) that if these states are not logically adjacent, little grouping or minimization can occur. Figure 5.44(b) indicates that significant grouping of 1s or Xs can take place to minimize the gating expressions for the IFL. These maps resulted from applying Principle 1 to the state assignment problem.

Principle 2 can be demonstrated by the partial state diagram of Fig. 5.45. Again we will assume a three flip-flop system. The maps of Fig. 5.46 show two possible state assignments. Figure 5.47 includes the next state or D input maps for the two state assignments.

It may appear from a comparison of the maps in Fig. 5.47 that little minimization has been accomplished by applying Principle 2. We must recognize that to express a term involving a map-entered variable requires

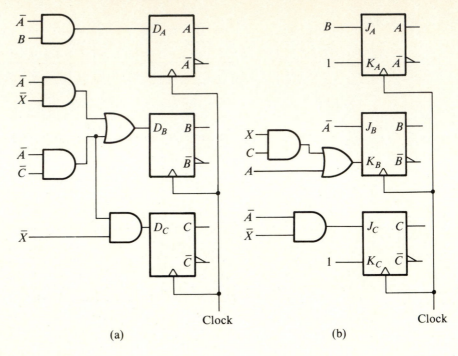

Figure 5.41 IFL design for Example 5.2: (a) using *D* flip-flops; (b) using *JK* flip-flops.

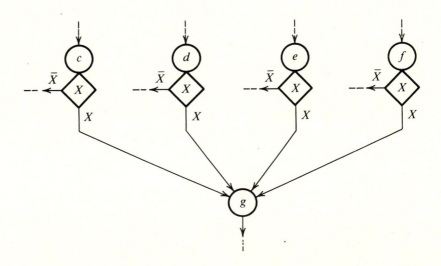

Figure 5.42 A partial state diagram.

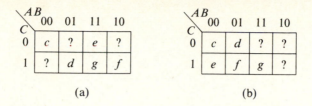

Figure 5.43 State assignments: (a) nonminimizing assignment; (b) minimizing assignment.

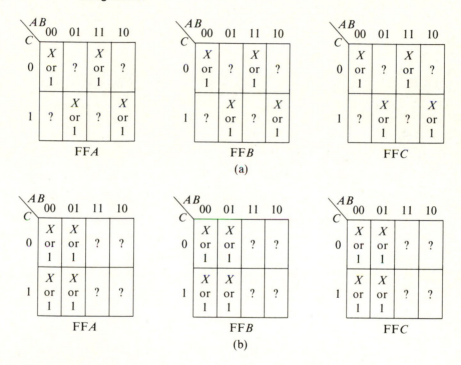

Figure 5.44 Next state or *D* input maps: (a) for nonminimizing assignment; (b) for minimizing assignment.

one more variable than required to express those terms involving a 1. To implement such an expression requires one more input to a gate. The maps of Fig. 5.47(b) then represent a slightly simpler logic system than the maps of Fig. 5.47(a) represent since one less MEV is included.

Intuitively, we might expect that observance of Principle 1 does more to minimize the IFL than does observance of Principle 2. This has been shown to be true [2] and dictates how conflicts between the two principles should be resolved. Not all desired adjacencies can be obtained in a given state assignment; thus Principle 1 is given higher priority for system minimization.

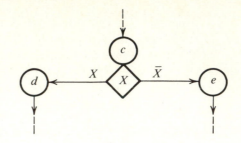

Figure 5.45 Partial state diagram to demonstrate the validity of Principle 2.

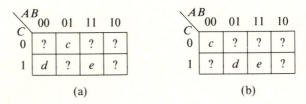

Figure 5.46 State assignments: (a) arbitrary; (b) chosen according to Principle 2.

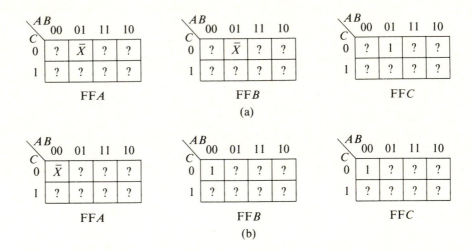

Figure 5.47 Next state or D input maps: (a) for arbitrary state assignments; (b) for assignment according to Principle 2.

Once this principle has been satisfied, Principle 2 is applied as far as possible without disturbing the assignments resulting from Principle 1.

IFL Using Direct-Addressed MUXs

The IFL of the state machine can be designed with MUXs instead of gates. Although component cost is generally higher for MUX design, the chip count can be minimized and design time is decreased. Furthermore, no restrictions are placed on state assignment for circuit minimization, allowing more flexibility in the overall design. We will see in the following sections that state assignment may affect the output forming logic design and system reliability. When IFL is designed with MUXs, state assignment can then be used for these purposes rather than for IFL minimization. One last advantage of MUX design lies in the ease of troubleshooting the state machine. It is very easy to isolate problems of faulty wiring or chips with this type of system.

The general arrangement of the direct-addressed MUX system is shown in Fig. 5.48. Although the IFL shown consists of gates and MUXs, the gates often are either very simple or not required at all. In this scheme the high asserted outputs of the state flip-flops are connected to the select lines of all MUXs in parallel. Each state then addresses a different set of MUX inputs.

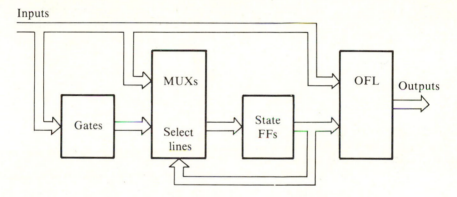

Figure 5.48 Architecture of the direct-addressed MUX system.

One MUX is required for each state flip-flop and each MUX must have n select lines, where n is the number of state flip-flops. A direct-addressed MUX design requires $n(2^n{:}1)$ MUXs to design the IFL. For example, a six-state system utilizes three flip-flops and would use three 8:1 MUXs. This type of design always uses D flip-flops.

In order to demonstrate the use of direct-addressed MUXs for IFL, we will redesign the systems of Examples 5.1 and 5.2. For the first system the D input maps of Fig. 5.34(b) will be used. We emphasize that these maps, while labeled as corresponding to incorrectly chosen states, are quite appropriate for MUX design. The present state of the system selects the input lines of the MUXs that connects to the flip-flop inputs. If we number the minterm locations in the maps for D_A and D_B, we then connect to the corresponding MUX input lines the values contained in the locations. Figure 5.49 shows the MUX realization of IFL for Example 5.1.

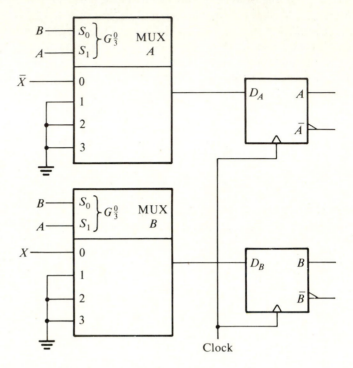

Figure 5.49 IFL using MUXs.

Example 5.2 is redesigned using the D input maps of Fig. 5.40(b). Again the minterm numbers determine which values to connect to the MUX input lines. The redesigned system is shown in Fig. 5.50.

We see the ease with which the design is carried out with MUXs for IFL. It might be noted though that the 8:1 MUXs of Fig. 5.50 are not used very efficiently. Several of the inputs connect to the same value. It is possible to use smaller MUXs, but additional circuitry must be added. When the MUX size is reduced, the system is an indirect-addressed MUX arrangement. We will consider this approach in the next chapter.

It might be instructive to consider the component cost for equivalent systems using gates and MUXs for IFL. Figure 5.41(b) shows the minimized gate realization of IFL for the system of Example 5.2 which uses three 2-input gates. This IFL could be realized with one chip. In small quantities these chips cost approximately 17 cents each, leading to a 17 cent component cost for IFL. The IFL for the MUX system would require three chips. Each of these 8:1 MUX chips would cost about 65 cents, bringing the total IFL cost to $1.95.

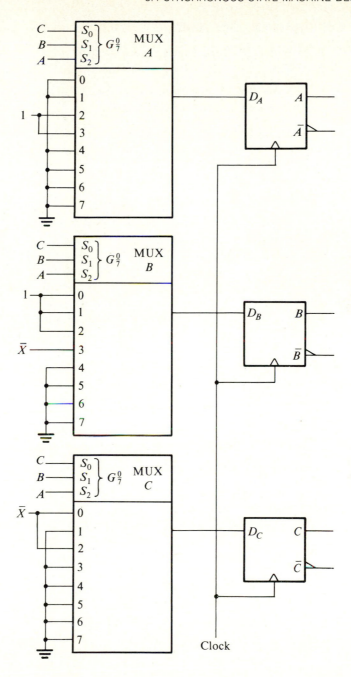

Figure 5.50 IFL for Example 5.2 using MUXs.

Drill Problems: Sec. 5.4D

1. How many flip-flops would be required to implement the state diagram of Fig. 5.16?

2. Design the IFL for the diagram of Fig. 5.16 using D flip-flops and gates, minimizing the number of gates used.

3. Repeat problem 2 using JK flip-flops.

4. Repeat problem 2 using D flip-flops and MUXs for the IFL.

E. Output Forming Logic

When an output is to be generated by a state machine, there are several choices of assertion time. Figure 5.51 demonstrates six different possibilities. The first output is asserted during state c and deasserted at all other times. The second output persists for one state time also, but both assertion and deassertion are delayed for one-half clock period compared with the first case. The assertion of the third output is delayed one-half clock period while deassertion corresponds to the ending of state c. The fourth output is not as common as the others. It is asserted at the state beginning and deasserted one-half clock period later. Outputs 5 and 6 are similar, both becoming asserted during state c and presumably deasserted during some later state. The particular output selected will often depend on output specifications, and we must design the circuit to achieve this signal. Other times, there may be enough flexibility in output signal that the circuit configuration will be chosen with little regard to the resulting output signal shape.

Unconditional and Conditional Outputs

An unconditional output (sometimes called an immediate output) is one that depends only on the state of the system. When a particular state is entered, an AND gate or IC decoder reflects this fact by asserting an output line. We will later see that this output information may be delayed by as much as one-half clock period, but its existence is a function only of the state of the system.

A conditional output not only depends on the system state but also requires certain input conditions before the output is asserted. For example, in state c the system may be required to assert output W only if the input X is asserted.

Returning to Fig. 5.16, we can see both conditional and unconditional outputs represented by this state diagram. Output P is conditional while all other outputs are unconditional. P will be asserted only when the system is in state f and the input signal X is not asserted. If X is asserted while the system is in state f, output P will not be asserted. It is important to note here that data changes take place on alternate clock transitions to state

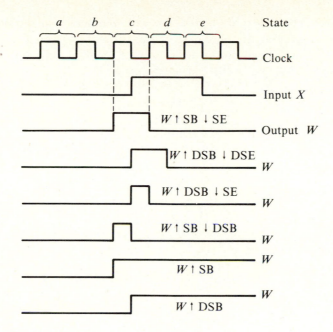

Figure 5.51 Six possible output assertion choices.

changes. When a state is entered, we must wait one-half clock period to allow data to be updated before generating the conditional output. This problem can be demonstrated by considering a system that is to sample a data line X to locate the 3-bit word 101. When this word has been located, the system is to generate a signal R and resume checking the next word. We will assume data changes occur on the positive-going clock transition which dictates that state changes take place on the negative-going transition. The state diagram of Fig. 5.52 is appropriate for this system. States c and e are timing states inserted to result in a three-clock time repetition period. If the first bit of the word is zero, we know the word cannot be 101, but we must delay two more clock periods before beginning the next word check. The timing chart of Fig. 5.53 corresponds to the word 101 appearing on X. When the system is in state a, a value of $X = 1$ causes a state change to state b. The value of X persists for one-half clock time before being updated. If X changes to zero during state b, the system then moves to state d. An output cannot be generated in this third state until the third data value occurs. X is again updated halfway through the state time. The output for this system is conditional and must not occur until the second half of the state time. We might consider the use of the inverted clock to enable the output R in this situation. Figure 5.54 shows a circuit that could be considered to generate a conditional output and one used to generate an unconditional output. These methods may lead to output glitches; consequently we must consider this problem further. Before considering the problem of output glitches, we will comment that the output W in Fig. 5.54 is asserted at the beginning of state b and deasserted at the end of this state. The conditional output R is asserted one-half state time after entering state c and deasserted at the end of this state.

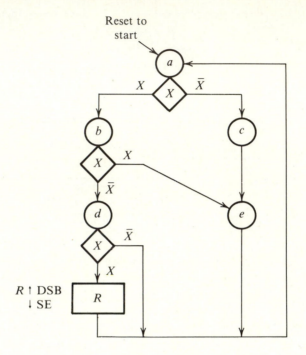

Figure 5.52 State diagram with conditional output.

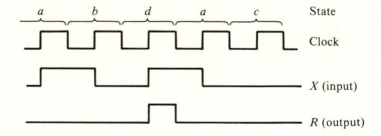

Figure 5.53 Timing diagram.

State Assignment for Minimization of Output Forming Logic

In OFL design, there are two bases upon which state assignment may be made. The first relates to minimization of gates required to implement the

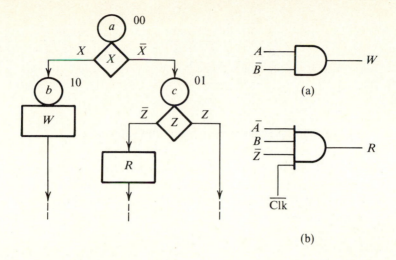

Figure 5.54 Partial state diagram and output decoders for (a) unconditional output; (b) conditional output.

design. Of course, if an IC decoder is used, this consideration is unnecessary. The second basis for state assignment is elimination of output glitches that might be troublesome. The relative importance of these considerations depends on the particular application at hand. As we discuss OFL design, we will mention the constraints that will influence the designer's choice of states.

> Principle 3. To minimize OFL design when using gates, those states with identical outputs should be assigned logically adjacent states on the present state map.

This will allow reduction of the expression for the output equation. If possible, we should use "don't care" conditions of unused states to further minimize the output equation. Let us consider the map shown in Fig. 5.55. The W in a map location indicates that output W is to occur for this state.

The present state map shows that states d and f are the only ones in which output W may occur. This output may not occur even in these states unless input X is asserted as shown in the output map. The expression for the output is

$$W = \bar{B}CX + ACX$$

If we rearrange the states to make d and f adjacent and adjacent to the "don't care" conditions, as shown in Fig. 5.56, the output expression becomes

$$W = AX$$

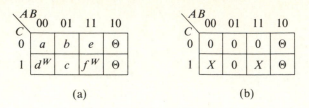

(a)   (b)

Figure 5.55 (a) Present state map. (b) Output map.

We will emphasize that the state assignment may have been chosen to mini-mize input forming logic, and a state rearrangement may not be possible. Generally, IFL minimization results in more significant savings than does OFL minimization [2].

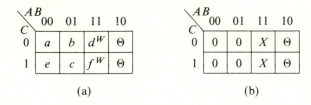

(a)   (b)

Figure 5.56 (a) Present state map. (b) Output map for adjacent states.

Although it is often more costly, a decoder chip can be used to generate state machine outputs. This approach may minimize chip count especially if several output signals must be generated. The flip-flop outputs are connected directly to the decoder inputs leading to the assertion of a particular line in each state. The decoder output lines corresponding to all states producing the same output signal are ORed to generate the output.

State Assignment to Eliminate Output Glitches

The elimination of output glitches will be more important than circuit mini-mization in some cases. If the output signal drives circuits such as RS latches or counter inputs, an output glitch cannot be allowed. In many applications the output may be important only after state change transients have died out. When this is true, glitches that occur during state change can be ignored. It is not always possible to eliminate output glitches by state assignment and other methods may then be required. We will discuss these methods later.

In order to eliminate output glitches, we must first understand the origin of these very narrow pulses. The circuit of Fig. 5.57 shows two flip-flops whose outputs drive an AND gate to create an unconditional output. The output X will occur whenever $A = B = 1$. If we happen to be in the 01 state and then a transition is made to the 10 state, a glitch may result if flip-flop A has a shorter propagation delay time than does flip-flop B. In this instance, A changes to 1 before B drops to 0. Actually, the AND gate levels, V_{IHmin} and V_{ILmax}, also affect this situation. For a very short time, both AND gate inputs appear to have 1s applied. A short sliver or glitch may be produced at the gate output. If B drops below V_{IHmin} before A rises to V_{ILmax}, then no glitch could occur. Unfortunately, all flip-flops and gates possess different propagation delays, and these sometimes troublesome glitches tend to be rather prevalent in this type of circuit.

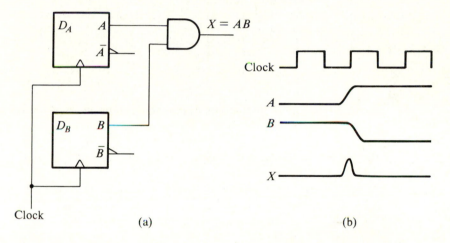

Figure 5.57 (a) A circuit that may produce a glitch. (b) Timing diagram.

It is often possible to avoid output glitches in state machines by proper assignment of states. In this method, state transition paths must not cross a state that produces an output on the way to the next state. The possible transition paths can be drawn on a present state map. Those states that should produce an output are identified by an asterisk. If a path crosses a state marked with an asterisk, a glitch may result. For example, consider the diagram and maps of Fig. 5.58. Since four states are present, two flip-flops are required. From this state diagram we see that a transition from state a to c is possible. If the map of Fig. 5.58(b) is chosen, the transition from a to c can follow one of two paths. If A is faster than B, the path 00-10-11 will be followed. If B switches faster than A, the path 00-01-11 will be traversed. This latter possibility crosses state d and may produce a glitch on the output. Choosing the present state map of Fig. 5.58(c) eliminates any possibility of glitches. As the system moves from state a to b or c, the

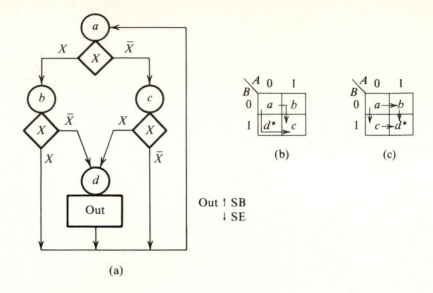

Figure 5.58 (a) State diagram. (b) Map leading to possible glitch. (c) Map that guarantees no output glitch.

paths do not cross state d. This output state is reached only when the system conditions dictate entry.

We will note here that if output glitches must be avoided, the selection of state assignment may be based entirely on this criterion. If so, neither the IFL nor the OFL will be minimal (if gates are used), but this is the price paid for a glitch-free output.

A more complex situation is represented in Fig. 5.59. From the map of Fig. 5.59(b) we see that a transition from state b to e may take a path that crosses state d, that is, 001-011-010-110. The map of Fig. 5.59(c) solves this problem. As shown in Fig. 5.60, all possible allowed state transitions indicated by the state diagram have possible paths that do not cross output states. From this table we see that no output state is crossed by any possible path; thus an output glitch will never occur. Although trial-and-error methods are required to select proper states, it is generally appropriate to start by selecting output states as far removed from each other as possible. In this case d is 001, while each flip-flop must change states to reach state e which is 110.

Before leaving the discussion of state assignment to avoid output glitches, a word of caution is appropriate. We noted previously that unused or "don't care" states can be placed adjacent to output states to minimize OFL. This implies that when the unused state is entered, an output will be generated. If our states are chosen to avoid output glitches, we must be certain that any "don't care" state used to reduce OFL is not crossed during a state transition.

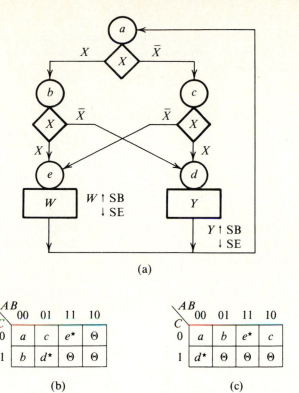

(a)

$$\begin{array}{c|c|c|c|c} \overline{AB} & & & & \\ C & 00 & 01 & 11 & 10 \\ \hline 0 & a & c & e^\star & \Theta \\ \hline 1 & b & d^\star & \Theta & \Theta \\ \end{array}$$

(b)

$$\begin{array}{c|c|c|c|c} \overline{AB} & & & & \\ C & 00 & 01 & 11 & 10 \\ \hline 0 & a & b & e^\star & c \\ \hline 1 & d^\star & \Theta & \Theta & \Theta \\ \end{array}$$

(c)

Figure 5.59 (a) State diagram. (b) State assignment with possible glitch. (c) Glitch-free assignment.

Transition From	To		Possible paths
a	b	....	000 → 010
a	c	....	000 → 100
b	d	....	010 → 011 → 001*
			010 → 000 → 001*
b	e	....	010 → 110*
c	d	....	100 → 000 → 001*
			100 → 101 → 001*
c	e	....	100 → 110*
d	a	....	001 → 000
e	a	....	110 → 100 → 000
			110 → 010 → 000

Figure 5.60 Transition paths for map of Fig. 5.59(c).

There are times when it is impossible to avoid glitches by state assignment. Furthermore, conditional outputs are produced by circuits that can lead to glitches without crossing output states; therefore other methods must be considered.

Glitches Due to Conditional Outputs

A conditional output is not generated when a particular output state is entered. Any output must be delayed one-half state time until the input data is updated. Figure 5.53 shows a timing chart for such a conditional output. In order to produce a conditional output, the gate of Fig. 5.54(b) might be used. This particular configuration can lead to glitches unless certain precautions are taken. The problem here is that as the gate is enabled by the inverted clock, the data line $\overline{Z}$ may also be changing. Furthermore, as the inverted clock disables the gate, the state could be changing to cause other gate inputs to be in transition as the gate closes. Whenever several inputs of a gate may change at the same time, the possibility of a glitch exists.

Clock Suppression to Avoid Glitches

In order to avoid the glitches due to conditional outputs, we should allow the input data to settle before the AND gate is enabled and we should disable the gate before the state change takes place. We can do this by delaying the inverted clock signal with respect to the data changes and delaying the state changes even more than we delay the inverted clock. Figure 5.61 shows a means of accomplishing these delays.

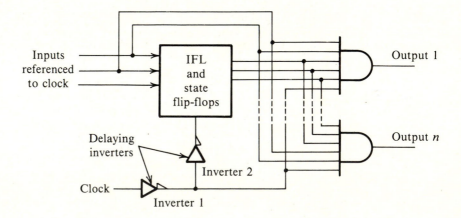

Figure 5.61 Clock suppression.

The outputs are formed by ANDing the appropriate state flip-flops and input signals with a delayed, inverted clock signal. State changes are delayed by an additional delaying inverter resulting in the timing chart of Fig. 5.62.

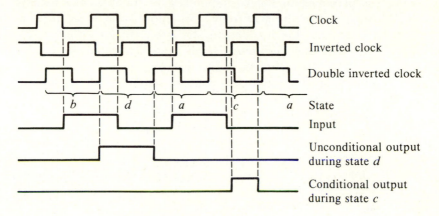

Figure 5.62 Timing diagram for clock suppression.

In Fig. 5.62 input data changes on the negative-going clock transition. Delaying inverter 1 allows any changing inputs to become stable before the output AND gates are enabled. Delaying inverter 2 delays the state changes even further, and so the output AND gates are disabled just prior to state changes. With this arrangement, during the time the output gates are enabled, no other gate inputs will be changing.

If the data changes on the positive clock transition and the state changes take place on the negative transition, a different arrangement of inverters is required. In this case, the gate again must be opened after one-half state time. The positive clock will then enable the output gate, but this signal must be delayed to avoid data change conflicts. Consequently, two inverters are used to delay the clock signal to the gate. Two more inverters are then inserted that connect to the clock inputs of the flip-flops. Figure 5.61 would be modified by using negative transition flip-flops and replacing each of the inverters by a pair of inverters.

While the clock suppression method is relatively simple, the output lasts for one-half of the state time rather than the full state time. This may be a problem in a small number of applications, but will generally not be a significant factor.

Output Holding Register

A component that can be added to eliminate output glitches is an output holding register. This register, generally consisting of D flip-flops, can be loaded with the output signals at or just after the midpoint of the state change. Figure 5.63 shows this scheme. One flip-flop is required for each output of the system. For unconditional outputs the decoder gates generate an output signal at the beginning of any output-producing state. This signal is presented to the corresponding D input of the holding register. At the midpoint of the state time, this information is set into the holding register. In general, the holding register produces an output that is asserted at the delayed state beginning ($\uparrow$DSB) and deasserted at the delayed state ending ($\downarrow$DSE). Any glitches at the decoding gates would occur at the state beginning and would disappear prior to setting data into the holding register.

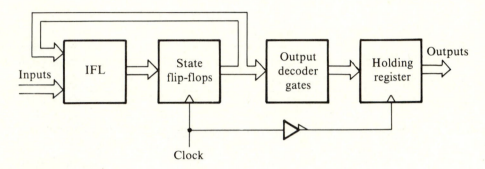

Figure 5.63 Use of an output holding register.

For conditional outputs, the data changes at the midpoint of the state time; thus we need to allow the data line time to settle before loading the holding register. In this case the inverter shown between clock signal and holding register in Fig. 5.63 must serve as a delaying inverter. The output register is then loaded long enough after the midpoint of the state time to allow the data line to become stable.

When a holding register is used, the output glitches appearing at the gate outputs are ignored. Consequently, state assignments need not be made to avoid output glitches. Instead, the IFL or the OFL or both can be minimized by state choice. Reduction in this circuitry will offset a portion of the cost of the output register.

Creating Delays

There are several methods of creating the needed delay time in logic circuits. In the TTL series, several different families can be mixed. The propagation delay time for an inverter might be 4 ns for Schottky TTL, 10 ns for high-speed TTL, 15 ns for low-power Schottky, and 22 ns for standard TTL. Thus, some difference in delay time can be effected by using a different TTL family. A second method uses a series of cascaded inverters to increase delay time. As many as six or eight inverters are sometimes used in logic circuits to achieve longer delays. A third method to control delay time consists of adding a capacitor from gate input to ground to delay a signal. The output impedance of the driving stage forms an RC time constant with the capacitor to control the total delay time of the circuit. While additional capacitors can be costly, this method allows delay time control over a very wide range of values.

Output Forming Logic Design

We will now return to Examples 5.1 and 5.2 to discuss the design of the OFL for these systems. The state diagram of Fig. 5.27 indicates that outputs Y and Z are unconditional outputs with Y asserted in state b and Z asserted in state c. If we consider the state assignment chosen to minimize IFL as shown in Fig. 5.29(a), we note that output glitches due to transient crossing of output states can result. In switching from state a to c or from c to a, state b may be traversed. If we insist on using this assignment, we must either accept the output glitches (if glitches can be tolerated at the output) or we must use holding registers or clock suppression to solve this problem. The arrangement of Fig. 5.64(a) will lead to possible glitches on output Y. This may be acceptable in some designs, but often these glitches must be removed. Figure 5.64(b) uses a holding register that sets the output data into the flip-flops at midstate time. The outputs in this case lag the outputs of the gates by one-half the period of the clock. In some rare cases, this time lag may be unacceptable. If so, the arrangement of Fig. 5.64(c) can be used. In this circuit, the clock is delayed long enough by the inverters to allow the glitches to die out before the information is set into the holding register. Although the outputs Y and Z will be slightly delayed from the state beginning, this delay is in the nanosecond range with TTL circuits; therefore we consider the output to be asserted at the state beginning.

Figure 5.64(d) shows a clock suppression scheme that delays the outputs again by two inverter delays. In this circuit, the outputs are asserted for only one-half clock period. Since the gates are disabled before the state changes take place, no delay of state change is required here.

Now let us consider the same system without any IFL constraints on state assignment. We can now assign states to remove output glitches or to minimize OFL. The assignment of Fig. 5.29(b) is then appropriate for an efficient output design. The system can switch from a to b or from a to c without crossing an output state. It can also move from b to a or from c to a without crossing an output state. In addition to this, the "don't care" condition can be placed next to both output states. With this state assignment, the output maps are shown in Fig. 5.65. The expression for Y is reduced to $Y = B$ and Z becomes $Z = A$. In this situation no gates are required for the OFL, and the outputs are taken directly from the state flip-flops.

We will now design the system of Example 5.2 with OFL assuming MUXs are to be used for IFL and output glitches cannot be tolerated. Since MUXs will make up the IFL, the state assignment shown in Fig. 5.66 is an

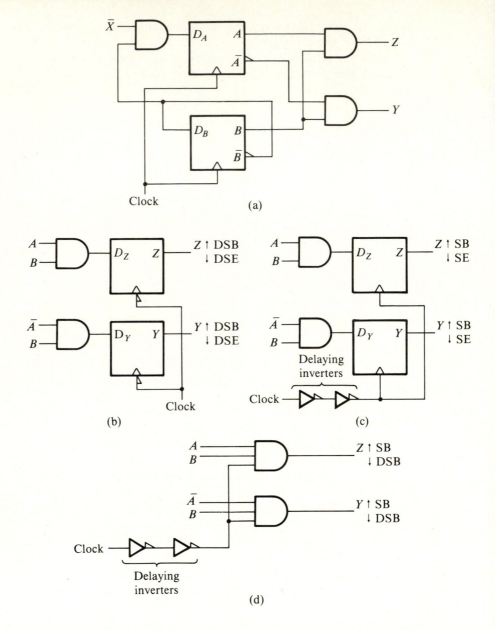

Figure 5.64 Unconditional outputs. (a) State machine with gates for OFL. (b) Holding register arrangement. (c) Alternate holding register arrangement. (c) Clock suppression.

B \ A	0	1
0	0	0
1	1	Θ

Y

B \ A	0	1
0	0	1
1	0	Θ

Z

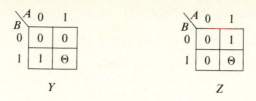

Figure 5.65 Output maps for Example 5.1.

acceptable one. No allowed transitions from the state diagram cross the output states d or f, nor does any transition cross the "don't care" location 001. In moving from state d to state a, location 100 is crossed; thus we must not group this location with output Z. We need not use clock suppression or holding registers, and we can minimize the OFL (except for output Z). The next state maps or D input maps and the output maps are given in Fig. 5.67. Application of these maps leads to the system of Fig. 5.68.

C \ AB	00	01	11	10
0	d^Y	e	a	Θ
1	Θ	b	c	f^Z

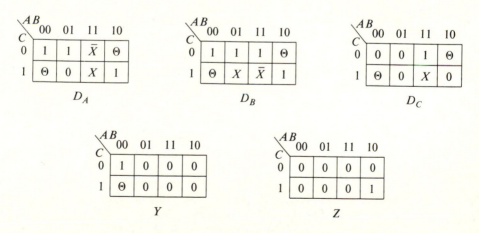

Figure 5.66 A present state map for Example 5.2.

C \ AB	00	01	11	10
0	1	1	$\bar{X}$	Θ
1	Θ	0	X	1

D_A

C \ AB	00	01	11	10
0	1	1	1	Θ
1	Θ	X	$\bar{X}$	1

D_B

C \ AB	00	01	11	10
0	0	0	1	Θ
1	Θ	0	X	0

D_C

C \ AB	00	01	11	10
0	1	0	0	0
1	Θ	0	0	0

Y

C \ AB	00	01	11	10
0	0	0	0	0
1	0	0	0	1

Z

Figure 5.67 Pertinent maps for system design of Example 5.2.

Drill Problems: Sec. 5.4E

1. Design the OFL for the system of Fig. 5.58(a) using the assignment of Fig. 5.58(c). Use a gate for the OFL.

2. Design the OFL for the system of Fig. 5.59(a) using the assignment of Fig. 5.59(c). Use gates for the OFL.

3. Repeat problem 2 using a decoder for the OFL.

4. Design the OFL for the system of Fig. 5.59(a) using the assignment of Fig. 5.59(b). Use gates for the OFL along with a holding register to eliminate possible glitches.

5. Design the OFL for Fig. 5.52 if state *d* is 011. Use clock suppression to eliminate possible glitches, and assume data changes on the positive clock transition.

F. Redundant States

The starting point of any engineering design is a set of specifications. These specs might be expressed in words, equations, input and output waveforms, or in some other appropriate form. The designer must then develop a system that operates within the requirements specified. We will consider the digital system design problem in more detail in Chap. 6.

In state machine design, one of the first tools used is the state diagram. There may be several state diagrams that can satisfy the original specifications of a given problem. Some of these diagrams may contain extra or redundant states that could be eliminated to decrease the design complexity. We must consider methods of locating and eliminating these redundant states before we implement the state diagram.

In a state machine two major tasks are completed during each state. Outputs are generated, if necessary, and the signals producing the correct next state must be generated. These output signals may depend on input signals (conditional outputs) as also may the next state signals. If two states generate identical output signals and identical next state signals for all possible input conditions, the two states are said to be equivalent. One of these equivalent states is considered to be redundant and can be removed.

As an example of redundant states we will again consider a system that is to sample a synchronous data line *X*. This system should take four samples before returning to its initial state to begin another 4-bit check. If the 4-bit sequence (word) 0111 is detected, an output is to be generated and the system should then begin the next check without missing a bit. One possible state diagram is shown in Fig. 5.69.

If the sequence 0111 occurs, the system will proceed through states *a*, *b*, *c*, and *d* and generate an output prior to returning to state *a*. Any other sequence will not generate an output, but the state diagram provides the proper delay of 4 bits before returning to state *a*. We note that a

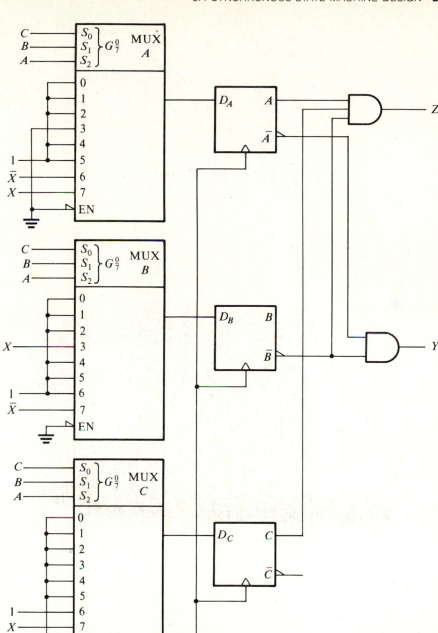

Figure 5.68 Complete system for Example 5.2.

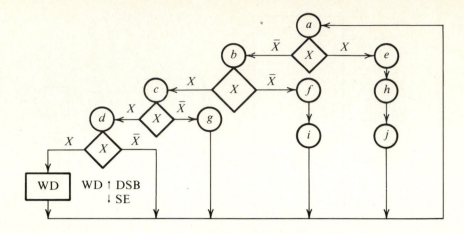

Figure 5.69 A state diagram for a word detector.

Present State	Input X	Next State	Output WD
a	0	b	0
	1	e	0
b	0	f	0
	1	c	0
c	0	g	0
	1	d	0
d	0	a	0
	1	a	1
e	0	h̷f	0
	1	h̷f	0
f	0	i̷g	0
	1	i̷g	0
g	0	a	0
	1	a	0
h̷f	0	j̷g	0
	1	j̷g	0
i̷g	0	a	0
	1	a	0
j̷g	0	a	0
	1	a	0

Equivalent states

Equivalent states

Figure 5.70 State table for a word detector.

conditional output is required in this instance. An alert designer might note that there are several redundant states in the diagram of Fig. 5.69. For example, states f and h are provided as timing states to ensure that if an incorrect bit has been detected in bit 1 or bit 2 of the word, two more samples will be required before returning to state a. States f and h could be combined into a single state. The same argument applies to states g, i, and j; that is, only one of these states is required.

Present State	Input X	Next State	Output WD
a	0 1	b e	0 0
b	0 1	f c	0 0
c	0 1	g d	0 0
d	0 1	a a	0 1
e	0 1	f f	0 0
f	0 1	g g	0 0
g	0 1	a a	0 0

Figure 5.71 Reduced state table.

There is an orderly method available to identify and eliminate redundant states. First, the next state table for the state diagram is constructed. This table consists of the present states, the various possible inputs, the next states, and the outputs of the system. Obviously, the state table for a given system may differ from one designer to another. Equivalent states are then identified. As equivalent states are identified, all but one are considered redundant and can thus be consolidated into a single state. This method will be demonstrated in Fig. 5.70. This is the state table for the word detector. We note from the table that states g, i, and j are equivalent. While state d is similar to states g, i, and j, it has a different output specification and is therefore not equivalent. We next choose names for these equivalent states. We will designate states g, i, and j by state g, replacing i and j by g in the table. This results in the modifications shown in Fig. 5.70. Before we construct the reduced state table for the system, we check the chart to see if the modified states lead to any further equivalent states. States f and h are now seen to be equivalent. We will select state f as the name for the states f

and h and replace the letter h with an f each time it appears in the table. We continue this process until no equivalent states exist and then develop the reduced state table. For this example, Fig. 5.71 shows the reduced state table while Fig. 5.72 indicates the reduced state diagram.

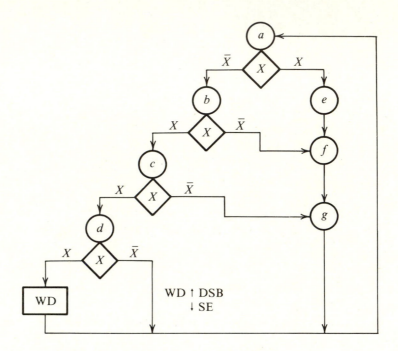

Figure 5.72 Reduced state diagram for word detector.

5.5 MAINLY SYNCHRONOUS SYSTEMS

Section overview: The mainly synchronous system is introduced in this section. This very important type of state machine requires special treatment of the asynchronous inputs. After establishing the problem of erroneous state transitions, three methods of solving this problem are discussed. The go–no go configuration is the best method if possible to use this configuration. Conversion of asynchronous to synchronous inputs is considered. State assignment to avoid erroneous glitches is the third method suggested.

The systems considered to this point were assumed to have input data that changed in synchronism with the system clock. These synchronous systems actually occur in practice, allowing the principles discussed to be applied to their design. Perhaps the state machine most often encountered in practice though is the mainly synchronous system. This system has clock-driven state changes and one or more asynchronous inputs. An asynchronous input has no fixed relationship to the clock signal and consequently may change values at any point of the clock period.

Most systems requiring human interaction are mainly synchronous systems. An operator must communicate with the system by switches, the assertion of which will have no relation to the clock signal. Two digital systems exchanging information over connecting lines will transmit or receive asynchronous signals. Even if the system clocks are adjusted to the same frequency, they cannot be controlled precisely enough to result in simultaneous clock transitions. Hence, data sent by one system in synchronism with that system's clock is asynchronous with respect to the other system's clock. Most terminals communicating serially with computers are examples of this type of system.

In this section we will consider the problems that can arise in asynchronous systems along with some possible solutions.

A. Erroneous State Transitions

When an asynchronous input changes just prior to the state-changing clock transition the system may move to a state not indicated by the state diagram. In order to demonstrate this problem, we will consider the diagrams of Fig. 5.73. If the present state of the system is a and $X = 0$, the IFL generates $D_A = 0$ and $D_B = 1$ to cause the next state to be 01. Now suppose that X changes to 1 just prior to the positive clock transition. The IFL will respond by generating $D_A = 1$ and $D_B = 0$ to set up a next state of 10 as indicated by the state diagram. Unfortunately, D_A and D_B will not change instantaneously due to propagation delays in the input logic. Perhaps D_A will move from 0 to 1 before D_B moves from 1 to 0. If the state change occurs at this instant, the next state will be 11 instead of 10. On the other hand, the propagation delays may be such that D_B responds to a change in X more rapidly than does D_A. If a state change occurs and D_B has changed from 1 to 0, but D_A has not completed the transition from 0 to 1, the next state will be 00 instead of 10.

If these states of 11 or 00 are reached instead of the proper state, an erroneous state transition has occurred. Although the probability of X producing an erroneous state transition may be very low, perhaps 1 in 1000, the one time that such a problem occurs may lead to serious consequences. Therefore, we must consider methods to deal with this problem if we intend to produce a reliable system.

It is possible to predict the code of the erroneous states by using a K-map. The system of Fig. 5.74 is a mainly synchronous system that can lead to erroneous states. The present state map is shown in Fig. 5.75. The state machine switches from state a to b or c depending on the asynchronous variable X. From the time state a is entered, the flip-flop inputs will contain the signals to result in a next state of b or c. If the variable X changes, the gating conditions change toward the proper next state but may not reach the

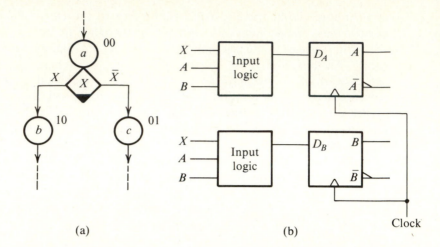

Figure 5.73 (a) A partial state diagram. (b) Block diagram of system.

proper levels before the state-changing clock transition occurs. If not, the next state will be one that lies on the transition paths between states b and c. From the map we see that the possible erroneous states are g and f, or 010 and 111.

Now let us assume a second state assignment as shown in Fig. 5.76. In this map there are six possible transition paths between b and c which traverse all six remaining states of the map. Any of these states are possible erroneous states which the system might reach when switching from state a to b or c. We note here that the state code of a is unimportant in determining the possible erroneous states.

Drill Problems: Sec. 5.5A

1. State 001 is exited via an asynchronous variable X. The two following states are 101 and 011. List all possible erroneous states to which the system might switch from 001.

2. Explain why the code of the present state (001) does not influence the code of the possible erroneous states in problem 1 (or in any mainly synchronous system).

B. Asynchronous to Synchronous Conversion

It is possible to convert all asynchronous input variables to synchronous variables by means of flip-flops. The system shown in Fig. 5.77 indicates the conversion of an asynchronous variable G to a synchronous variable called T driven by the negative clock transition. From a design standpoint,

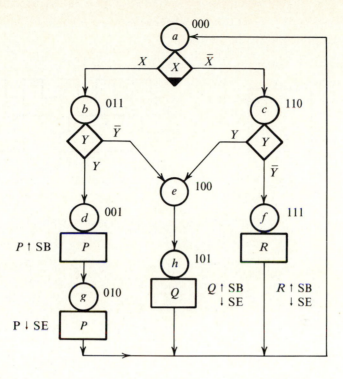

Figure 5.74 A mainly synchronous system.

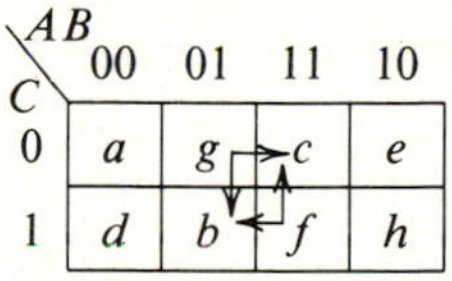

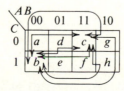

Figure 5.75 Present state map.

Figure 5.76 Another present state map.

conversion to synchronous signals is very simple, allowing synchronous design principles to be utilized.

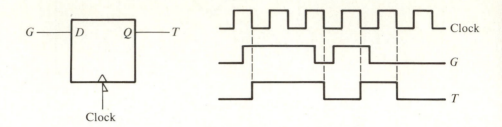

Figure 5.77 Asynchronous to synchronous signal conversion.

There are two problems related to the use of a flip-flop in converting asynchronous inputs to synchronous. The first is the obvious increase in cost and complexity. One flip-flop is required for each asynchronous input variable. In addition to this problem, there is a possibility of unreliable operation in the D flip-flop. This problem has been well-documented in the literature [2] and occurs when the asynchronous input changes value very near the clock transition that loads the flip-flop. Normally, the output of the flip-flop sets at either the high or low voltage level, but never at the midpoint of the two levels. If the input changes very near the clock transition, it is possible for the flip-flop output to move toward the opposite level, then stop moving somewhere between the levels, then move to either of the two normal levels. This midpoint level is called a metastable state and leads to an ambiguous output. The occurrence of this metastable state is quite rare, and many systems can tolerate the occasional problem introduced by the ambiguous output. The resulting effect of the metastable state can generally be minimized if the asynchronous input is guaranteed to change at a lower frequency than the system clock. This means the input will always exist at a particular level for over a full clock period. If it changes near the clock transition and causes an ambiguous output, the correct output will definitely exist one cycle later after a second clock transition takes place.

Although the synchronizing register is a reasonable solution in many applications, there is one situation that warrants closer attention. If the system state changes take place on the positive-going clock transitions, the input synchronizing register should be updated prior to this time. This register update should be far enough in advance of the state change to allow the input forming logic outputs to settle before the state-changing clock transition. In some instances, the synchronizing register will be updated on the opposite clock transition. If the asynchronous variable changes just after the register is updated, this change does not affect the next state transition. It is not until the next update of the synchronizing register plus a half-period of the clock that this variable has an effect on the system.

If it is important to register the change of the asynchronous variables as soon as possible, the register update should take place just prior to the state change. This can be accomplished by changing the synchronizing register flip-flops and state flip-flops on the same clock transition. In this case, the state change must be slightly delayed to allow the IFL outputs to settle prior to this state change. Delaying inverters are often used to achieve this delay. This minimizes the delay between input variable assertion and effect on the system. Unfortunately, a maximum delay of almost one state time is still possible.

Figure 5.78(a) shows the use of an input synchronizing register which is updated on the transition that is alternate to the state-changing transition. Figure 5.78(b) indicates the method of registering input changes nearer the state-changing transition. Either additional inverters or slower inverters may be required to achieve the proper delay.

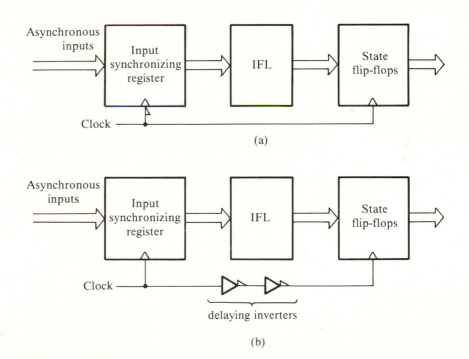

Figure 5.78 Asynchronous to synchronous conversion: (a) on alternate clock transition; (b) on same transition.

C. The Go—No Go Configuration

In several instances, branching from a given state to only a single next state is controlled by an asynchronous variable. When this is true, the go–no go arrangement of Fig. 5.79 is appropriate. In this configuration states a and b must be logically adjacent to avoid erroneous state transitions.

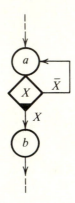

Figure 5.79 The go–no go configuration.

We will assume state a corresponds to $AB = 10$ and state b is 11. A change in X from 0 to 1 near the clock transition will either occur early enough to change flip-flop B to 1 or it will be late enough that B will remain at 0. If B does not change, the system remains in state a. The next state-changing clock transition will lead to a change, assuming X lasts at least one clock period. Thus, the system will always move to state b from state a when X is asserted. In this method all asynchronous variables must last more than one clock period to ensure proper switching. A shorter input pulse must be stretched to guarantee reliable operation.

The go–no go configuration is the most reliable scheme available in mainly synchronous system design. Although modification of a state diagram to achieve this configuration may result in additional states, the increased reliability is well worth the investment for many applications.

D. Logical Adjacency

Another configuration that can be used with asynchronous inputs is shown in Fig. 5.80 when a branch to one of two states is required. In the arrangement of Fig. 5.80, states b and c must be chosen to be logically adjacent. If state b is 01 and state c is 11, the gating of only one flip-flop changes as X changes. If X arrives early enough, the change is reflected by switching to the proper state. The disadvantage with this configuration is that if X changes from 0 to 1 near the state change time, the system may move to state c. If X remains equal to 1, the system never reaches state b and the assertion of X is ignored. If X had been asserted just slightly sooner, state b would have been reached. When it is important to register a change in X before moving to another state, the go–no go arrangement of Fig. 5.79 should be used.

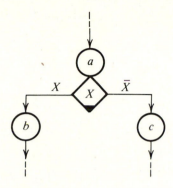

Figure 5.80 Branching to two states.

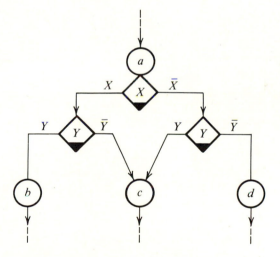

Figure 5.81 State diagram leading to erroneous state transitions.

An important conclusion can be drawn from the preceding paragraphs. We have found that in order to avoid erroneous state transitions, the states following an asynchronous variable must be logically adjacent. Therefore, a maximum of two states may follow a state that is exited via asynchronous variables. Figure 5.81 shows an improper arrangement since three states b, c, and d cannot be mutually adjacent.

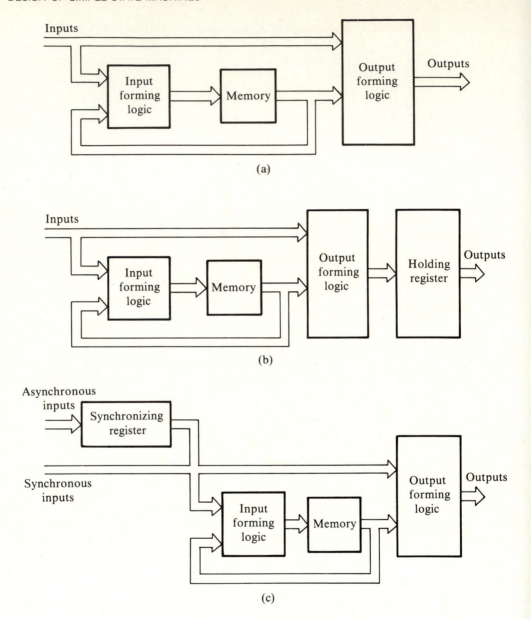

Figure 5.82 (a) General state machine. (b) State machine with output holding register. (c) State machine with input synchronizing register.

5.6 GENERAL STATE MACHINE ARCHITECTURE

Section overview: This section reviews various possible arrangements for state machines.

Figure 5.82 shows the general block diagram of the state machine along with two possible variations. The output holding register of Fig. 5.82(b) is added to latch the outputs into this register at the midpoint of the state time or after the state flip-flops settle. As explained earlier, this arrangement is used to filter out the glitches from the output decoder. The information shifted into the register is unaffected by any glitches since these transients would occur only when state changes take place. One point that is significant here is that if any outputs are conditional, that is, dependent on input variables, the transition that fills the holding register must be delayed slightly from the midpoint of the state time. We recall that when a single phase clock is used, the synchronous data must change on the alternate half-cycle to that which causes the state change. Thus, the midpoint of the state change corresponds to the time that the synchronous inputs are changing. In order to allow these inputs time to change and to allow the output decoder to become stable before loading the register, the loading clock pulse must be delayed. Delaying inverters are used for this purpose as shown in Fig. 5.63. The amount of delay achieved can be controlled by the number of inverters cascaded or by the inverter propagation delay time. In TTL there are several versions of compatible inverters giving a variety of delay times that can be used.

Figure 5.82(c) shows a synchronizing register that converts the asynchronous inputs to synchronous variables. The state machine is then controlled by only synchronous variables. Although no output holding register is shown in this figure, this register may be added to the system if needed.

We have discussed the realization of all architectures of Fig. 5.82 using rather simple components. We have considered the use of gates to realize both IFL and the output decoder. We have also considered decoders for these sections, emphasizing the use of this component for output decoder design. The MUX has been used for IFL design in a direct-addressed mode. That is, the state flip-flops are connected directly to the MUX select line, allowing a unique input line to connect to the MUX output in each state. Generally, this method is used to minimize design time rather than component cost. It is possible to spend more time in design to yield a smaller MUX size. This system is referred to as the indirect-addressed MUX state machine. We will discuss this configuration and also the ROM and PLA configurations in Chap. 7, considering advantages and shortcomings of each.

In an effort to demonstrate various types of design and compare differences, we will design actual systems in Chap. 6 using several approaches. As we consider each approach, we will develop pertinent design principles and indicate why such an approach would be used.

REFERENCES AND SUGGESTED READING

1. T.L. Booth, *Sequential Machines and Automata Theory*. New York: John Wiley and Sons, 1967, chap. 3.
2. W.I. Fletcher, *An Engineering Approach to Digital Design*. Englewood Cliffs, N.J.: Prentice-Hall, 1980, chaps. 5 and 6.
3. D. Winkel and F. Prosser, *The Art of Digital Design*. Englewood Cliffs, N.J.: Prentice-Hall, 1980, chaps. 5 and 6.

CHAP. 5 PROBLEMS

Sec. 5.2

5.1   Completely define the meaning of a synchronous state machine.

* **5.2**   What is a mainly synchronous system?

Sec. 5.3A

* **5.3**   Develop the excitation table for the circuit shown in Fig. P5.3 assuming positive logic for inputs A and B.

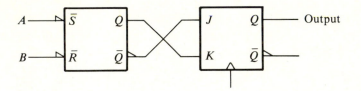

Figure P5.3

5.4   Repeat Prob. 5.3 if the connections between flip-flops are reversed.

5.5   An output sequence of

	t_1	t_2	t_3	t_4	t_5
Q1	—	0	1	1	0
Q2	—	1	0	1	1

is to be produced with two D flip-flops. What is the sequence of values that $D1$ and $D2$ must take on?

5.6   Repeat Prob. 5.5 if JK flip-flops are used.

* **5.7**   Repeat Prob. 5.5 if T flip-flops are used.

5.8   If the input signals to a gating circuit are as shown, design the circuit to drive two D flip-flops to create the sequence of outputs, Q_1 and Q_2.

	t_1	t_2	t_3	t_4	t_5
$I1$	0	0	1	1	—
$I2$	0	1	0	1	—
$Q1$	—	0	1	1	0
$Q2$	—	1	0	1	1

5.9 Repeat Prob. 5.8 for *JK* flip-flops.

* **5.10** Repeat Prob. 5.8 for *T* flip-flops.

Sec. 5.3C

5.11 Explain why input data changes should generally take place on alternate clock transitions to those causing the state change.

5.12 Analyze the circuit of Fig. 5.5 if $\overline{X}$ is applied to the IFL rather than X.

5.13 Analyze the circuit of Fig. 5.10 if $\overline{X}$ is applied to the IFL rather than X.

5.14 Plot the next state maps of Fig. 5.12(b) as variable entered maps with X as the MEV.

Sec. 5.3D

5.15 Draw the state diagram for the circuit of Fig. 5.5.

* **5.16** Draw the state diagram for the circuit of Fig. 5.10.

5.17 Draw the state diagram for a system that checks an incoming data line X that changes on the positive clock. This system is reset to start the process of checking of 3-bit words. If exactly two 1 bits are detected in a word, an output Y is to be generated and the system resumes checking the next word. You may assume that 1 bit time separates each word.

5.18 Repeat Prob. 5.17 if there is no separation between words, that is, the first bit of a word immediately follows the last bit of the previous word.

Section 5.4A

* **5.19** Design a counter to generate the repetitive sequence 000, 001, 010, 100, 011, 110, 000, ... The counter should be self-starting.

5.20 Repeat problem 5.19 for a sequence of 0010, 0011, 0111, 1110, 0000, 1010, 1001, 0100, 1100, 0110, 0010, ...

Sec. 5.4D

5.21 Design the IFL for a state machine to implement the waveform shown in Fig. P5.21. Minimize the gates for the IFL. When the input is asserted, it will always have a duration of four clock periods. *Hint:* Use trial-and-error methods to generate a state diagram with four total states and two output states.

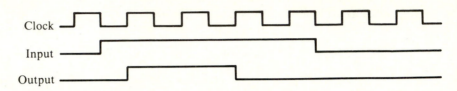

Figure P5.21

* **5.22** Repeat Problem 5.21 using MUXs for the IFL.

Sec. 5.4E

* **5.23** Select states to eliminate output glitches for the diagram shown in Fig. P5.23.

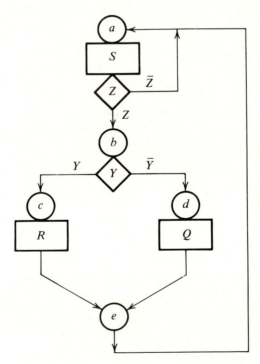

Figure P5.23

5.24 Implement the system shown in Fig. P5.24. Show timing chart. Use JK flip-flops and gates for IFL.

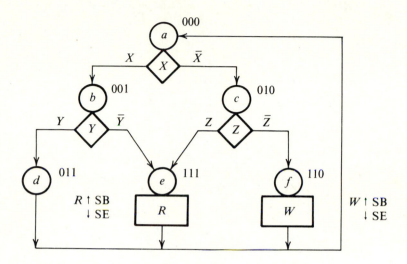

Figure P5.24

5.25 Repeat Prob. 5.24 using D flip-flops and MUXs for IFL.

5.26 Implement the circuit described in Prob. 5.17. Show state diagram and timing chart. Use JK flip-flops and gates for IFL.

5.27 Repeat Prob. 5.26 using D flip-flops and MUXs for IFL.

5.28 Implement the circuit described in Prob. 5.18. Show state diagram and timing chart. Use D flip-flops and gates.

5.29 Assuming output glitches are unimportant, realize the state machine corresponding to the state diagram (Fig. P5.29). Use MUXs for the IFL.

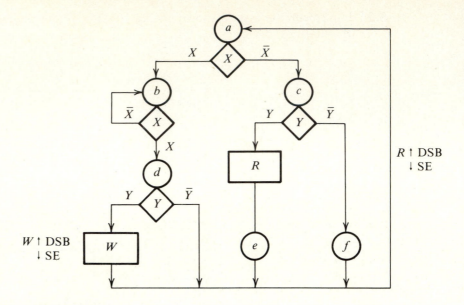

Figure P5.29

5.30 Realize the state machine represented by the diagram shown in Fig. P5.30. Use MUXs for the IFL, and use clock suppression to eliminate output glitches.

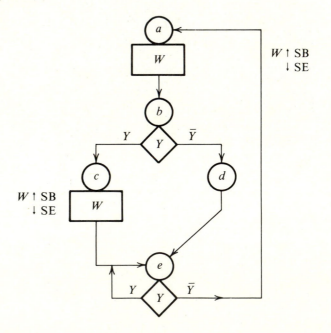

Figure P5.30

5.31 Repeat Prob. 5.29 using a holding register to eliminate output glitches. Assume R and W can remain asserted for a full clock period.

5.32 Ignoring output glitches, design a minimal state machine for the diagram of Prob. 5.30 using gates for the IFL and OFL.

Sec. 5.4F

5.33 Construct a reduced state diagram for a system that checks an input line to find the word 101101. The check repeats at 6-bit intervals.

Sec. 5.5

5.34 An erroneous state transition from state a to state d can occur in the system of Fig. P5.34 if X is asserted near the time of state change. What are the possible state codes for state d?

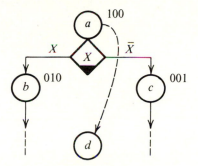

Figure P5.34

* **5.35** What are the possible codes for the erroneous state e in Fig. P5.35?

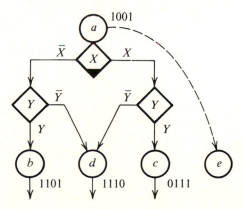

Figure P5.35

5.36 Design a state machine using gates for the IFL and OFL along with *JK* flip-flops to implement the state diagram shown in Fig. P5.36. Be as economical as possible. Eliminate output glitches.

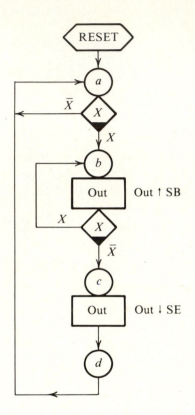

Figure P5.36

5.37 Design the state machine to implement the state diagram shown in Fig. P5.37. Assume *Y* changes on positive-going clock transitions and output glitches must not occur.

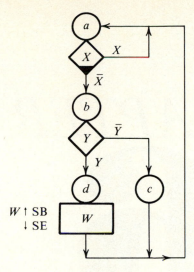

Figure P5.37

CHAPTER
6
Design
of Digital
Systems

6.1 INTRODUCTION TO DESIGN

Section overview: This section introduces the concept of design. Design and analysis are compared and the role of trial-and-error methods in design are discussed.

Since we intend to discuss the design of digital systems, we must first consider the meaning of the word "design."

A design is an underlying scheme or plan that governs the development of some entity.

This definition can be expanded to apply to digital system design [1]:

> A digital design is a plan, often expressed in schematic or block diagram form, that governs the development of a digital system.

The digital area is notorious for products designed by some "fly-by-night" company that fail to meet specifications. This results from the fact that many powerful logic circuits are available as building blocks for digital systems. The adage that "a little knowledge can be dangerous" often applies to this area. A poorly trained designer can manipulate building blocks, generally using trial-and-error methods only, and produce a system that functions. Unfortunately, it may not function with the necessary reliability. When used under different circumstances from those of the manufacturer's test setup, the system may fail to perform according to specs. Any design method used then should allow the designer to produce a reliable digital system while optimizing those parameters most important to that specific system.

A. Analysis vs Design

In elementary school we are given analysis problems such as the following:

> Mother buys three cans of corn at 27 cents per can, two loaves of bread at 58 cents per loaf, four quarts of milk at 30 cents per quart, and 2 pounds of steak at $2.65 per pound. How much money did Mother spend on groceries?

This type of problem serves a useful purpose in teaching a child how math can be used. It has a single answer that immediately measures a person's ability to apply the required mathematics. Furthermore, design methods are generally based on analysis; thus it is logical to teach analysis prior to design.

A design problem based on this same example might read:

> Mother has $10 with which to purchase groceries. She must buy enough food to serve three people and she desires to include items from the four basic food groups. Given the following price list for food, construct an appropriate shopping list for Mother.

Obviously, this problem has many answers depending on the length of the price list, the rate of sales tax, the appetites of the people involved, and other factors. This design problem is more closely related to a real world problem than is the analysis problem. It is a situation that Mother may face daily, and its solution requires an ability to analyze. Generally, it is solved by a trial-and-error approach and may require several trials to reach an acceptable answer.

Analysis can be considered to be an important tool in solving a design problem. A designer must fully understand analysis before he or she can effectively apply design methods. Unfortunately, the use of this tool is often emphasized over the design procedure even in digital system classes.

B. Trial-and-Error Methods in Design

As mentioned previously, a trial-and-error approach might be used to construct the desired shopping list. A trial solution (shopping list) is proposed, and then the proposal is analyzed. By comparing the results for analysis to the specifications, an error can be evaluated. If this error is too large, the trial solution is modified in a way that decreases the error. This procedure continues until acceptable results are achieved. If the list of food in each group is extensive, there may be hundreds of possible combinations that satisfy the constraints imposed. There may also be hundreds of unacceptable combinations.

This time consumed in the trial-and-error method is usually excessive. In addition, certain problems that decrease reliability of digital system operation are not always considered when this approach is used. While it is perhaps impossible to eliminate trial-and-error methods, a good design procedure should minimize time required for such methods and also produce a reliable system.

6.2 TOP-DOWN DESIGN

Section overview: This section defines top-down design and proposes a method to apply this type of design to digital systems. An example of the design of a simple digital system is discussed.

A. Definition

The term "top-down design" has grown out of the programming area and relates closely to the concept of structured programming. Niklaus Wirth, the developer of the PASCAL programming language, provides the following definition of structured programming [3]:

> Structured programming is the formulation of programs as hierarchical, nested structures of statements and objects of computation.

Implied in this definition is a decomposition or breaking down of a large problem into component parts. These parts are then decomposed into smaller problems with this successive refinement continuing until each remaining task can be implemented by simple program statements or a series of statements. As each statement or structure is executed, a part of the overall objective is accomplished. The program is decomposed into units called modules. These modules are decomposed into control structures or series of statements. The statement is the basic unit of the program.

A very significant point in applying top-down design is that the modules should be selected to result in minimum interaction between these units. In general, complexity can be reduced with the weakest possible coupling between modules. Minimizing connections between modules also minimizes the paths along which changes and errors can propagate into other parts of the system [4]. While complete independence is impossible since all modules must be harnessed to perform the overall program objective, interaction is minimal in a well-designed program.

B. Top-Down Design of Digital Systems

When applied to digital systems top-down design proposes a broad, abstract unit to satisfy the given system function. This function will be characterized by a set of specifications. The overall system is then decomposed into modules. These modules are partitioned in such a way to be as independent of each other as possible, but working together will satisfy the overall system function. Just as in programming where higher levels of structure contain the major decision-making or control elements of the program [4], one of the partitioned modules will be the control unit of the system. Successive refinement of the modules leads to decomposition into operational units. These units will often be MSI circuits or groups of circuits. Refinement continues, with an accompanying increase in detail, until all sections can be implemented by known devices or circuits.

There are certain characteristics of the digital field that make top-down design an easy method to apply. For example, the large number of SSI and MSI circuits available allows operational units and even modules to be realized directly with chips. These chips are generally designed to require a minimal number of control inputs, and thus minimal interaction between modules is automatically achieved, at least to an extent.

Another advantage of the top-down method in digital design is that the control module can be realized as a state machine. This allows well-known design procedures to be used in designing a reliable control unit.

C. An Example

In order to demonstrate the use of the top-down approach, the design of a 2½ digit digital voltmeter (DVM) will be considered [1]. The problem statement would begin with a set of specifications that the DVM should meet. After a study of these specifications, the DVM can be decomposed into a set of modules as shown in Fig. 6.1. These modules are chosen according to function performed. The DVM must perform the measurement function and display the results. A control module is required to direct the operation of the measurement and display modules.

Figure 6.1 reflects a high level of abstraction. At this step of the design, almost no detail appears. Modules indicate only broad functions to be performed with no indication as to how those functions should be carried out. Before proceeding to the next step, the designer must choose an approach. In this example, the designer might choose to use a dual-slope technique for the measurement module, three 7-segment LED displays for the display module, and a state machine for the control module.

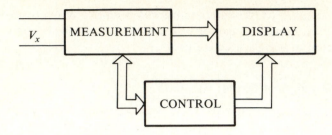

Figure 6.1 Functional modules.

The modules are then decomposed into operational units as shown in Fig. 6.2. It is at this point that the inexperienced designer or student experiences the most frustration. There is no plug-in procedure that can be used here. This step requires creativity and experience along with trial-and-error methods to complete. A knowledge of available MSI chips is also very helpful.

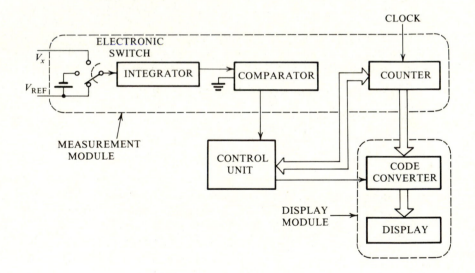

Figure 6.2 Operational units.

Figure 6.2 represents a more detailed picture of the system. Modules have now been broken down into units that perform specific operations—hence the name operational units. Each of these units can be implemented by a circuit or series of circuits using well-known design procedures or presently available IC chips. The comparator, counter, code converter/driver, and 7-segment displays are off-the-shelf items. The electronic switch and integrator may also be purchased or designed specifically to meet the given specifications. State machine design procedures will be used to implement the control unit.

Figure 6.3 shows the basic circuits that may be used to make up the counting, code converting, and display units. The system is now completely specified with the exception of the control unit.

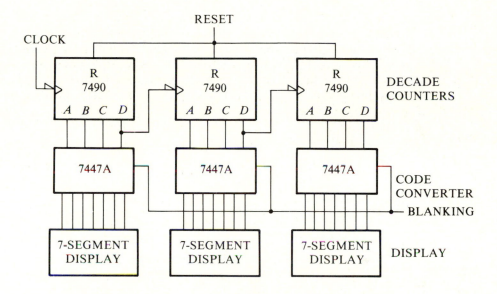

Figure 6.3 Basic circuits.

A timing chart is constructed, showing the relationship between all necessary control signals, inputs, and outputs. The controlling state machine is then designed to produce the control signals to drive the basic circuits. This completes the design process. The following section will formalize these ideas in terms of a suggested design procedure.

6.3 DESIGN PROCEDURES

Section overview: This section proposes two design procedures. The first procedure is for the design of an overall digital system. Among other things this procedure will lead to a control unit module. The second procedure proposes a method of designing this control unit using a state machine.

A. Digital System Design

An important component in most digital systems is the control unit that oversees the operation of all other units in the system. This component accepts input information from an external source along with information from other internal units and generates appropriate output signals for controlling other units of the system. As mentioned earlier this control unit can consist of a state machine. If a state machine is used, all design principles developed over the years for such devices can be applied to the controller design.

The overall system design can be broken down into two procedures. The first applies top-down design to decompose the system into a controlling state machine module and other sections easily implemented with digital circuits. The second procedure consists of designing the state machine.

We will now consider the accomplishment of the first procedure. Using the idea of top-down design, this procedure might consist of the following:

DD1. Determine, from a thorough study of the system specifications, the major goals of the system.

DD2. Decompose the system into functional modules required to implement the major goals. One module will be the controlling state machine.

DD3. Determine the next lower level of tasks that must be accomplished by each module.

DD4. Decompose all modules except the controlling state machine into operational units designed to accomplish these tasks. Often these subsections will correspond to MSI circuits.

DD5. Decompose the operational units until the entire system is reduced to components that can be realized with basic circuits.

DD6. Determine all control signals necessary to make the circuits operate.

DD7. Generate a timing chart to show the time relationship of all signals referenced to the clock signal.

DD8. Design the controlling state machine to implement the system.

DD9. Document your work.

We will emphasize that trial-and-error methods may play a part in several steps of this procedure. Furthermore, two or more different designs may be equally good; there is generally no single "best" design. Experience improves the efficiency of applying this or any design method. Some examples in the next section will demonstrate the application of the proposed procedure.

One of the major parts of the proposed design procedure is the development of the controlling state machine which will now be considered.

B. Overview of State Machine Design

As indicated previously, a state machine is a sequential circuit that controls the behavior of a digital system. Although there may be several different sequences of states that can be executed, the operation of the system is cyclic. The particular sequence executed depends on external conditions or signals applied along with internal status of the system. At a given state, the next state of the sequence always depends on the present state and often depends on the input signals. Outputs can be generated at appropriate states to drive or control other circuits or systems in order to perform useful functions. As a result of this control capability the system is often called a controlling state machine.

The design of a state machine consists of

1. Developing an appropriate state diagram

2. Selecting the proper number of states

3. Designing the input forming logic

4. Designing the output forming logic

While these four steps appear to be reasonably straightforward, there is a tremendous latitude of choice involved in each step that requires a creative designer to be very effective. There are, for example, many different state diagrams that generally can satisfy a given set of specifications. As we shall later see, states can often be exchanged for counter circuits to reduce the total number of states. In IFL and OFL design, the highest priority goes to developing a reliable system. Elimination of erroneous state problems and troublesome glitches must be accomplished before other matters such as minimization can be considered.

There are various quantities that a designer might attempt to minimize, depending on the circumstances related to the design. A state machine may be needed for a research or development project with no intention of marketing the resulting system. In such a case, minimizing the required design and construction time would be appropriate. Off-the-shelf SSI and MSI circuits would probably be used for this project with little consideration given to component cost. On the other hand, if a controller with a high-volume sales potential is to be marketed, production cost is the quantity that must be minimized. If SSI and MSI circuits are to be used, chip count and cost will determine overall system cost. Thus, attention to these factors must be given. If the system is to be integrated, the design should follow guidelines imposed by the IC fabricator to minimize cost.

While the design process of a state machine is not difficult, such factors as component cost, fabrication cost, company policy, market potential, design flexibility, labor cost, and other considerations tend to drastically influence the process.

C. State Machine Design

The design of the controlling state machine is the second task that must be accomplished to complete the major goal of the system. This task can be broken down into a series of steps also. Because of the many degrees of freedom in designing a controlling state machine, no single procedure can be considered to be the best available. To provide a framework for solving this problem, the following steps are suggested.

SM1. Construct a state diagram for the state machine.

SM2. Eliminate all redundant states.

SM3. Construct the reduced state diagram.

SM4. Determine how to handle asynchronous inputs.

SM5. Determine whether output glitches are significant.

SM6. Assign proper states.

SM7. Construct next state table or maps along with input information for flip-flops.

SM8. Design input forming logic.

SM9. Design output forming logic.

One of the more difficult steps in this procedure is that of constructing a state diagram. The timing chart is very useful at this point and may be developed prior to or in conjunction with the construction of the state diagram. The following section will consider several examples of design demonstrating the application of these procedures.

6.4 DESIGN EXAMPLES

Section overview: This section demonstrates the two design procedures of the previous section. The first example applies the state machine design procedure. The next example is a simple digital system that applies both procedures. The last example involves a slightly more complex digital system that again requires the application of both procedures.

A. A Tester Control

A device receives input data from a set of sensors as shown in Fig. 6.4. The tester generates a synchronous signal lasting exactly three clock periods long. When this signal is applied to device 1, lines X and Y may become asserted depending on the sensor inputs. These lines will contain synchronous, NRZ signals. If either of the combinations of X and Y shown in Fig. 6.4 occur, the corresponding output should be generated on the Out line. Any combination of X and Y assertions other than that shown should result in no output. The tester cannot tolerate glitches on the T_{in} line.

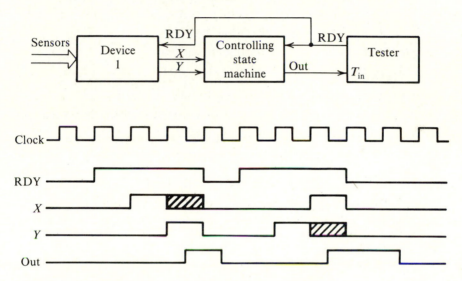

Figure 6.4 A test system controlled by a state machine.

This digital system is not completely specified, but the requirements on the controlling state machine are given. We have only to apply the state machine design procedure of Sec. 6.3C to complete the problem.

Step SM1   Construct a state diagram for the state machine.

Using a trial-and-error approach, we ultimately arrive at the diagram shown in Fig. 6.5. The system will remain in state a until RDY is asserted. When RDY is asserted, the system checks inputs X and Y. If either of these signals are asserted during the first clock period following RDY assertion, the state machine should produce no output regardless of future values of X and Y. Since RDY lasts exactly three clock periods, delay state d is required to force the deassertion of RDY before the system returns to state a. If neither X nor Y is asserted during the initial clock period, the system moves to state

b. This state checks the values of X and Y during the second clock period. If X is 1 and Y is 0, the system moves to state c. If X is 0 and Y is 1, the system moves to state e. Any other combination of X and Y moves the system to state d. In state c the system checks Y during the third clock period. If $Y = 1$, the system moves to output state f. If $Y = 0$, the system moves to state d. If the system had reached state e instead of c, X would also be checked during the third clock period. If $X = 1$, output state g is reached. For this condition the output should be asserted for two clock periods. An output state could follow g to produce a longer output, but this state would be redundant. It would be equivalent to output state f. Therefore, it is eliminated and the system proceeds from state g to state f to produce the required length of output assertion.

Step SM2   Eliminate all redundant states.

This was accomplished as part of step SM1.

Step SM3   Construct the reduced state diagram.

Removing the dashed lines in Fig. 6.5 results in the reduced diagram.

Step SM4   Determine how to handle asynchronous inputs.

There are no asynchronous inputs in this system.

Step SM5   Determine whether output glitches are important.

The original problem statement indicates that glitches should not occur. There are two possible sources of glitches in this system. One source is that of transient crossing of an output state. This source could be eliminated by state assignment. The second source lies in the OFL gates which could produce a static zero hazard when the system moves from state g to state f. We will choose states f and g to be adjacent and group both states together to avoid the hazard.

Step SM6   Assign proper states.

With seven states, three flip-flops are required. If we want to minimize design time, we could use MUXs for IFL which would not impose any particular constraints on the state assignment. We would then assign states g and f to be adjacent to minimize output gates. If we wanted to minimize component cost, we would choose to use gates for IFL and attempt to select the following adjacencies: a, f, and d adjacent—same next state a; c, d, and e adjacent—next state of state b; a, b, and d adjacent—next state of state a. Figure 6.6 shows one possible state assignment. With this assignment no state transition crosses an output state.

Step SM7   Construct next state maps and flip-flop input maps.

The next state maps are shown in Fig. 6.7. We will consider only a part of the development of these maps. From the state diagram the system moves from state a, 001, to b, 011, if RDY and $\overline{X}\overline{Y}$ are asserted. If RDY and any other combination of X and Y occur, the system moves to state d, 010. If

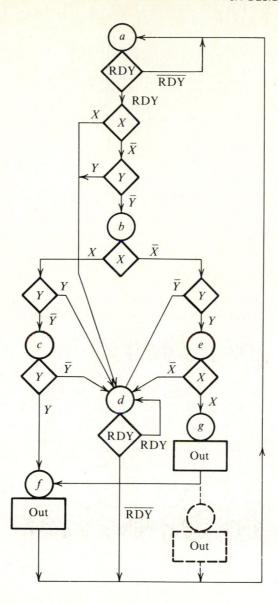

Figure 6.5 State diagram for system of Fig. 6.4.

RDY is never asserted, the system remains in state a. Flip-flop A will become 0 for any input condition. Flip-flop B will become 1 if state b or d is to be entered, and flip-flop C becomes 1 if the next state is a or b. The entries in the 001 location of the next state maps reflect these conditions. Each location of the map is completed using similar considerations.

	AB 00	01	11	10
C				
0	c	d	e	f^*
1	a	b	Θ	g^*

Figure 6.6 State assignment to minimize IFL and OFL.

	AB 00	01	11	10
C				
0	Y	0	X	0
1	0	$\bar{X}Y$	Θ	1

$$D_A = \bar{A}\bar{B}\bar{C}Y + ABX + BC\bar{X}Y + AC$$

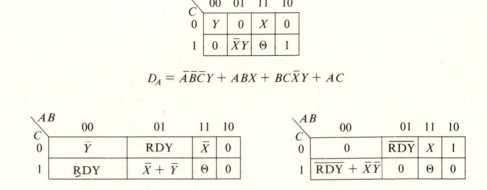

	AB 00	01	11	10
C				
0	$\bar{Y}$	RDY	$\bar{X}$	0
1	R̰DY	$\bar{X} + \bar{Y}$	Θ	0

$$D_B = \bar{A}\bar{B}C\bar{Y} + \bar{A}BC\,\mathrm{RDY} + \bar{A}B\bar{C}\,\mathrm{RDY}$$
$$+ BC\bar{X} + BCY + AB\bar{X}$$

	AB 00	01	11	10
C				
0	0	$\overline{\mathrm{RDY}}$	X	1
1	$\overline{\mathrm{RDY} + \bar{X}\bar{Y}}$	0	Θ	0

$$D_C = \bar{A}B\bar{C}\,\overline{\mathrm{RDY}} + \bar{A}BC\bar{X}\bar{Y}$$
$$+ \bar{A}\bar{B}C\,\mathrm{RDY} + A\bar{C}X + A\bar{B}C$$

Figure 6.7 Next state and D input maps.

Step SM8 Design the IFL.

We recall that the next state maps also correspond to the input maps for D flip-flops. The reduced expressions for D_A, D_B, and D_C are shown in Fig. 6.7. When using gates for IFL, it is sometimes more economical to use *JK* flip-flops. Accordingly, the input maps for this type of device are given in Fig. 6.8. These apply the excitation table for a *JK* flip-flop in the map development. The resulting expressions are no simpler than those for the D flip-flops, and so D flip-flops will be used.

Step SM9 Design the OFL.

The output map is shown in Fig. 6.9 along with the expression for Out. The final circuit is shown in Fig. 6.10.

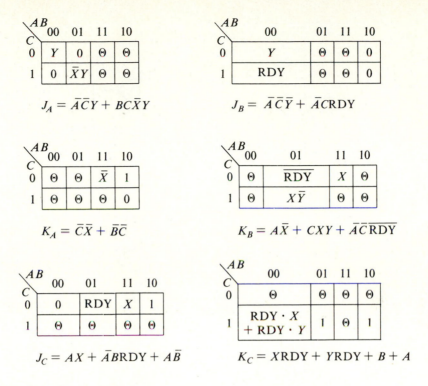

$$J_A = \bar{A}\bar{C}Y + BC\bar{X}Y \qquad\qquad J_B = \bar{A}\bar{C}\bar{Y} + \bar{A}C\text{RDY}$$

$$K_A = \bar{C}\bar{X} + \bar{B}\bar{C} \qquad\qquad K_B = A\bar{X} + CXY + \bar{A}\bar{C}\overline{\text{RDY}}$$

$$J_C = AX + \bar{A}B\text{RDY} + A\bar{B} \qquad\qquad K_C = X\text{RDY} + Y\text{RDY} + B + A$$

Figure 6.8 *JK* input maps.

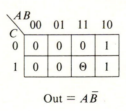

$$\text{Out} = A\bar{B}$$

Figure 6.9 Output map.

B. A Frame Counter

We will now demonstrate the design of a digital system using the procedures given in the preceding section. We will consider a rather simple frame counter design. This system monitors a serial line for a particular word. Each time this word is detected, a counter is incremented to record the total number of occurrences of the word.

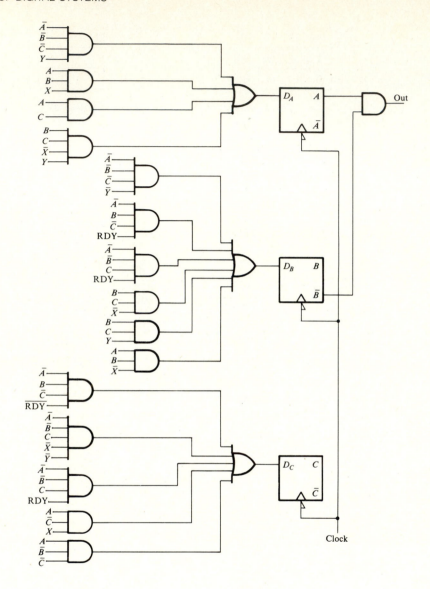

Figure 6.10 Controlling state machine for system of Fig. 6.4.

Specifications: A data acquisition system sends serial data at a 1 kHz bit rate to a recording system. The data is in the form of 4-bit words. Several records, each consisting of a variable number of words, will be transmitted each time the data acquisition unit becomes active. Each record is separated from the following record by the frame separator word 1101 which will never be contained in a data record. A count line from the acquisition unit to the recording system will be asserted to signal that counting should begin and will be deasserted to signal the end of record transmission. A clock signal from the data acquisition unit is available, and transmitted data changes on the negative transition of this clock. The count signal will always be asserted as the first bit of the first record occurs. The maximum number of records to be counted is 200.

A block diagram of the overall system is shown in Fig. 6.11. We first apply the system design procedure of Section 6.3A.

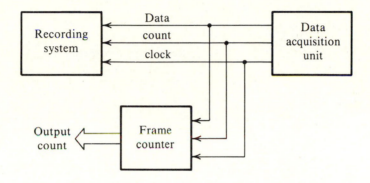

Figure 6.11 A data acquisition/recording system.

Steps DD1 and DD2

After studying the specifications, we draw a block diagram of the frame counter with one module representing the state machine as indicated in Fig. 6.12. The controlling state machine is also used to detect the frame separator word, producing an output each time 1101 occurs. This output is used to increment a counter having a reset assumed to be driven by the recording system.

Steps DD3, DD4, and DD5

We now consider the next lower level of task that must be accomplished by the controlled system module, that is, the counter. This must be at least an 8-bit counter to reach a count of 200. We could propose to cascade two 4-bit binary counter chips such as the 7493 to construct this counter. We can now decompose this block into the more detailed operational units of Fig. 6.13 where each unit corresponds to an IC chip.

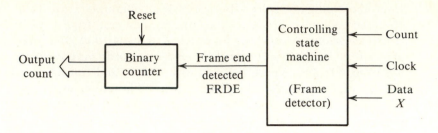

Figure 6.12 Block diagram of frame counter modules.

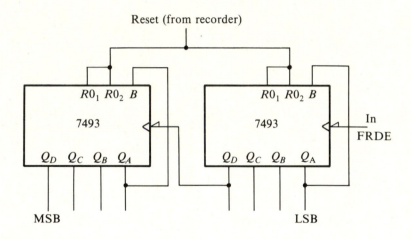

Figure 6.13 An 8-bit binary counter.

Step DD6

The only control signal produced by the state machine is the frame end detected signal FRDE. Each time the frame separator word is detected, a pulse should appear at the input of the counter.

Step DD7

A timing chart may not be necessary for this simple system, but one is shown in Fig. 6.14.

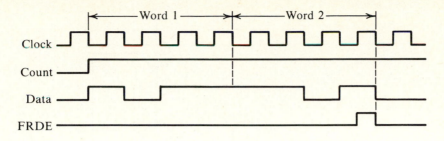

Figure 6.14 Timing chart for frame counter.

Step DD8

Here we must design the controlling state machine by applying the procedure of Sec. 6.3C.

Step SM1

We recognize that four bit checks are required to identify the FS word. Once a series of four checks are started at a word boundary, the next set of four checks must not begin until the next word boundary. Even when the state machine discovers a bit in the first position that indicates the FS word is not present, the next word check must begin at the next word boundary. The state diagram of Fig. 6.15 includes delay states to make the check cycle always equal 4 bits in duration.

If the FS word is detected, the state machine will proceed through states a, b, c, and d with an output generated during the latter half of state d. When any other word is being received, the system starts in state a but will move into the sequence of states e, f, or g depending on when the first bit occurs that differs from the FS word. Assertion of the count line deasserts the reset, allowing the system to leave state a.

Steps SM2 and SM3

There are no redundant states in this diagram.

Step SM5

Output glitches on FRDE will increment the counter causing a false count to result. We will eliminate output glitches by clock suppression.

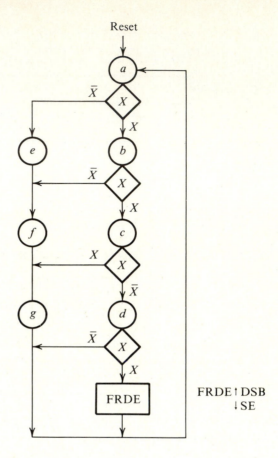

Figure 6.15 State diagram for frame detector.

Step SM6

We will choose MUXs for the IFL and a gate for the OFL. Since clock suppression will be used to eliminate glitches, state assignment is arbitrary. We could decrease the number of inputs to the OFL gate by arranging state d next to the "don't care" state. One possible assignment is shown in Fig. 6.16.

Step SM7

Since MUXs will be used, D flip-flops are appropriate. The flip-flop input (and next state) maps are shown in Fig. 6.17.

Step SM8

We will use three 8:1 MUXs for the IFL connecting the state flip-flop outputs to the select lines. The maps of Fig. 6.17 will be used to determine input connections.

$$
\begin{array}{c|c|c|c|c|}
 & \multicolumn{4}{c}{AB} \\
C & 00 & 01 & 11 & 10 \\
\hline
0 & a & b & c & d\star \\
\hline
1 & e & f & g & \Theta \\
\hline
\end{array}
$$

Figure 6.16 State assignment for frame detector.

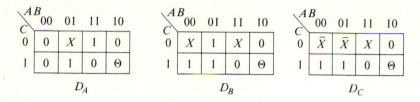

Figure 6.17 Flip-flop inputs for frame detector.

Step SM9

The output map is shown in Fig. 6.18 along with the expression for FRDE. Figure 6.19 indicates the final state machine.

$$
\begin{array}{c|c|c|c|c|}
 & \multicolumn{4}{c}{AB} \\
C & 00 & 01 & 11 & 10 \\
\hline
0 & 0 & 0 & 0 & X \\
\hline
1 & 0 & 0 & 0 & \Theta \\
\hline
\end{array}
$$

$$\mathrm{FRDE} = A\bar{B}X$$

Figure 6.18 Output map for frame detector.

Step DD9

Although it is important to document your work in terms of wiring diagrams, schematics, explanatory notes, or other paper work, we will skip this step due to space limitations.

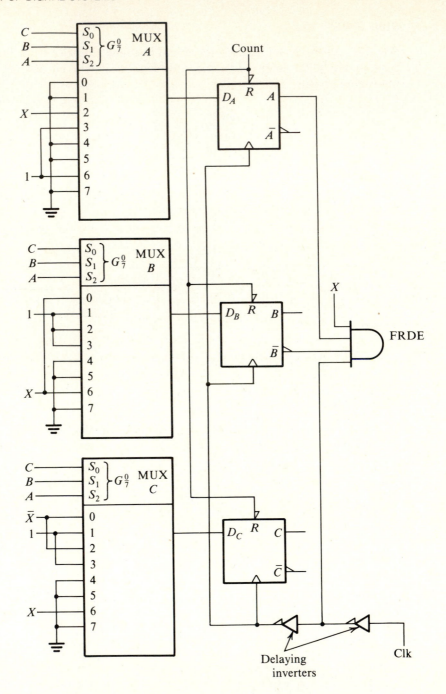

Figure 6.19 State machine for frame counter.

C. A Reaction Timer

In this example we will assume that our company is approached by a prominent agent for athletes. This individual makes a living by negotiating contracts between professional sports organizations and his clients. Over the years he has been retained by some highly successful athletes and has recently become very selective in the clients he agrees to represent. He now requires each prospective client to complete a series of physical tests. From these tests the agent hopes to evaluate the probability of success in professional sports for each athlete. Of course he will represent only those athletes with the greatest potential.

Among other devices, the agent would like our company to build a reaction timer to measure human reflexes. After a few meetings between the agent and one of our engineers, the following set of specifications are agreed upon:

1. The reaction timer will have a RESET-READY toggle switch. In one position the system is reset. When the switch is set to the READY position, the system will initiate the measurement process.

2. Some random time between 1 and 9 seconds after READY is asserted, the GO LED (light-emitting diode) is asserted. The assertion of the GO LED signals the start of the reaction timing process. The random time period before lighting the GO LED removes the possibility of anticipating the start of the timed period.

3. A pushbutton STOP switch is depressed to end the reaction measurement by stopping the timing process.

4. The reaction time in milliseconds must be displayed by three 7-segment displays. The error should be less than or equal to ± 2 ms.

5. If an individual depresses the STOP switch prior to assertion of the GO LED, a CHEAT LED should be asserted.

6. The 7-segment displays are to be blanked until a valid measurement is completed.

7. An LED display is to be available that classifies various ranges of reaction time. When reaction time falls within a given range, the corresponding LED should be asserted. These ranges are given in Fig. 6.20 along with the display layout.

In this situation we again do not have the system block diagram. We must first do the system design after which the controlling state machine will be designed. We will apply the procedure of Sec. 6.3A to produce the system, and then we will use the procedure of Sec. 6.3C to design the state machine.

Reaction time (ms)	LED	Classification
000–149	○	Tennis player
150–179	○	Basketball player
180–199	○	Baseball player
200–299	○	Quarterback
300–399	○	Offensive lineman
400–499	○	Referee
500–599	○	Athlete's spouse
600–699	○	Golfer
700–799	○	Head coach
800–899	○	Assistant coach
900–998	○	Team owner
999 or over	○	Dummy

Figure 6.20 Range LEDs.

Digital System Design

Step DD1 Determine the major goals of the system.

The major goals are to measure the reaction time, display the results, check for cheating, and activate the proper LED to identify the range of results.

Step DD2 Decompose the system into functional modules.

A block diagram that implements the broad functions of the system is shown in Fig. 6.21. The controlling state machine will exchange signals with the timing section and will transmit to the display section to indicate when the display should be active. The classifier is a combinational circuit that is driven by the timing section and drives a portion of the display.

Step DD3 Determine the next lower level of tasks.

This step considers the functions that each block should accomplish. The timing section must generate a random time period between the time that the READY switch is depressed and the GO LED assertion. At this point it must begin timing the period taken for the person to press the STOP switch. The display section must contain means for indicating the number of milliseconds taken to respond to the GO LED assertion. The range here is 0 to 999 ms. This section must also contain the GO LED along with 12 range LEDs. The classifier section must determine in which range the reaction time falls.

Step DD4 Decompose the modules into operational units.

This is a significant step that requires creativity and experience to perform effectively. A knowledge of available SSI and MSI circuits is necessary if these devices are to be used. The timing section can use three decade

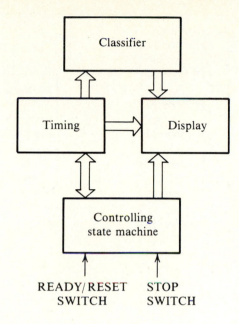

Figure 6.21 Block diagram of reaction timer.

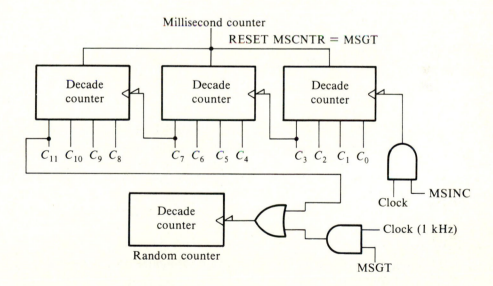

Figure 6.22 Timing module.

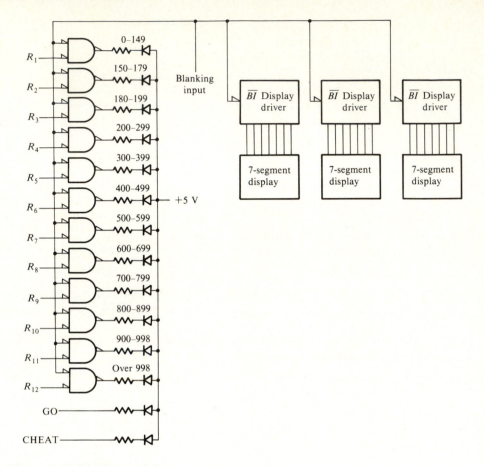

Figure 6.23 Display module.

counters driven by a 1 kHz clock to perform the timing function. The random time can be generated with a fourth decade counter. The display section can be made up of three 7-segment displays and drivers with blanking. Fourteen LEDs and twelve 2-input AND gates are also required. The classifier section uses two 4-bit comparators, gates, and a decoder to identify the range of results. Each module broken down into operational units is shown in Figs. 6.22 through 6.24.

Step DD5 Continue to refine the operational units until they can be realized with basic circuits.

This step has essentially been completed as step 4 was carried out. The decade counters can be realized with 7490 chips. The 7447A chip is appropriate for the display drivers including the blanking function. A 7442 chip can be used for the BCD-to-decimal decoder, while two 7485 comparator chips will complete the classifier module.

This module uses a decimal decoder to determine the number of hundreds of milliseconds contained by the millisecond counter. The most significant 4 bits of the counter drives this decoder. Comparator 1 generates a 1 on its $A > B$ line if the number of hundredths of a second (tens of milliseconds) is greater than 4. Comparator 2 generates a 1 on its $A > B$ line if this number is greater than 7. If the millisecond counter stops below 150 ms, either output line 0 or 1 of the decoder will be asserted. When one of these lines is asserted and comparator 1 is not asserted, the result must be less than 150 ms. These conditions assert the line R_1 to activate the correct LED. The assertion of R_2 requires that output line 1 of the decoder is asserted, the output of comparator 1 is asserted, and the output of comparator 2 is not asserted. This condition is true only when the counter contains over 149 and less than 180 ms. The conditions for assertion of R_3 through R_{11} can be found in a similar manner. Since the counter stops if 999 is reached, R_{12} will be asserted for all times equal to or greater than this value.

Step DD6 Determine all necessary control signals.

This step requires that we not only consider the control signals needed but also consider the proper assertion levels related to the actual circuits selected in the previous step.

The first signal needed is the RESET which can be derived from the READY/RESET switch. Although the RESET signal need not be debounced since it will connect to the direct sets of the state flip-flops, the READY signal should be. We will debounce this switch and also convert READY to a synchronous variable with a *JK* flip-flop. Prior to the assertion of READY, the signal MSGT must be asserted to increment the random counter at 1 ms intervals. When the READY switch is asserted, MSGT is deasserted. This results in a random number being contained in the random counter when the READY switch is asserted. At this time the signal incrementing this counter changes from a period of 1 ms to 1 s. The signal MSGT is also applied to the direct reset of the millisecond counter and will be removed from this counter when READY is asserted. MSINC is now asserted to apply the input clock signal. This 1 kHz input signal to the millisecond counter produces an output signal that increments the random counter once per second.

When the random counter reaches a count of 8 or 9 and the millisecond counter reaches a count of 999, the START signal is asserted. This signal will be created by an AND gate asserted by the correct counter conditions. This activates the GO LED and begins the timing process. We note that the next clock pulse received by the millisecond counter causes a count of 000 to appear. Thus, we need not be concerned with resetting this counter as the START signal is asserted.

The next signal required is the STOP signal produced by closure of the STOP switch. This signal will be debounced and synchronized by a *JK* flip-flop. A CHEAT signal must be generated if the STOP switch was asserted prior to assertion of the START signal. After the STOP signal is asserted, or if the millisecond counter reaches 999 prior to STOP assertion, the results

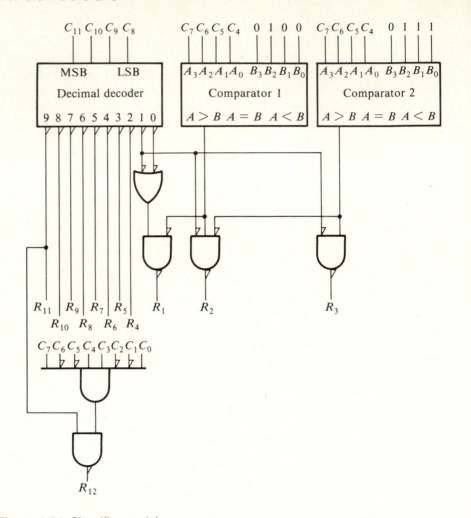

Figure 6.24 Classifier module.

can be displayed. The MSINC signal must be deasserted at this point to store the reaction time in the millisecond counter. The displays are now unblanked by deasserting the BLANK signal. The system will remain in this condition, displaying the results, until the RESET signal is again asserted.

Step DD7 Generate a timing chart.

The timing chart for the reaction timer is shown in Fig. 6.25. Since the counters are incremented on the negative clock transition, we choose the positive transition for state changes. The START signal is an input to the state

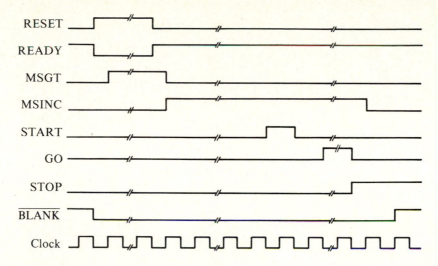

Figure 6.25 Timing chart for reaction timer.

machine and will be satisfied in synchronism with the millisecond counter change on the negative clock transition. The conversion of READY and STOP to synchronous signals is also done on the negative clock to avoid conflicts with state changes.

The timing chart should help resolve any timing problems that might exist as a result of our choice of times of signal generation. As an example of this, we note that as MSINC is deasserted to deactivate the millisecond counter input, the clock signal is simultaneously going positive. Depending on circuit delays, this could allow a sliver to pass on to the counter input. To avoid this, we might consider delaying the clock signal to the MSINC gate to allow MSINC to drop prior to the positive transition of the clock.

Step DD8 Implement the state machine.

This step will actually be broken down into the nine steps required to design the state machine. We will not cover these in detail, but will propose a possible state diagram as shown in Fig. 6.26.

The asynchronous inputs for READY and STOP need not have been converted to synchronous variables in this system. Branching from all states controlled by these variables appears in the go–no go configuration, and logically adjacent assignments could be made. The switch signals must be debounced in either instance; thus *JK* flip-flops can be used to synchronize the signals as shown in Fig. 6.27. Since manual assertion of these switches will last for several clock periods, no problem from ambiguous metastable states will result.

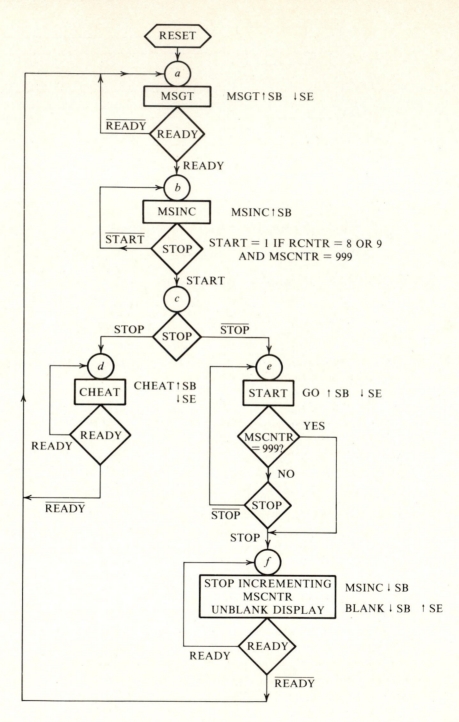

Figure 6.26 State diagram for reaction timer.

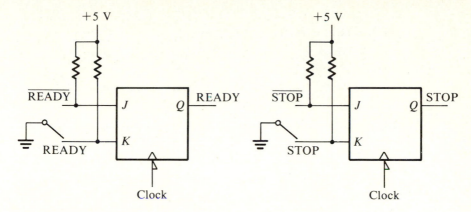

Figure 6.27 Debouncing and synchronizing with a flip-flop.

6.5 MISCELLANEOUS ASPECTS OF STATE MACHINE DESIGN

Section overview: This section discusses the reliability of state machines and then proceeds to a consideration of timing problems in these systems.

A. State Machine Reliability

There are two major sources of error in any digital system; extraneous noise and timing inaccuracies. The theory of noise is complex and will not be developed here. We will briefly mention some practical noise sources that might be encountered in actual systems.

One source of noise comes from within the circuit itself and is a result of ground loops. Figure 6.28(a) indicates schematically the general problem of allowing too many ground points in a circuit. As a result of the different currents through each section of the circuit, the voltages V_1, V_2, V_3, and V_4 are not quite equal. Although these points are connected to ground via wires or printed circuit conductors, these conductors each possess a small amount of ohmic resistance. Small voltages exist at the points indicated by V_1, V_2, V_3, and V_4 in Fig. 6.28(a). As gates switch on and off, the currents and voltages at these points change. These small changes can be amplified and may ultimately affect the overall circuit operation adversely. This problem can be minimized by the common point ground shown in Fig. 6.28(b). Although the voltage of this point may vary slightly as current into the node varies, the reference voltages for all sections of the circuit are identical. There are now no fluctuations of reference voltage between sections, and ground loop problems are minimized. While this method need not be applied to the positive voltage feed, it is good practice to use a common point feed also. It is also significant not to allow loops to be completed in the grounding circuit. Line a in Fig. 6.28(a) is removed as the common point ground is established in part (b).

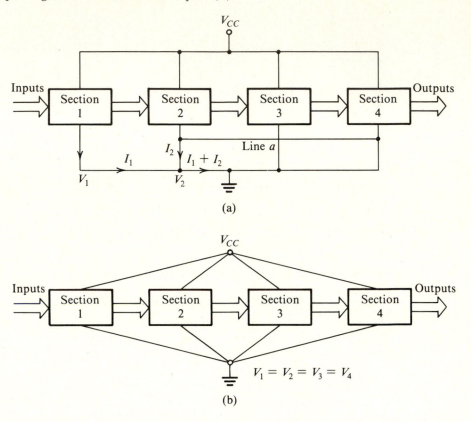

(a)

(b)

Figure 6.28 Grounding circuits: (a) with possible ground loops; (b) common-point grounding.

Another source of noise within the circuit is introduced by resistors and transistors. This is one source that can be treated on a theoretical basis and related fairly accurately to practice. We will here say only that the larger the circuit bandwidth or the higher the temperature, the greater is the resulting noise.

External radiation sources can also lead to noise in a given circuit. There are many radio and TV stations transmitting continuous energy into the environment. Closing or opening solenoids radiate electromagnetic waves as do spark plug wires in an automobile when the motor is running. Even though the Federal Communications Commission imposed the high-resistance carbon spark plug wire on the auto industry to minimize radiation, a significant amount of energy continues to radiate from a running engine. Long wires in a circuit may act as antennas to pick up radiated energy from these sources. In general, the longer the wires, the more extraneous noise is received and conducted to the circuit elements. Coaxial conductors with an external shield are often used to reduce this type of noise pickup when longer conductor lengths are required.

The preceding brief description of noise sources is given to emphasize the fact that a digital system is not immune to noise problems. On the other hand, certain methods of design lead to circuits that are less susceptible to noise than other circuit configurations. As mentioned in Chap. 4, the one-shot multivibrator is quite susceptible to noise and therefore is not often used in circuits that must be highly reliable. This is particularly true if the trigger input must connect to a long transmission line. The SR flip-flop is likewise not a good element to receive signals over a potentially noisy line.

State machines are inherently more reliable than asynchronous devices such as one-shots or SR flip-flops. Not only must a noise pulse with large amplitude occur, it must do so precisely at a time that just precedes the state changing clock transition. Otherwise, the clocked flip-flop will not respond to the noise pulse. It should be obvious at this point why a "ones-catching" device (master-slave FF) is inappropriate for state machine design.

There is another reason why the master-slave flip-flop cannot be used in state machine design. When input data to the IFL changes, hazards may be generated at the IFL output. Since we force the state-changing clock transition to be a half-period later than data changes, these short hazards do not affect the edge-triggered device, but they would cause erroneous state changes with the master-slave device. Since hazards are ignored by the edge-triggered flip-flop, hazard covers need not be used for the IFL.

We conclude that the state machine is more immune to extraneous noise than other types of system and also ignores hazards that are generated by the IFL.

B. Timing Considerations

We found in Chap. 5 that output glitches may result from differences in flip-flop or gate switching times. A holding register or clock suppression methods can be used to eliminate these problems. The necessary delays can be introduced in various ways to avoid timing conflicts. At higher frequencies, timing problems becoming more critical. A 1 MHz clock frequency results in 500 ns half-periods which allow reasonable delay times to be introduced with ease. A 20 MHz clock leads to 25 ns half-periods; thus delay times now make up a significant part of the clock period. In such a case, higher frequency devices such as ECL or Schottky TTL must be used to minimize required delays.

In calculating maximum clock frequency for the state machine, we must consider the IFL and the state FFs. Since there is one-half clock period between data change and flip-flop change, the IFL must settle in less than a half-period. The equation relating the clock frequency to IFL is

$$t_{cd} + t_{pIFL} + t_{su} < 1/2f$$

where t_{cd} is clock to data stable time, t_{pIFL} is propagation delay time through the IFL, t_{su} is the set-up time for the flip-flop, and f is the clock frequency. Isolating f in this inequality results in

$$f < \frac{1}{2(t_{cd} + t_{pIFL} + t_{su})}$$

It would be possible to increase this upper limit by allowing less than one half-period between state change and data change. A period greater than one-half period is then allowed for data changes to set up the proper flip-flop inputs. In the general case, however, the upper limit of f is calculated from the above inequality. For ECL circuits, this figure can exceed 100 MHz; thus the state machine can operate at very high frequencies.

SUMMARY OF SIGNIFICANT POINTS IN CHAP. 6

1. Design is a complex process requiring a knowledge of analysis.
2. Top-down design can be applied to digital system design. In this method successive refinement is used to decompose the system, first into modules and then into operational units. These units are realized with familiar circuits.
3. One module of the digital system is always a controlling state machine.
4. A design procedure for digital systems is listed.
5. A design procedure for state machines is listed.
6. The reliability of state machines and maximum frequency of operation are considered.

REFERENCES AND SUGGESTED READING

1. D.J. Comer, "Application of top-down principles to digital system design," *IEEE Transactions on Education*, November 1983.
2. W.I. Fletcher, *An Engineering Approach to Digital Design*. Englewood Cliffs, N.J.: Prentice-Hall, 1980, chap. 7.
3. D. Winkel and F. Prosser, *The Art of Digital Design*. Englewood Cliffs, N.J.: Prentice-Hall, 1980, chap. 6.
4. N. Wirth, "On the composition of well-structured programs," *Computing Surveys*, vol. 6, no. 4, December 1974.

CHAP. 6 PROBLEMS

6.1 Reduce the state diagram shown in Fig. P6.1 if possible. Then implement the state machine using MUXs for the IFL. Output A should be asserted for only one-half clock period, while B is asserted for a full period.

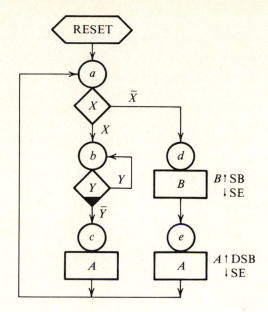

Figure P6.1

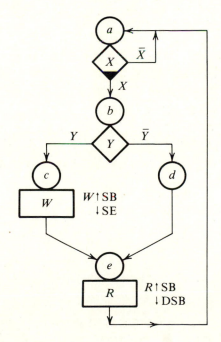

Figure P6.2

6.2 Design a minimal state machine for the diagram shown in Fig. P6.2 that eliminates output glitches and possible erroneous states. Explain all assumptions made.

* **6.3** Two separate data lines are changing synchronously on the rising edge of the clock. Design a system to examine these lines, and find the sequence 100 on line X coincidentally with the sequence 110 on line Y. When these combinations occur simultaneously anywhere in the data stream, an output strobe lasting one-half clock period should be generated. The system should then resume checking without missing a bit. Show the state diagram and timing chart. Use D flip-flops and gates for the IFL being as economical as possible. Output glitches should be eliminated.

6.4 Repeat Prob. 6.3 using JK flip-flops and gates for IFL.

6.5 Repeat Prob. 6.3 using D flip-flops and MUXs for the IFL.

6.6 Using positive clock transition D flip-flops, design a register and controlling state machine that receives a 4-bit parallel word on lines I_1, I_2, I_3, and I_4 and then transmits this word serially onto a line. The transmitted word should be preceded by a start bit equal to 1. When a transmit button is pushed, the register should accept the parallel data, then transmit serially. The line should be low at all times except when transmitting the NRZ code. Each of the 5 bits should last exactly one clock period. Assume a 1 kHz clock is present.

6.7 Design the state machine to control the adder described in Sec. 5.2C. State all assumptions made.

CHAPTER
7
Alternate Configurations for State Machines

In Chaps. 5 and 6 we introduced the basic concept of the state machine and developed principles and examples of design. With the information presented in these chapters, a great deal of practical design of sequential logic circuits can be accomplished. This chapter is intended to expand on the possible configurations that can be used with the state machine. These configurations offer more options such as programmable state machines and smaller chip-count systems. Prior to this discussion of alternate configurations of conventional state machines, we will consider the factors that determine whether a microprocessor controller will be more cost-effective than the state machine approach.

7.1 STATE MACHINE VS MICROPROCESSOR-BASED DIGITAL SYSTEMS

Section overview: This section considers the use of the microprocessor as a controller. The strengths and weaknesses of this device compared with a conventional state machine controller are discussed. The definition of a microprocessor state is then developed. The section is concluded with a discussion indicating how a designer might decide whether or not to use a microprocessor in digital system design.

In systems requiring complex controllers at moderate operation frequencies, the microprocessor is often a more cost-effective solution than the state machine. Although this text does not consider microprocessor design, it is important to understand when conventional state machine design should give way to microprocessor design. It is also significant to recognize what design principles are directly transferable from the state machine area to the microprocessor area.

A. Strengths and Weaknesses of the Microprocessor

There are those who feel the microprocessor has made the state machine obsolete. While the microprocessor has taken over several jobs once performed solely by state machines, there are other areas they have not been able to touch. The strength of the microprocessor lies in the fact that it can act not only as a state machine controller, but it can also perform many functions once requiring special purpose modules. In a digital system requiring addition or subtraction, for example, the microprocessor can accomplish this function with its internal arithmetic unit. An application calling for variable state times can be performed by microprocessor timing circuits and software. In conventional digital systems, extra circuitry must be provided for these purposes. Thus, the microprocessor can consolidate the control function and other component functions of a digital system into a single unit.

A second strength of a microprocessor is the very large number of equivalent states that can be generated without increasing system complexity. In a conventional state machine, additional states require more flip-flops. More state flip-flops lead to additional complexity in IFL and OFL design. While five state flip-flops may be handled without great difficulty, six or more flip-flops lead to a very difficult design procedure. The microprocessor can produce a number of states limited only by the size of the program memory area.

A third strength of the microprocessor is the flexibility due to its programmability. A single microprocessor can function as multiple state machines. This device can contain several separate program sequences each of which controls a different device. Alternatively, two programs may control the same device, but each program may lead to a different sequence of operations of the device. Furthermore, controlling programs stored in read/write memory (RAM) can be modified to change a control sequence. While this chapter considers programmed state machines in a later section, the microprocessor flexibility cannot be approached with conventional state machines.

The preceding list of strengths of the microprocessor might lead one to conclude that this device should replace the state machine in all controlling applications. Such a conclusion would be incorrect, for there are several weaknesses possessed by this device.

One major weakness of the microprocessor design compared with conventional state machines is the cost for small systems. The component cost for a conventional system with four state flip-flops can be under $2.00 using TTL chips. A typical microprocessor system could easily cost 10 times this figure for components. The older Intel 8080A (which is rather obsolete for design purposes) costs around $4.00, but it requires several support chips to create a functioning system. More modern stand-alone microprocessor chips with on-board ROM generally exceed a $20.00 cost figure.

While the component cost differential itself is enough to dictate the use of conventional state machine design for small controllers, there is an additional cost when using the microprocessor. The microprocessor system requires programming. At best, this means paying a manufacturer to program the ROM for the system. Of course, additional cost is incurred in this step. If the designer is to program the ROM, it must be done with a ROM burner that is quite expensive and the labor required also adds cost. Regardless of the method used, programming the microprocessor system leads to a significant expense over and above the component cost. For smaller systems, certainly for systems requiring less than 32 states, the conventional state machine can be constructed at a considerably lower cost.

Above 32 states, the design time for the conventional machine becomes more significant. Generally, if speed requirements can be met, the microprocessor is more appropriate for systems having more than 32 states. There are other factors that influence this decision which may make a 16-state system a candidate for microprocessor implementation.

A second weakness of the microprocessor is its relatively slow speed of operation. A 100 MHz state machine using ECL or Schottky TTL is not a particularly difficult design although care must certainly be taken at these higher clock frequencies. A state time of 10 ns results from a 100 MHz clock. A 100 MHz clock is unheard of for presently available microprocessors. Furthermore, several machine cycles will be required at each state for this device. We will shortly see that a state for the microprocessor may require a read operation to test system inputs, some manipulative operations, and a write operation to generate outputs. The microprocessor controller is then limited to much lower frequencies of operation than the conventional state machine controller. Typically, this advantage exceeds two orders of magnitude for the conventional state machine.

We see from this discussion that the microprocessor exhibits some serious limitations that disqualify its use in certain areas. For high-frequency design above a few MHz, the microprocessor cannot be used. For simpler systems having less than 32 states and requiring no programmability, the cost of the microprocessor system is prohibitive. We would expect to see the microprocessor controller used in low-frequency applications requiring either a large number of states (perhaps 64 or more) or programmability. We will consider other factors that relate to the choice of controlling device in the following paragraphs.

B. The Microprocessor as a State Machine

The microprocessor can be used as a controller without using state machine design techniques. There are certain concepts that are transferable from the well-developed state machine theory, thus it can be useful to treat the microprocessor as a state machine. This subsection will pursue this approach.

When using a microprocessor as a state machine, we must recognize several differences between this system and a conventional state machine. A conventional state machine allows input information to directly determine the next state to which the system branches. The microprocessor must read the external inputs into the accumulator then check the accumulator contents to determine the next state. Thus, at least one data transfer is involved in this case. In some microprocessors there are one or more testable pins that cause branching based on the voltage levels of these pins. In this case, the inputs can directly dictate the next state, but, in general, the input data must be shifted into the accumulator to be checked.

A second difference is the variable time required for each state. A conventional state machine remains in a given state for one clock period and then moves to a new state or remains in this same state. With a microprocessor, we have some states that may be present for timing purposes only. Another state may produce an output. Some states may have to consider an input, generate a conditional output, and branch to one of several next states depending on the input data. The number of machine cycles taken to accomplish the tasks required at each state may vary greatly from one state to another. Of course, if a fixed state time in each state is required, no-op codes can be added to the shorter states to equalize all state times. This procedure is generally avoided unless absolutely necessary to keep the design simple.

A third difference between conventional and microprocessor state machines lies in the data manipulation or processing capabilities of the microprocessor. This system can perform arithmetic or logical operations during any state, producing results that can be stored or used for branching purposes. The conventional system has no manipulative capabilities and, if such are needed, external circuitry must be added. This circuitry then requires control by the state machine to accomplish the required tasks. The microprocessor integrates data manipulation into the list of tasks performed by the state machine.

With these features of the microprocessor state machine we must provide for several possible operations during each state time. The following list tabulates the various types of machine cycles or tasks that may be accomplished at a given state.

1. Machine cycles for timing purposes

2. Data transfer from I/O bus (input)

3. Branching based on accumulator or flag data

4. Internal data transfer

5. Data manipulation

6. Data transfer to I/O bus (output)

Item 1 is often used when the microprocessor must be synchronized with an external system. If operations must be performed at specific time intervals, no-op cycles may be inserted to effect this requirement.

Item 2 may include the read in of synchronous or asynchronous signals. If the input signals are synchronous, it is only necessary to read this data at the proper time. Item 1 will be used for this purpose. For asynchronous inputs, an interrupt signal is normally used to indicate to the microprocessor that input data is valid.

Item 3 allows the microprocessor to check input data and branch to the next proper state.

If internal data manipulation is to take place before leaving a state, items 4 and 5 allow this function to take place.

Item 6 allows communication from the microprocessor to an external device.

We may now define a state as a series of instructions that accomplishes from one to six of the items of the preceding list. On this basis the state time will be variable, depending on the number of items that must be completed at each state. If fixed state times are required, additional no-op codes can be inserted in the shorter states requiring extended times.

It is beyond the scope of this text to consider the details of microprocessor design. With the preliminary information of this section, we can develop some design guidelines to determine whether or not a microprocessor controller should be used.

C. Top-Down Design Using the Microprocessor

The basic idea of microprocessor-based digital system design is identical to state machine design. We recall from Chap. 6 that a digital system, characterized by a set of specifications, is first decomposed into modules. These modules are basically functional modules that are as independent as possible of other modules. One of these modules is the controlling state machine. Each module is now decomposed into operational units which often consist of MSI circuits. These operational units are circuits designed to perform specific functions. Binary counters, shift registers, and ALUs are typical examples of operational units.

Once the system is decomposed into operational units, the comparison between conventional state machine and microprocessor-based design can be made. The steps suggested for this comparison are as follows:

1. Determine the number of states required for the controller.

2. Determine if the speed of the microprocessor is sufficient to meet specifications.

3. Determine the number of operational units that can be implemented by the microprocessor. In addition to the ability to act as a system controller, the microprocessor can also perform many operations. Timing, counting, arithmetic, and logical operations are examples of functions that may be implemented by the microprocessor. These capabilities can lead to the elimination of MSI circuits in the overall design.

4. Consider any possible enhancements to system features that could result from the presence of a microprocessor.

At this point the major factors dictating a choice can be considered. Component cost for the conventional state machine system can be approximated and compared with component plus programming cost of the microprocessor-based system. The predicted design complexity can be compared, as can the chip count and wiring cost. Any other factors of importance can be considered at this time in arriving at a final decision.

An important general conclusion can be drawn from this discussion. The top-down method of Chap. 6 is applicable to both conventional and microprocessor-based digital system design. Furthermore, this approach allows a comparison of the two designs to be made in order to make a reasonable choice of system type.

We might now return to the digital voltmeter considered in Sec. 6.2C. This system is composed of the operational units shown in Fig. 6.2. These operational units can each be decomposed into basic circuits as done for the counter, code converter, and display units shown in Fig. 6.3. After some rough design of the controlling state machine, a component cost can be calculated.

If we want to consider a microprocessor-based DVM, we can return to Fig. 6.2 and evaluate all operations that can be performed with a microprocessor. The controlling state machine is the first function that can be absorbed by the microprocessor. The counting operation and the code conversion can also be performed by this device. A register must be added to drive the LED display and the other operational units; the electronic switch, integrator, comparator, and display will remain in this design. The component cost can now be approximated and compared with that of the conventional design. This comparison is an important factor that will influence the final decision.

7.2 ALTERNATE CIRCUIT CONFIGURATIONS

Section overview: This section introduces other device configurations for the construction of state machines. The indirect-addressed MUX, the direct-addressed ROM, and the indirect-addressed ROM state machines are considered.

In Chaps. 5 and 6 we discussed only a few basic circuit configurations for the state machine. We considered the use of gates or direct-addressed MUXs for IFL and gates or decoders for OFL. There are several other types of device that can be used for both IFL and OFL. We will now discuss some of these devices and attempt to point out the kinds of application in which a particular device would be useful.

Another configuration that can be used for IFL is the indirect-addressed MUX arrangement. This scheme is used to reduce the MUX size at the expense of an additional code converter. If flip-flops are used for a system having more than 16 states, direct-addressed MUXs cannot easily be used. The largest MUX available on a chip is a 16:1 device. Indirect addressing can then be used in order to reduce the MUX size to a realistic value.

In situations where the sequence of states must be varied occasionally, a ROM or PLA controlled state machine may be used. In such a system, one ROM or PLA can be replaced by another to cause a different control sequence to be carried out. A direct-addressed ROM is conceptually simple, but it is rather wasteful in terms of ROM size. The indirect-addressed ROM reduces ROM size by again using a code converter. The following sections will consider these configurations.

A. Indirect-Addressed MUXs

Figure 7.1 indicates the general configuration of the indirect-addressed MUX system. This arrangement is very similar to the direct-addressed MUX system except for the addition of the code converter. This device allows the reduction in size of the MUXs. While this method is often used to reduce larger MUXs to practical sizes, we will demonstrate the method of design using a smaller system. We will redesign the IFL of the system of Fig. 5.50, shown again in Fig. 7.2.

We note that MUX A has only two signals connected to the inputs: ground and $+5$. MUX B has three signals connecting to the inputs, while MUX C also has two. Since these MUXs are 8-input devices, we conclude that they are not being used very efficiently. Smaller MUXs could be used to multiplex the required inputs to the flip-flop inputs. A 2:1 MUX could be used for MUX A and MUX C, while a 4:1 MUX could be used for MUX B, as shown in Fig. 7.3.

Inputs

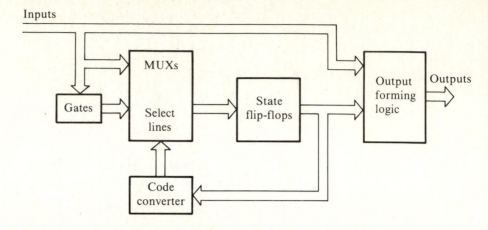

Figure 7.1 Indirect-addressed MUXs for IFL.

The input assignment is somewhat arbitrary. Extra inputs can be connected to logically adjacent inputs to reduce the code converter complexity. For example, MUX B has one extra input. By connecting two logically adjacent inputs to ground, this value can be addressed with $S_{1B} = 0$ and $S_{0B} = \Theta$. It is also appropriate if the assignment can be made so that MUXs of the same size require the same select line addressing. A dual 4:1 MUX chip has the select lines connected in parallel for each chip as does a quad 2:1 MUX chip. If the same size MUXs require the same addressing, a single chip can be used. Otherwise, two separate chips are required. Although it is not a trivial task to achieve the proper assignment, we will not discuss it further here.

The code converter must be designed to select the correct inputs for all three MUXs in each state of the system. Figure 7.4 contains the necessary information to develop the code converter. This table is derived from a consideration of the pertinent next state maps of Fig. 5.40(b) or from the direct-addressed MUX schematic. The former source is preferred since "don't care" conditions have not been chosen at this point and can help reduce the code converter. Using maps to reduce the expressions for select lines gives

$$S_{0A} = \overline{A}B; \quad S_{1B} = \overline{A}, \quad S_{0B} = \overline{C}; \quad S_{0C} = \overline{A}\overline{C}$$

The IFL using indirect-addressed MUXs is shown in Fig. 7.5.

Actually MUX A could easily be eliminated since the equation for $S_{0A'}$ also equals the equation for D_A. Instead of this gate driving a select line with MUX A driving D_A, the MUX could be removed allowing the gate to drive D_A. The indirect-addressed MUX scheme is generally applied to larger systems which will not result in this situation.

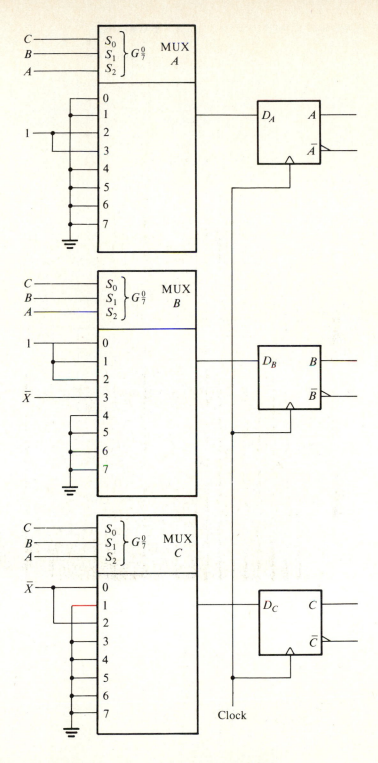

Figure 7.2 IFL for Example 5.2 using direct-addressed MUXs.

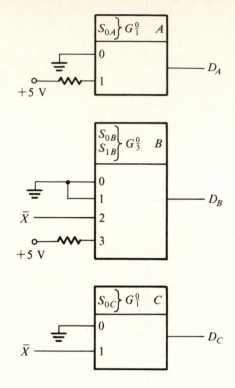

Figure 7.3 Input assignment for indirect-addressed MUXs.

Present State			MUX select lines			
			MUX A	MUX B		MUX C
A	B	C	S_{0A}	S_{1B}	S_{0B}	S_{0C}
0	0	0	0	1	1	1
0	0	1	Θ	Θ	Θ	Θ
0	1	0	1	1	1	1
0	1	1	1	1	0	0
1	0	0	0	0	Θ	0
1	0	1	Θ	Θ	Θ	Θ
1	1	0	0	0	Θ	0
1	1	1	0	0	Θ	0

Figure 7.4 MUX select lines as a function of present state.

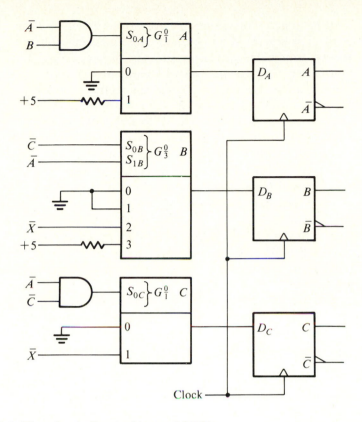

Figure 7.5 IFL using indirect-addressed MUXs.

Drill Problems: Sec. 7.2A

1. Realize the IFL for the system represented by the diagram of Fig. 5.59(a) using indirect-addressed MUXs. Use the state assignment of Fig. 5.59(b).

2. Repeat problem 1 using the state assignment of Fig. 5.59(c).

B. Direct-Addressed ROM

In some applications, the sequences executed by a state machine may need to be modified. This requires a change in the IFL in order to control a new sequence. In such a case, it is appropriate to use programmable devices for the IFL. A ROM or PLA can be programmed to create the correct signals to drive the state flip-flops. A change in sequence of the state machine can be effected simply by replacing the ROM or PLA with a new chip containing a different program. Changing the machine with conventional IFL would require a rewiring job rather than a chip change.

The concept of changeable ROM cartridges has gained great acceptance in the last several years, both in state machine and microprocessor control. Several hand-held calculators provide the capability of canned programs or subroutines that can be executed under ROM control. One of several programs can be used, simply by plugging the proper ROM into the program socket. Video games use a similar approach. The game to be played depends on the ROM plugged into the program socket. Control sequences are different for each ROM, providing a quick method of changing games.

A second advantage of ROM-controlled state machines occurs in system development work. A complex digital system will often require a large amount of debugging and modification during initial development. A hard-wired system may lead to many necessary wiring changes during this phase. Wiring changes are costly and can be time consuming in a larger system. Some firms use the EPROM to avoid this problem.

An EPROM (erasable programmable read-only memory) is a ROM that can be programmed upon application of the correct electrical signals. Once programmed, the EPROM retains this information indefinitely or until the chip is exposed to intense ultraviolet radiation for approximately 15 min. The package contains a quartz window through which the ultraviolet rays can impinge on the chip. After erasure, the EPROM can be reprogrammed. There is no upper limit on the number of program-erase cycles for the EPROM.

Using this device to design a ROM-controlled state machine allows a system to be debugged and modified under software control. The EPROM can initially be written into by a programming device under control of a computer. The state machine system is then exercised to locate any problems in operation. As problems are encountered, the system is modified simply by rewriting the contents of the EPROM. This procedure continues until the system operates according to specs. A production system can be created simply by replacing the more expensive EPROM with a mask-programmable device containing a duplicate program. Most EPROMs have a companion ROM that is pin-compatible.

The configuration for a state machine using the direct-addressed ROM is shown in Fig. 7.6. The next state and the outputs of a state machine are completely determined at any time by the present state of the system and the values of the input variables. The ROM uses the system input variables and present state (flip-flop outputs) as input addresses. The device is programmed to produce the correct next state conditions and correct system output variables as ROM outputs. The ROM then becomes both IFL and OFL for the system. The number of system inputs plus the number of state flip-flops determine the width of the input address, and hence the number of ROM locations needed. The number of outputs plus the number of state flip-flops determine the width of the output or the word size. A 5-input, 4-output system with three state flip-flops would require an 8-input ROM.

This leads to a 2^8 or 256 word capacity ROM. Each word must be 7 bits wide. If a standard ROM is used, a 256 × 8 device would be selected.

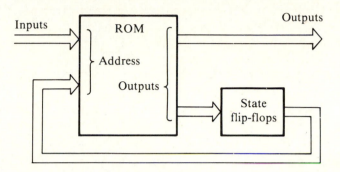

Figure 7.6 A direct-addressed ROM state machine.

The two state diagrams of Fig. 7.7 represent systems with the same possible sequences of states, but with different output states. We will realize these systems with direct-addressed ROMs indicating the ROM programming for each system. One system can then be converted to another simply by exchanging ROMs. We emphasize that the modification of a nonprogrammable state machine would require wiring changes.

With six states in the systems three flip-flops are required. Two system inputs, X and Y, bring the total number of ROM inputs needed to five. Since there are three system outputs, the number of ROM outputs is six. The minimum ROM size is then 32 × 6. A standard 32 × 8 ROM can be used. Figure 7.8 shows the realization of either of these systems using clock suppression to eliminate output glitches on the conditional lines R and W.

In order that the system perform properly, the ROM must be programmed correctly. While not every location must be programmed, those locations corresponding to each state and every possible input combination require programming. Figure 7.9 contains the programming for both systems.

The total delay required by the inverters is based on the address to output delay time of the ROM. A data change occurring on the negative clock transition represents an address change on the ROM. After a delay specified as address to data time, the output of the ROM will be stable. The clock suppression gates must not open until this delay time has elapsed. Bipolar transistors are used for smaller ROMs, leading to relatively small delay times. An address to data time of 30 ns might be typical of such a device. The delaying inverter propagation delay time must exceed this value which may require that a series of inverters be used.

The direct-addressed ROM is a simple system, but of course the ROM

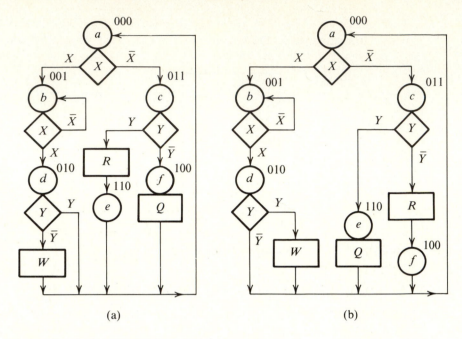

Figure 7.7 State diagrams of similar systems.

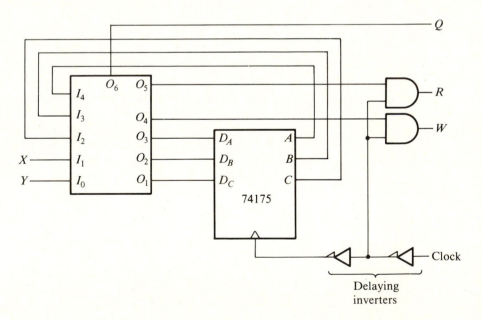

Figure 7.8 Direct-addressed ROM controlled state machine.

	ROM Address					ROM Outputs						ROM Outputs					
	Present State			System Inputs		System Outputs			Next State			System Outputs			Next State		
	A	B	C	X	Y	Q	R	W	A	B	C	Q	R	W	A	B	C
	I_4	I_3	I_2	I_1	I_0	O_6	O_5	O_4	O_3	O_2	O_1	O_6	O_5	O_4	O_3	O_2	O_1
a	0	0	0	0	0	0	0	0	0	1	1	0	0	0	0	1	1
	0	0	0	0	1	0	0	0	0	1	1	0	0	0	0	1	1
	0	0	0	1	0	0	0	0	0	0	1	0	0	0	0	0	1
	0	0	0	1	1	0	0	0	0	0	1	0	0	0	0	0	1
b	0	0	1	0	0	0	0	0	0	0	1	0	0	0	0	0	1
	0	0	1	0	1	0	0	0	0	0	1	0	0	0	0	0	1
	0	0	1	1	0	0	0	0	0	1	0	0	0	0	0	1	0
	0	0	1	1	1	0	0	0	0	1	0	0	0	0	0	1	0
d	0	1	0	0	0	0	0	1	0	0	0	0	0	0	0	0	0
	0	1	0	0	1	0	0	0	0	0	0	0	0	1	0	0	0
	0	1	0	1	0	0	0	1	0	0	0	0	0	0	0	0	0
	0	1	0	1	1	0	0	0	0	0	0	0	0	1	0	0	0
c	0	1	1	0	0	0	0	0	1	0	0	0	1	0	1	0	0
	0	1	1	0	1	0	1	0	1	1	0	0	0	0	1	1	0
	0	1	1	1	0	0	0	0	1	0	0	0	1	0	1	0	0
	0	1	1	1	1	0	1	0	1	1	0	0	0	0	1	1	0
f	1	0	0	0	0	1	0	0	0	0	0	0	0	0	0	0	0
	1	0	0	0	1	1	0	0	0	0	0	0	0	0	0	0	0
	1	0	0	1	0	1	0	0	0	0	0	0	0	0	0	0	0
	1	0	0	1	1	1	0	0	0	0	0	0	0	0	0	0	0
e	1	1	0	0	0	0	0	0	0	0	0	1	0	0	0	0	0
	1	1	0	0	1	0	0	0	0	0	0	1	0	0	0	0	0
	1	1	0	1	0	0	0	0	0	0	0	1	0	0	0	0	0
	1	1	0	1	1	0	0	0	0	0	0	1	0	0	0	0	0

(a)   (b)

Figure 7.9 ROM contents: (a) for system of Fig. 7.7(a); (b) for system of Fig. 7.7(b).

and programming requirement lead to a higher component cost than does a more conventional design. This scheme is rather inefficient in its usage of storage locations. For example, in states e and f, the next state is a regardless of the values of inputs X and Y. Since each of the four input combinations leads to a different address, four locations corresponding to each of these states must be programmed. Furthermore, each of the four locations have the same contents.

Another disadvantage of the direct-addressed ROM is that the size is determined by the number of inputs rather than by the number of unique words that must be developed at the ROM output.

These disadvantages can be overcome by using an indirect-addressed ROM or a PLA. The indirect-addressed ROM results in very efficient ROM usage at the expense of an additional code converter and a more complex design procedure. The PLA can reduce the number of storage locations required since this number is not a function of the number of inputs as is the case with the ROM. Flexibility is lost with these methods, however, and this explains why system development applies the direct-addressed ROM (EPROM) configuration.

Drill Problems: Sec. 7.2B

1. Realize the system represented by the diagram of Fig. 5.59(a) using a direct-addressed ROM. Use the state assignment of Fig. 5.59(b). Specify the contents of the ROM.

2. Repeat problem 1 using the state assignment of Fig. 5.59(c).

C. Indirect-Addressed ROM

An indirect-addressed ROM-controlled state machine is shown in Fig. 7.10. The code converter provides a unique address to access each unique ROM output required.

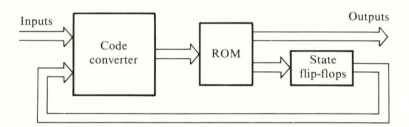

Figure 7.10 An indirect-addressed ROM controlled state machine.

An examination of Fig. 7.9(a) shows a total of eight unique output combinations required from the ROM. The same output combination may result from several different inputs. We can produce these eight combinations of outputs from a 3-input ROM. A code converter is then required to drive the ROM inputs. The inputs to this converter consist of the system inputs and the state flip-flop outputs. The design of the code converter becomes more difficult as the number of inputs increases. Even in this small system there are five inputs to the converter. It may be necessary to involve a computer in the design of larger converters to achieve a minimum code converter.

In order to design the code converter, we arbitrarily assign one unique address to each unique output combination required. All present state and system input combinations resulting in a particular output must produce a

Present State			System Inputs		ROM Outputs								
					System Outputs			Next State			ROM Address		
A	B	C	X	Y	Q	R	W	A	B	C	I_2	I_1	I_0
					O_6	O_5	O_4	O_3	O_2	O_1			
0	0	0	0	0	0	0	0	0	1	1	0	0	0
0	0	0	0	1	0	0	0	0	1	1	0	0	0
0	0	0	1	0	0	0	0	0	0	1	0	0	1
0	0	0	1	1	0	0	0	0	0	1	0	0	1
0	0	1	0	0	0	0	0	0	0	1	0	0	1
0	0	1	0	1	0	0	0	0	0	1	0	0	1
0	0	1	1	0	0	0	0	0	1	0	0	1	0
0	0	1	1	1	0	0	0	0	1	0	0	1	0
0	1	0	0	0	0	0	1	0	0	0	0	1	1
0	1	0	0	1	0	0	0	0	0	0	1	0	0
0	1	0	1	0	0	0	1	0	0	0	0	1	1
0	1	0	1	1	0	0	0	0	0	0	1	0	0
0	1	1	0	0	0	0	0	1	0	0	1	0	1
0	1	1	0	1	0	1	0	1	1	0	1	1	0
0	1	1	1	0	0	0	0	1	0	0	1	0	1
0	1	1	1	1	0	1	0	1	1	0	1	1	0
1	0	0	0	0	1	0	0	0	0	0	1	1	1
1	0	0	0	1	1	0	0	0	0	0	1	1	1
1	0	0	1	0	1	0	0	0	0	0	1	1	1
1	0	0	1	1	1	0	0	0	0	0	1	1	1
1	0	1	0	0	Θ	Θ	Θ	Θ	Θ	Θ	Θ	Θ	Θ
1	0	1	0	1	Θ	Θ	Θ	Θ	Θ	Θ	Θ	Θ	Θ
1	0	1	1	0	Θ	Θ	Θ	Θ	Θ	Θ	Θ	Θ	Θ
1	0	1	1	1	Θ	Θ	Θ	Θ	Θ	Θ	Θ	Θ	Θ
1	1	0	0	0	0	0	0	0	0	0	1	0	0
1	1	0	0	1	0	0	0	0	0	0	1	0	0
1	1	0	1	0	0	0	0	0	0	0	1	0	0
1	1	0	1	1	0	0	0	0	0	0	1	0	0
1	1	1	0	0	Θ	Θ	Θ	Θ	Θ	Θ	Θ	Θ	Θ
1	1	1	0	1	Θ	Θ	Θ	Θ	Θ	Θ	Θ	Θ	Θ
1	1	1	1	0	Θ	Θ	Θ	Θ	Θ	Θ	Θ	Θ	Θ
1	1	1	1	1	Θ	Θ	Θ	Θ	Θ	Θ	Θ	Θ	Θ

Figure 7.11 Information required for code converter design.

unique 3-bit code with which to address the ROM. Using the table of Fig. 7.9(a), we generate the information necessary to design the code converter in Fig. 7.11. Using maps to reduce the expressions for I_2, I_1, and I_0 results in

$$I_2 = A + BC + BY$$

$$I_1 = A\bar{B} + \bar{B}CX + BCY + \overline{AB}\bar{C}\bar{Y}$$

$$I_0 = \bar{B}C\bar{X} + A\bar{B} + \overline{AB}\bar{Y} + \bar{B}CX$$

The indirect-addressed ROM with code converter is shown in Fig. 7.12.

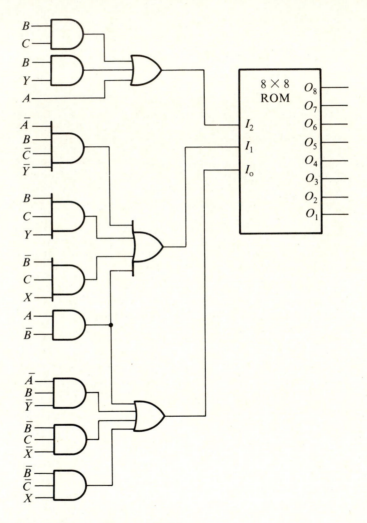

Figure 7.12 An indirect-addressed ROM.

While the ROM size and programming requirements have been reduced, the additional cost and effort involved in the code converter design may offset this advantage. There is another solution that often represents a reasonable compromise between the direct-addressed ROM and the indirect-addressed ROM. This will be referred to as the reduced-input ROM-

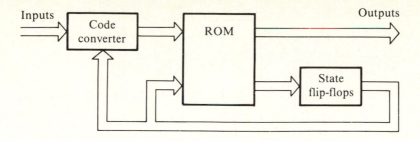

Figure 7.13 Reduced-input ROM controlled state machine.

controlled state machine. In this configuration, the present state variables are connected directly to ROM address lines, but the system inputs connect to a code converter whose outputs connect to the remaining ROM address lines. Figure 7.13 indicates this configuration.

The reduced-input system does not generally reduce the ROM size as much as the indirect-addressed system does. On the other hand, the code converter design is simpler and less costly. Relatively large systems can be designed with this scheme without resulting in a complex code converter. In order to demonstrate this type of design, the system of Fig. 7.7(a) will again be implemented.

Since the present state variables make up a portion of the ROM address, we recognize that each state will lead to a unique set of addresses. Within each state, the code converter must provide outputs that address the number of required locations. We again examine the table of Fig. 7.9(a) containing the direct-addressed information. The greatest number of locations that must be addressed within a given state is two. This situation occurs in states a, b, c, and d. In states e and f, only one set of outputs is required and hence only one location must be addressed. In order to address two locations within a state, a single binary variable is required. The code converter need only drive one address line to provide the required addressing capability. The ROM will then include three inputs from the state flip-flops and one input from the code converter. This leads to the requirement for a 16×8 ROM compared with a 32×8 for the direct-addressed scheme and an 8×8 for the indirect-addressed system.

We will use a MUX for the code converter. Since one code converter output is required and two system inputs are present, a 2:1 MUX is needed. We will arbitrarily connect X to the 0 line of the MUX and Y to the 1 line as shown in Fig. 7.14. We must next determine what to connect to the select line. From Fig. 7.9(a) we see that in states a or b, the inputs $X = 0$ and $Y = 0$ produce one set of ROM outputs while $X = 1$ and $Y = 0$ produce a second set. We can produce these two sets of outputs by allowing the ROM address to change with X. We can accomplish this by letting $S_0 = 0$ when the system is in states a or b. In states c and d, the two sets of

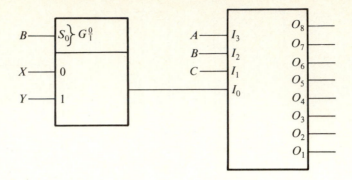

Figure 7.14 Code converter for reduced-input scheme.

	System Inputs	MUX Select Line	Present State			MUX Output	ROM Outputs					
							System Outputs			Next State		
	X Y	S_0	A I_3	B I_2	C I_1	Out I_0	Q O_6	R O_5	W O_4	A O_3	B O_2	C O_1
$a\begin{cases} \\ \\ \end{cases}$	0 Θ	0	0	0	0	0	0	0	0	0	1	1
	1 Θ	0	0	0	0	1	0	0	0	0	0	1
$b\begin{cases} \\ \\ \end{cases}$	0 Θ	0	0	0	1	0	0	0	0	0	0	1
	1 Θ	0	0	0	1	1	0	0	0	0	1	0
$d\begin{cases} \\ \\ \end{cases}$	Θ 0	1	0	1	0	0	0	0	1	0	0	0
	Θ 1	1	0	1	0	1	0	0	0	0	0	0
$c\begin{cases} \\ \\ \end{cases}$	Θ 0	1	0	1	1	0	0	0	0	1	0	0
	Θ 1	1	0	1	1	1	0	1	0	1	1	0
$f\begin{cases} \\ \\ \end{cases}$	Θ Θ	Θ	1	0	0	Θ	1	0	0	0	0	0
	Θ Θ	Θ	1	0	0	Θ	1	0	0	0	0	0
$e\begin{cases} \\ \\ \end{cases}$	Θ Θ	Θ	1	1	0	Θ	0	0	0	0	0	0
	Θ Θ	Θ	1	1	0	Θ	0	0	0	0	0	0

Figure 7.15 Conditions for reduced-input scheme.

outputs are a function of Y; that is, when $Y = 0$ one set of outputs is required and when $Y = 1$ the other set should be present. In these states, the value of S_0 should equal 1. Figure 7.15 reflects this information.

A K-map of S_0 as a function of A, B, and C is shown in Fig. 7.16. This leads to an expression $S_0 = B$ which completes the design.

Although 12 locations must be programmed in this ROM, the code con-

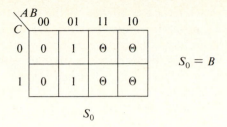

Figure 7.16 K-map of S_0.

verter is much simpler than the indirect-addressed system. This advantage becomes more significant as the total number of system inputs increases.

There is a loss of flexibility with the indirect-addressed ROM compared with the direct-addressed device. Since the code converter depends on the ROM contents, a change in ROM may require a change in code converter. When this is the case, the advantage of programmability is offset by the need for wiring changes.

Drill Problems: Sec. 7.2C

1. Realize the system represented by the diagram of Fig. 5.59(a) using an indirect-addressed ROM. Use the state assignment of Fig. 5.59(c).

2. Repeat problem 1 using a reduced-input ROM addressing scheme.

D. The PLA

A PLA can replace the ROM in the direct-addressed state machine with little effort. In systems that require a small ratio of programmed locations to total number of input combinations, the cost of the PLA can be significantly less than that of the ROM. In order to compare these devices, we will consider a system requiring a 10-input ROM needing only 48 programmed locations of 8-bit width. The ROM size would be 1024×8, whereas a PLA size of $10 \times 48 \times 8$ could be used. The cost of a 48-location PLA may be 30 percent of the cost of a 1024-location ROM. While the savings may not always be this great, the PLA offers a reasonable alternative to the ROM. Unfortunately, the choice of PLA size is rather limited, and the size needed for a given application may not be available.

Although mentioned in a previous chapter, we will emphasize an advantage of the PLA for rapid production of finished systems. The PLA can be used to realize combinational circuits such as the IFL and OFL for the state machine. Other related circuits, called gate arrays or master slice chips, include additional logic elements on a chip. In some cases the flip-flops for the state machine are available on the chip, and a state machine can be constructed by properly interconnecting the existing elements. Very short turn-around times from design to production result since the deposition of metal conductors can be done rather easily. In cases where a firm needs to beat a competitor to market with a finished system, the PLA or related chips can be used to advantage.

7.3 EXCHANGING CIRCUITRY FOR STATES

Section overview: This section considers the use of additional circuits to reduce the number of states in a state machine to the minimum level.

The previous section discusses various circuit configurations used to implement a state machine after a state diagram has been selected. Of course, all redundant states would be eliminated prior to selecting the final state diagram. At this point of the design, the number of states can be reduced no further. If discrete gates and flip-flops were used to design the state machine, the final design would often be minimal when previously discussed techniques are applied.

With the tremendous variety of circuits and systems now available on IC chips, state machine design is never done using discrete logic elements. Registers, counters, and arrays of gates or flip-flops are available on single chips. Chip count, component cost, and design time are often important quantities to minimize rather than the number of discrete logic elements.

In many design situations the addition of a single chip containing an MSI circuit will allow a flip-flop to be eliminated from the state machine. This, in turn, eliminates gating circuitry and simplifies the design. In some situations, depending on the design parameters, this may be a better solution to the problem.

It should be emphasized that any attempt to apply these methods must be made during the selection of functional modules or operational units for the system, prior to development of the state diagram. Alternatively, after an initial state diagram is produced, it can be examined for possible change resulting from a different choice of functional modules or operational units with which to implement the system.

Quite often a counter chip can be used to reduce the total number of states in a system having several states that perform identical functions. The state diagram of Fig. 7.17 demonstrates such a case. This system contains 11 states which would require four state flip-flops to implement. States d, e, and f are similar states generating the same output as also are states h, i, and j. Instead of including two sets of three similar states, a counter can be used to eliminate four of these states. The diagram of Fig. 7.18 shows this revised system.

The state machine now remains in state d or h for exactly three state times. The counter will be incremented once each clock period and, after a

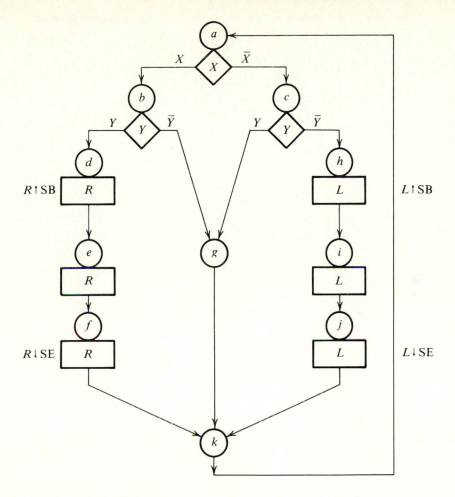

Figure 7.17 A state diagram.

count of 3 is reached, the system moves to state k. The count of 3 can be detected by a gate whose output serves as a state machine input. Thus, the counter must be incremented on the alternate clock transition to the state-changing transition. In this way, input data change does not conflict with state changes.

Only seven states and three state flip-flops are now required, but a binary counter with at least two flip-flops is necessary. Furthermore, additional gating to control the counter and to detect a count of 3 must also be used.

Counters are, by no means, the only circuit that can save states in a controlling state machine. Anytime a similar function must be performed at more than one state, a possibility for exchanging circuitry for states exists.

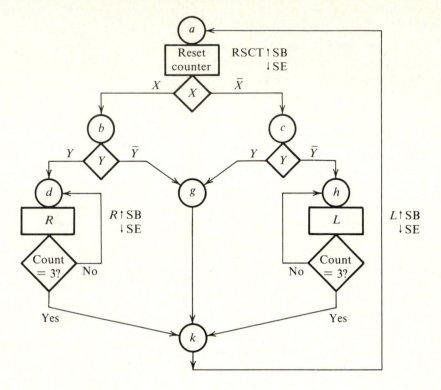

Figure 7.18 Revised state diagram.

Multiple circuits that can then perform the similar functions simultaneously may lead to a reduction in states.

We mention that a trade-off of states and circuitry is not always cost-effective. Each situation must be considered individually to determine if the reduction of states is worth more than the additional circuitry required.

7.4 A DRAG STRIP CONTROLLER

Section overview: This section demonstrates the various configurations discussed in the earlier sections of the chapter. These methods are applied to a practical digital system.

We will assume that our company has been awarded the contract to design, construct, and install the timing, control, and display mechanism for a drag strip. The company has assigned us the responsibility of overseeing the digital system design while another engineering group within the firm will do the sensor and display design. This group will make all interfaces TTL compatible, thus we need not be concerned with any interface circuitry.

A. Specifications

A drag strip consists of two straight tracks that provide a measured quarter-mile over which two dragsters race. The two dragsters position themselves at the start line where they remain until the start signal is given. The vehicles then accelerate to reach the quarter-mile mark as quickly as possible. The elapsed time from reception of the start signal to crossing the finish line must be measured for each dragster. In addition, the average speed of each vehicle must also be measured. Generally, the top speed is measured rather than the average speed, but this task would make our problem too difficult to be instructive.

Our specifications tell us that the average speed we must measure will never fall below 60 mph and never exceed 150 mph. The total elapsed time will then never be longer than 15 seconds or shorter than 6 seconds.

When the system is in the reset mode, a red light, located at the starting gate, must be activated. The operator throws a ready switch to indicate that the race will shortly begin. This switch closure will activate a yellow light and turn the red light off. A random time period between 1 and 4 seconds will begin at the closure of the switch. At the end of this time period, the yellow light shuts off as a green light is turned on. Simultaneous to the activation of the green light, a counter begins which will determine the times taken for the vehicles to reach the finish line.

Two electronic detectors are required in both lanes. One detector indicates when a vehicle crosses the start line, the other indicates the crossing of the finish line. The finish line is exactly one-quarter mile from the start line.

If either vehicle crosses the start line prior to the occurrence of the green light, that vehicle must be disqualified. The elapsed time from green light assertion to the crossing of the finish line must be recorded in milliseconds for each vehicle. The accuracy must be ± 3 ms. The average speed of each vehicle must also be registered to within ± 1 mph accuracy. The results are displayed after both vehicles have completed the race. A pointer should identify the dragster that achieved the shortest elapsed time as the winner.

B. Diagrams

After studying the specifications of the drag strip controller to identify the major goals, we proceed to draw the block diagram in terms of functional modules as shown in Fig. 7.19. Our major goal is to provide a system that will do the following:

1. Start the race some random time after the ready button is pushed.

2. Check for false starts and indicate disqualification when one occurs.

3. Time both vehicles over a quarter mile and display both times.

4. Indicate the winner.

5. Measure and display the average speeds of both vehicles.

The various functions that must be performed are separated as much as possible in the block diagram.

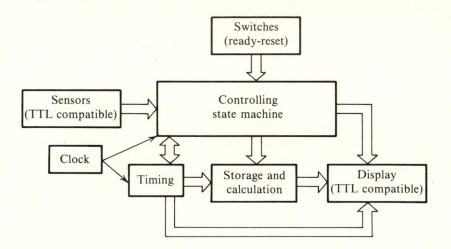

Figure 7.19 Block diagram for drag strip controller as functional modules.

We next consider each module individually to determine how each subfunction can be implemented. The timing circuitry must generate a random time ranging from 1 to 4 seconds and then must measure the times taken for each vehicle to reach the finish line. We will choose to include a millisecond counter driven by 1 kHz clock for each lane, capable of counting to 19999. These counters will begin counting when the start signal is generated, and each will stop when the corresponding vehicle reaches the finish line.

In order to generate the random time between depression of the READY button and the generation of the start signal, a 2-bit binary counter will be used. This counter will be driven by the 1 kHz clock when the RESET button is pushed. When the READY button is depressed, some random number between 0 and 3 will be contained by this counter. At this time, the 1 kHz clock is disconnected from the random counter input and connected to the input of the right lane millisecond counter. An input is also connected to the random counter that produces a pulse every second. This signal can be derived from the active millisecond counter. When a count of 3 is reached on the random counter along with a count of 999 on the active millisecond counter, the start condition will exist. The next clock pulse should lead to activation of the start light and allow both counters to begin counting the clock pulses starting at a zero count. These counters will be stopped as the corresponding vehicles cross the finish line.

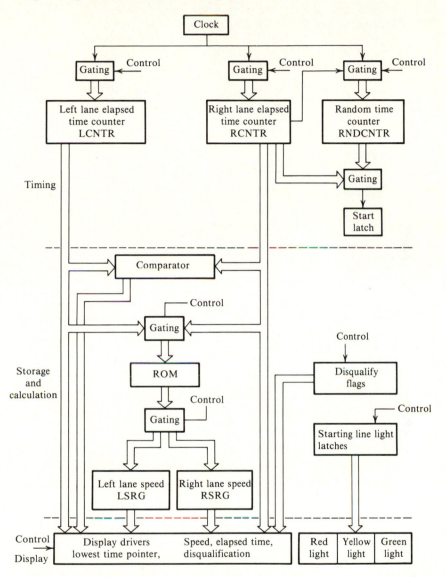

Figure 7.20 Operational units.

The calculation module will require two storage registers to hold the average speed figure for each vehicle. A ROM will be used to store "table look-up" information to convert from elapsed time to average speed. The ROM will be addressed by each elapsed time counter and the outputs, representing average speed, will be stored in two speed registers. A comparator will be required to determine the lowest elapsed time. Two latches, which will store disqualification information, are also included in this section.

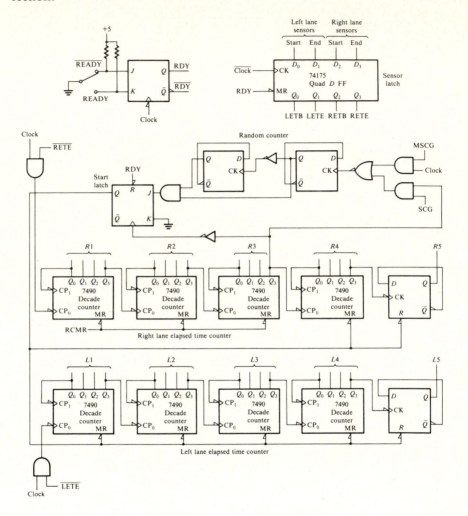

Figure 7.21 Timing subsection.

The display section must convert the information in various latches or registers to visual displays. To display actual numbers, BCD to 7-segment code converters can be used. Each elapsed time display will contain a decimal point occurring at a fixed point in the number. This point will always be active when the display is activated. A pointer associated with the lowest time should automatically be activated to identify the winner. At this point, we can propose the rough block diagrams for each subsection as shown in Fig. 7.20.

Although a particular system may ultimately be custom integrated, the original prototype will generally be constructed with discrete ICs for testing. An alternative approach that is becoming more popular is to use a simulation program to test the design before integrating the entire system. For this design problem, we will assume that a prototype is to be built and tested.

Obviously, a good knowledge of available ICs is important. The block diagrams of Fig. 7.20 must be implemented with ICs or refined further until such implementation is possible. As these circuits are chosen, the required control signals are also determined. While this step is an important one, we will not discuss it further. Rather we will propose the implementation of Figs. 7.21, 7.22, and 7.23 to realize the block diagrams of Fig. 7.20.

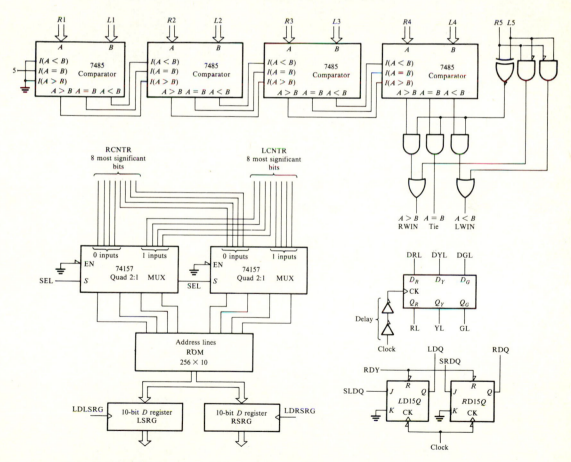

Figure 7.22 Storage and calculation subsection.

The lane sensors will generate asynchronous signals which will be converted to synchronous variables by the four D flip-flops. The RDY signal, which is initiated by a switch closure, will also be converted to a synchronous input. The RDY signal initializes all flip-flops to the appropriate states.

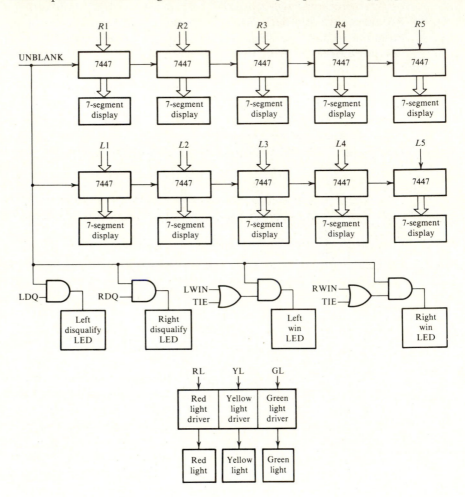

Figure 7.23 Display subsection.

The race is started by asserting the READY switch, causing the red light to go off and lighting the yellow light. The random counter provides an arbitrary time between 1 and 4 seconds before the START condition is met. The start latch sets at this time, and this action extinguishes the yellow light and activates the green. Both elapsed time counters now begin to count the 1 ms clock pulses. If either vehicle has crossed the starting line prior to the assertion of the START signal, the corresponding disqualify flag will be set.

When a vehicle reaches the finish line, the corresponding counter input is gated off; thus the decade counters contain elapsed times.

An accuracy of 1 mph will be obtained if only the 8 most significant bits of the elapsed time counters are used. This requires an 8-input ROM. To represent any speed between 60 and 200 mph in BCD requires a 10-bit word. The memory must then be a 256 × 10 ROM. This device must be programmed to output the correct speed for each of the 256 possible elapsed time inputs.

When both vehicles have finished the race, the 8 most significant bits of the right lane counter will access the ROM. The output is stored in a 10-bit register, RSRG. The left lane counter then accesses the ROM, and this output is stored in the LSRG.

All results are then displayed. Both counters drive displays, and both speed registers drive displays. The comparator output selects the winning time and activates the proper pointer. If either disqualify latch has been set, the disqualify indicator will be activated. All displays are blanked until the proper time. When unblanked, the display will remain until the READY switch is deasserted.

In order to provide the proper control signals, the state diagram of Fig. 7.24 can be implemented. Although this diagram implements the basic requirements of the drag-strip controller, there are certain shortcomings that must be solved for a practical system. For example, the results will only be displayed if both vehicles cross the finish line. Obviously, this presents a problem since one vehicle may experience mechanical or other difficulty and never reach the finish line. We will ignore this problem to keep the example simple enough to be instructive.

There are some signals that, once asserted, can remain asserted through several states. The signal SCG which enables an input to the random counter is asserted when state b is entered. This signal can be deasserted as the system exits state b, but it can also remain asserted up to the time state i is exited. Rather than specify a deassertion time at this point, the notation SCG ↓ ⊖ $cdefghi$ is used. This tells us deassertion can occur anywhere from state c to state i. We will see later that this flexibility allows us to design a smaller output decoder.

The last task remaining before the circuit design can be done is the construction of an accurate timing diagram. This chart should show all important inputs and outputs referenced to the system clock. Figure 7.25 includes this information. Note that the states are also indicated on the timing chart near the clock reference signal.

C. Direct-Addressed MUX Design

In this circuit one MUX is required for each state flip-flop. With this approach, the Q outputs of the state flip-flops are connected directly to the select lines of each MUX. The size of the MUX is then determined by the

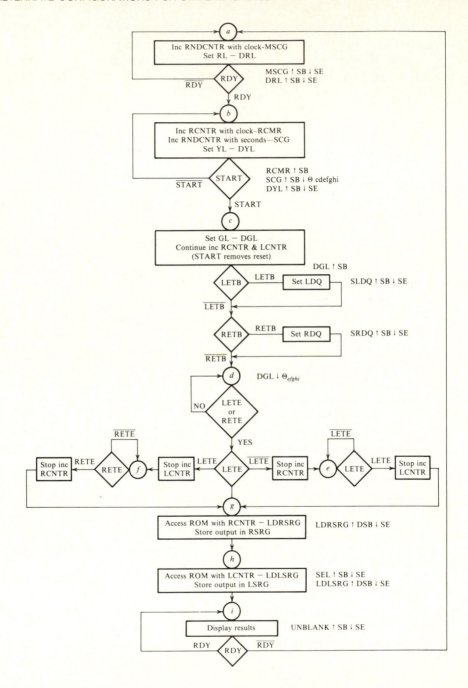

Figure 7.24 State diagram for drag strip controller.

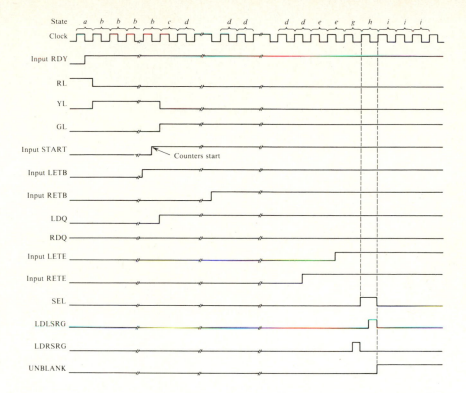

Figure 7.25 Timing diagram for drag strip controller.

number of state flip-flops. Each state directly addresses the MUX inputs to create the proper flip-flop input.

Although this scheme has been discussed before, we will implement the state diagram of Fig. 7.24 with direct-addressed MUXs for later comparison to other types of implementation. Since all inputs are synchronized, we have no state assignment constraints imposed by input signals. It also appears that output glitches are unimportant and can be ignored for all outputs except LDRSRG and LDLSRG. We will use clock suppression to remove these glitches. With 9 states in the system, four state flip-flops are needed. Four variables will then address each of the four MUXs, requiring 16:1 devices. State assignment is unimportant here since gates are not used for the IFL and no states have identical output specs. The present state map of Fig. 7.26 shows one possible state choice. The next state maps, or D flip-flop input maps are shown in Fig. 7.27. The output maps are now constructed as shown in Fig. 7.28.

We note that gates may precede the MUXs in this scheme as shown in Fig. 7.29, but the distinguishing feature is that the state flip-flops address the MUX input lines. Obviously this particular circuit does not use the MUXs

AB \ CD	00	01	11	10
00	a	e	i	Θ
01	b	f	Θ	Θ
11	c	g	Θ	Θ
10	d	h	Θ	Θ

Figure 7.26 State assignment for the drag strip controller.

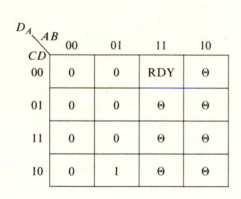

D_A

AB \ CD	00	01	11	10
00	0	0	RDY	Θ
01	0	0	Θ	Θ
11	0	0	Θ	Θ
10	0	1	Θ	Θ

D_B

AB \ CD	00	01	11	10
00	0	1	RDY	Θ
01	0	1	Θ	Θ
11	0	1	Θ	Θ
10	LETE + RETE	1	Θ	Θ

D_C

AB \ CD	00	01	11	10
00	0	LETE	0	Θ
01	START	RETE	Θ	Θ
11	1	1	Θ	Θ
10	$\overline{LETE} \cdot \overline{RETE}$	0	Θ	Θ

D_D

AB \ CD	00	01	11	10
00	RDY	LETE	0	Θ
01	1	1	Θ	Θ
11	0	0	Θ	Θ
10	LETE	0	Θ	Θ

Figure 7.27 Next state of D input maps.

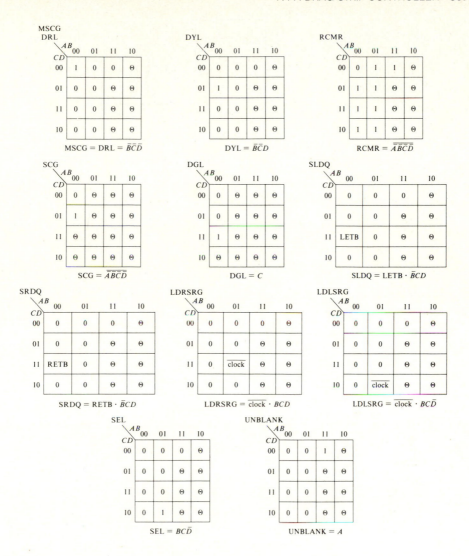

Figure 7.28 Output maps for drag strip controller.

efficiently. Of the 64 total MUX inputs, 44 are connected to ground. The first MUX only provides three different signals to the flip-flop input D_A. These signals are logic zero, logic one, and RDY. MUX B supplies only four signals to D_B. MUX C has five different input signals, and MUX D has four input signals. It is possible to reduce the size of the MUXs by using the indirect-addressed MUX approach.

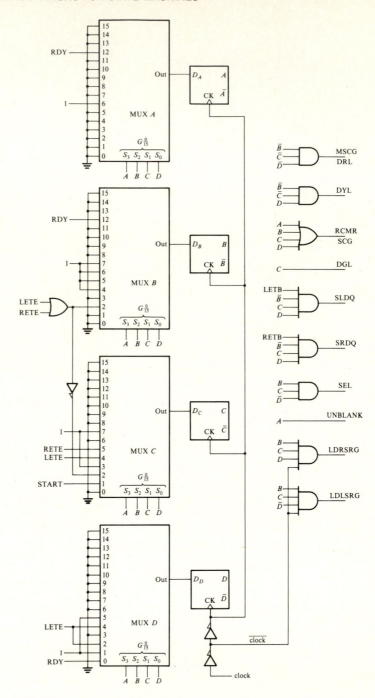

Figure 7.29 Drag strip controller using direct-addressed MUXs.

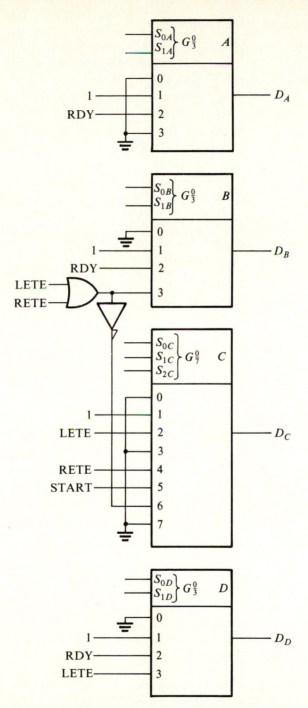

Figure 7.30 Input assignment for MUXs.

D. Indirect-Addressed MUX Design

The MUX size is reduced by adding the code converter to drive the MUX select lines. All states that require the same output for a given decoder can address a single MUX input line.

If we were to implement the drag strip controller using this scheme, we would first select the MUX sizes. MUXs A, B, and D can be 4:1 devices, while C must be an 8:1 device to handle the required inputs. We then assign the inputs, somewhat arbitrarily, as shown in Fig. 7.30.

Present State				MUX Select Lines								
				MUX A		MUX B		MUX C			MUX D	
A	B	C	D	S_{1A}	S_{0A}	S_{1B}	S_{0B}	S_{2C}	S_{1C}	S_{0C}	S_{1D}	S_{0D}
0	0	0	0	0	0	0	0	0	0	0	1	0
0	0	0	1	0	0	0	0	1	0	1	0	1
0	0	1	0	0	0	1	1	1	1	0	1	1
0	0	1	1	0	0	0	0	0	0	1	0	0
0	1	0	0	0	0	0	1	0	1	0	1	1
0	1	0	1	0	0	0	1	1	0	0	0	1
0	1	1	0	0	1	0	1	0	0	0	0	0
0	1	1	1	0	0	0	1	0	0	1	0	0
1	0	0	0	Θ	Θ	Θ	Θ	Θ	Θ	Θ	Θ	Θ
1	0	0	1	Θ	Θ	Θ	Θ	Θ	Θ	Θ	Θ	Θ
1	0	1	0	Θ	Θ	Θ	Θ	Θ	Θ	Θ	Θ	Θ
1	0	1	1	Θ	Θ	Θ	Θ	Θ	Θ	Θ	Θ	Θ
1	1	0	0	1	0	1	0	0	0	0	0	0
1	1	0	1	Θ	Θ	Θ	Θ	Θ	Θ	Θ	Θ	Θ
1	1	1	0	Θ	Θ	Θ	Θ	Θ	Θ	Θ	Θ	Θ
1	1	1	1	Θ	Θ	Θ	Θ	Θ	Θ	Θ	Θ	Θ

Figure 7.31 MUX select lines as a function of present state.

The code converter must be designed to select the correct inputs for all MUXs in each state of the system. Figure 7.31 contains the necessary information to develop the decoder. This table can be derived from the next state maps of Fig. 7.27 and the input assignments of Fig. 7.30. Using a 4-line to 16-line decoder could reduce the chip count in Fig. 7.32 somewhat, but we see that the price paid for reducing MUX inputs is added circuitry.

E. Direct-Addressed ROM Design

We will now realize the drag strip controller using a ROM. The system inputs and outputs are tabulated in Table 7.1.

Table 7.1 Tabulation of Inputs and Outputs

Inputs	Outputs
1. RDY	1. MSCG, DRL
2. START	2. DYL
3. LETB	3. RCMR, SCG
4. RETB	4. DGL
5. LETE	5. SLDQ
6. RETE	6. SRDQ
	7. LDRSRG
	8. LDLSRG
	9. SEL
	10. UNBLANK

Since there are four state flip-flops and six system inputs, the ROM is required to be at least a 1024 × 14 device. Two standard ROM chips, each 1024 × 8 devices, can be used to provide a 1024 × 16 ROM for this problem. Figure 7.33 shows the realization of the drag strip controller using the direct-addressed ROM.

In order that the circuit function properly, the ROM must be programmed correctly. Although not every location must be programmed, those locations corresponding to every possible input combination must be programmed. Figure 7.34 shows the contents of all locations that must be programmed.

An important point should be noted here. In states d, e, f, g, h, and i the inputs LETB and RETB have no effect on the outputs to be generated. Unfortunately, the four possible combinations of these two variables lead to different input addresses to the ROM. The contents of these four locations must all be programmed with the same output word. These "don't care" situations multiply the number of locations that must be programmed. For example, state d only has four possible sets of outputs, but the LETB and RETB combinations dictate that nine locations be programmed. This situation leads to additional programming when using the direct-addressed ROM. If we could somehow ignore the inputs LETB and RETB, except when the system is in state c, the number of locations that must be programmed could be reduced from 33 to 20.

A 1024-word ROM is more costly than a small ROM that need contain only 20 unique words. Furthermore, these additional inputs lead to the requirement of additional programming as explained previously. There are two methods that may be more cost-effective while still leading to the advantage of programmability. An indirect-addressed ROM or a PLA can be used for the IFL.

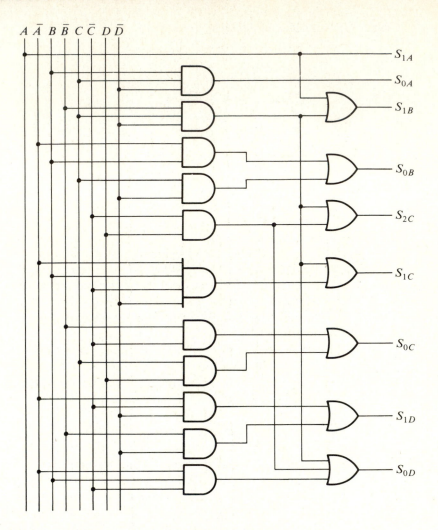

Figure 7.32 Decoder for indirect-addressed MUXs.

F. Reduced-Input ROM Design

The number of inputs to the ROM is reduced by the code converter logic. Programming flexibility is maintained by allowing the state flip-flops to connect directly to the ROM inputs. The combinational logic can be gates, decoders, or MUXs. Not only is the number of inputs to the ROM decreased, the "don't care" conditions can now be used to minimize the logic. With 10 inputs to the code converter, a computer would be required to minimize converter design. We will use the reduced-input scheme at this point; it does not truly minimize circuitry, but it allows a manual design procedure to be used. In this situation the inputs are decoded, but the state variables are not.

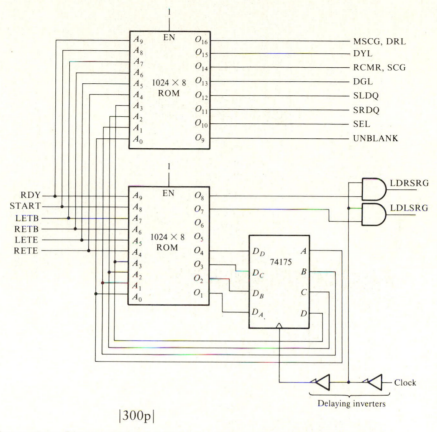

|300p|

Figure 7.33 Direct-addressed ROM controlled state machine.

We next consider how many outputs from the code converter are required. Examination of Fig. 7.34 indicates that state c must generate four outputs; therefore, the ROM must store these outputs in four different locations. Two variables are then required to address these four ROM locations. The address decoder provides these two outputs. At this point we could specify the ROM size. With six total inputs, four from the state flip-flops and two from the address decoder, the ROM must have a 64 × 14 capacity. Again, two standard ROMs could be used. Two 64 × 8 devices would be appropriate here.

We will refer to these code converter outputs as variables E and F. The assignment of values to these variables is somewhat difficult. On the other hand, E and F are primarily functions of the system inputs with only a slight dependence on state variables. Generally, the expressions for code converter

State		ROM Address — Present State				System Inputs						Next State				ROM Outputs — System Outputs											
		A_0 A	A_1 B	A_2 C	A_3 D	A_4 RETE	A_5 LETE	A_6 RETB	A_7 LETB	A_8 START	A_9 RDY	O_1	O_2	O_3	O_4	O_5 LDLSRG	O_6 LDRSRG	O_7 UNBLANK	O_8 SEL	O_9 SRDQ	O_{10} SRLQ	O_{11} SLDQ	O_{12} DGL	O_{13}	O_{14} RCMR&SCG	O_{15} DYL	O_{16} MSCG&DRL
a	{	0	0	0	0	0	0	0	0	0	0	0	0	0	0	Θ	Θ	0	0	0	0	0	0	0	0	0	1
		0	0	0	0	0	0	0	0	0	1	0	0	0	1	Θ	Θ	0	0	0	0	0	0	0	0	0	1
b	{	0	0	0	1	0	0	0	0	0	1	0	0	0	1	Θ	Θ	0	0	0	0	0	0	0	1	1	0
		0	0	0	1	0	0	0	1	0	1	0	0	0	1	Θ	Θ	0	0	0	0	0	0	0	1	1	0
		0	0	0	1	0	0	1	0	0	1	0	0	0	1	Θ	Θ	0	0	0	0	0	0	0	1	1	0
		0	0	0	1	0	0	1	1	0	1	0	0	0	1	Θ	Θ	0	0	0	0	0	0	0	1	1	0
		0	0	0	1	0	0	0	0	1	1	0	0	1	1	Θ	Θ	0	0	0	0	0	0	0	1	1	0
		0	0	0	1	0	0	0	1	1	1	0	0	1	1	Θ	Θ	0	0	0	0	0	0	0	1	1	0
		0	0	0	1	0	0	1	0	1	1	0	0	1	1	Θ	Θ	0	0	0	0	0	0	0	1	1	0
		0	0	0	1	0	0	1	1	1	1	0	0	1	1	Θ	Θ	0	0	0	0	0	0	0	1	1	0
c	{	0	0	1	1	0	0	0	0	1	1	0	0	1	0	Θ	Θ	0	0	0	0	0	0	1	1	0	0
		0	0	1	1	0	0	0	1	1	1	0	0	1	0	Θ	Θ	0	0	0	0	0	1	1	1	0	0
		0	0	1	1	0	0	1	0	1	1	0	0	1	0	Θ	Θ	0	0	0	0	1	0	1	1	0	0
		0	0	1	1	0	0	1	1	1	1	0	0	1	0	Θ	Θ	0	0	0	1	1	1	1	1	0	0
d	{	0	0	1	0	0	0	0	0	1	1	0	0	1	0	Θ	Θ	0	0	0	0	0	0	1	1	0	0
		0	0	1	0	0	0	0	1	1	1	0	0	1	0	Θ	Θ	0	0	0	0	0	0	1	1	0	0
		0	0	1	0	0	0	1	0	1	1	0	0	1	0	Θ	Θ	0	0	0	0	0	0	1	1	0	0
		0	0	1	0	0	0	1	1	1	1	0	0	1	0	Θ	Θ	0	0	0	0	0	0	1	1	0	0
		0	0	1	0	0	1	0	1	1	1	0	1	0	1	Θ	Θ	0	0	0	0	0	0	Θ	1	0	0
		0	0	1	0	0	1	1	1	1	1	0	1	0	1	Θ	Θ	0	0	0	0	0	0	Θ	1	0	0
		0	0	1	0	1	0	1	0	1	1	0	1	0	0	Θ	Θ	0	0	0	0	0	0	Θ	1	0	0
		0	0	1	0	1	0	1	1	1	1	0	1	0	0	Θ	Θ	0	0	0	0	0	0	Θ	1	0	0
		0	0	1	0	1	1	1	1	1	1	0	1	0	1	Θ	Θ	0	0	0	0	0	0	Θ	1	0	0
e	{	0	1	0	0	1	0	1	0	1	1	0	1	0	0	Θ	Θ	0	0	0	0	0	0	Θ	1	0	0
		0	1	0	0	1	0	1	1	1	1	0	1	0	0	Θ	Θ	0	0	0	0	0	0	Θ	1	0	0
		0	1	0	0	1	1	1	1	1	1	0	1	1	1	Θ	Θ	0	0	0	0	0	0	Θ	1	0	0
f	{	0	1	0	1	0	1	0	1	1	1	0	1	0	1	Θ	Θ	0	0	0	0	0	0	Θ	1	0	0
		0	1	0	1	0	1	1	1	1	1	0	1	0	1	Θ	Θ	0	0	0	0	0	0	Θ	1	0	0
		0	1	0	1	1	1	1	1	1	1	0	1	1	1	Θ	Θ	0	0	0	0	0	0	Θ	1	0	0
g	{	0	1	1	1	1	1	1	1	1	1	0	1	1	0	Θ	Θ	0	1	0	0	0	0	Θ	1	0	0
h	{	0	1	1	0	1	1	1	1	1	1	1	1	0	0	Θ	Θ	0	0	1	0	0	0	Θ	1	0	0
i	{	1	1	0	0	1	1	1	1	1	1	1	1	0	0	Θ	Θ	0	0	1	0	0	0	Θ	1	0	0
		1	1	0	0	0	0	0	0	0	0	0	0	0	0	Θ	Θ	0	0	1	0	0	0	Θ	1	0	0

Figure 7.34 ROM contents for drag strip controller.

	Present State				R E T E / E	L E T E / E	R E T B / B	L E T B / B	S T A R T	R D Y	Decoder Outputs	
	A	B	C	D							E	F
a	0	0	0	0	0	0	0	0	0	0	1	0
	0	0	0	0	0	0	0	0	0	1	0	1
b	0	0	0	1	0	0	0	0	0	1	0	1
	0	0	0	1	0	0	0	1	0	1	0	1
	0	0	0	1	0	0	1	0	0	1	0	1
	0	0	0	1	0	0	1	1	0	1	0	1
	0	0	0	1	0	0	0	0	1	1	1	0
	0	0	0	1	0	0	0	1	1	1	1	0
	0	0	0	1	0	0	1	0	1	1	1	0
	0	0	0	1	0	0	1	1	1	1	1	0
c	0	0	1	1	0	0	0	0	1	1	0	0
	0	0	1	1	0	0	0	1	1	1	0	1
	0	0	1	1	0	0	1	0	1	1	1	0
	0	0	1	1	0	0	1	1	1	1	1	1
d	0	0	1	0	0	0	0	0	1	1	1	0
	0	0	1	0	0	0	0	1	1	1	1	0
	0	0	1	0	0	0	1	0	1	1	1	0
	0	0	1	0	0	0	1	1	1	1	1	0
	0	0	1	0	0	1	0	1	1	1	0	1
	0	0	1	0	0	1	1	1	1	1	0	1
	0	0	1	0	1	0	1	0	1	1	1	1
	0	0	1	0	1	0	1	1	1	1	1	1
	0	0	1	0	1	1	1	1	1	1	0	0
e	0	1	0	0	1	0	1	0	1	1	1	1
	0	1	0	0	1	0	1	1	1	1	1	1
	0	1	0	0	1	1	1	1	1	1	0	0
f	0	1	0	1	0	1	0	1	1	1	0	1
	0	1	0	1	0	1	1	1	1	1	0	1
	0	1	0	1	1	1	1	1	1	1	0	0
g	0	1	1	1	1	1	1	1	1	1	0	0
h	0	1	1	0	1	1	1	1	1	1	0	0
i	1	1	0	0	1	1	1	1	1	1	0	0
	1	1	0	0	0	0	0	0	0	0	1	0

Figure 7.35 Address decoder outputs.

outputs can be developed by considering only the system inputs. If conflict-ing requirements on identical input conditions arise, an appropriate number of state variables are introduced as inputs to resolve this problem. Figure 7.35 indicates one possible assignment for the variables E and F. Each row of the table requires a different output which can be found by referring to Fig. 7.34. In all states except c the equations for E and F are

$$E = \overline{\text{LETE}} \text{ START} + \overline{\text{START}} \ \overline{\text{RDY}}$$

$$F = \text{RETE} \ \overline{\text{LETE}} + \overline{\text{RETE}} \text{ LETE} + \overline{\text{START}} \text{ RDY}$$

In state c the equations are

$$E = \text{RETB}$$

$$F = \text{LETB}$$

The address decoder is implemented in Fig. 7.36.

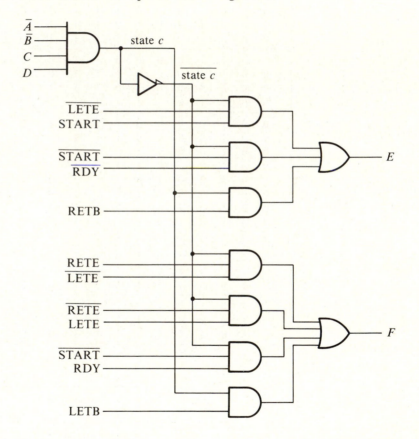

Figure 7.36 Address decoder.

We could also use MUXs in the reduced-input ROM circuit to form the code converter. Since two outputs, E and F, are required we must use two MUXs. The MUXs must have a total number of inputs that equals or exceeds the number of system inputs. Six system inputs dictate that each MUX have four inputs, resulting in eight total inputs.

Next we consider the inputs that are important in each state. We must determine from Fig. 7.35 which variables control the outputs E and F. In state a, RDY determines the values of E and F. In state b, START controls these values, while in state c RETB and LETB are the controlling variables. In states d through i, the outputs E and F are functions of RETE and LETE. We now associate these input variables with different MUX input lines. In this case, we will associate the RDY input with the zero input line of the two MUXs. START will be associated with the input lines 1. RETE and LETE will be assigned to input lines 2, and RETB and LETB will be assigned the input lines 3. Thus, when the system is in state a, the MUX select lines must address lines 0, requiring that the select lines are $S_1 = 0$ and $S_0 = 0$. State b must generate $S_1 = 0$ and $S_0 = 1$, while state c must generate $S_1 = 1$ and $S_0 = 1$. States d through i should produce $S_1 = 1$ and $S_0 = 0$. This information can be plotted as in Fig. 7.37 to develop the expressions for S_1 and S_0. The address decoder using MUXs is realized in Fig. 7.38 using a dual 4:1 MUX chip.

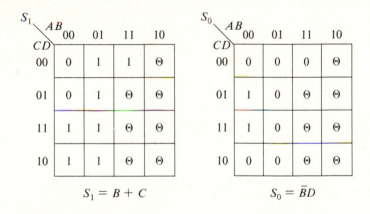

$$S_1 = B + C$$

$$S_0 = \bar{B}D$$

Figure 7.37 MUX select line maps.

The last step in this design is to specify the programming for the ROM. Using both Figs. 7.34 and 7.35, we generate Fig. 7.39. Note that only 20 of the 64 available locations are programmed. Ability to change operation by changing ROM is maintained, but less programming and a smaller ROM is required compared with the direct-addressed ROM state machine. A few extra gates plus more design time is the price paid for this reduction in circuitry and programming. Furthermore, some reduction in flexibility occurs in the indirect-addressed scheme since the code converter design may depend on the state variables.

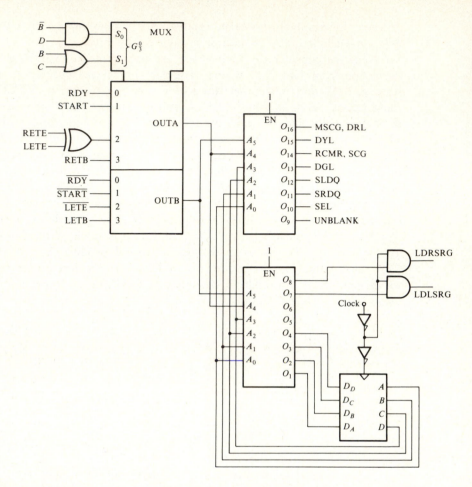

Figure 7.38 The indirect-addressed ROM state machine.

SUMMARY OF SIGNIFICANT POINTS IN CHAP. 7

1. A microprocessor can function as a state machine controller while performing other functions.
2. Top-down design methods can also be applied to microprocessor-based design.
3. The microprocessor has several strengths and weaknesses compared with the conventional state machine.
4. Other configurations for the conventional state machine are the indirect-addressed MUX, direct-addressed ROM, and indirect-addressed ROM systems.

	ROM Address						ROM Outputs															
	A_0	A_1	A_2	A_3	A_4	A_5	O_1	O_2	O_3	O_4	O_5	O_6	O_7	O_8	O_9	O_{10}	O_{11}	O_{12}	O_{13}	O_{14}	O_{15}	O_{16}
a	0	0	0	0	1	0	0	0	0	0	Θ	Θ	0	0	0	0	0	0	0	0	0	1
	0	0	0	0	0	1	0	0	0	1	Θ	Θ	0	0	0	0	0	0	0	0	0	1
b	0	0	0	1	0	1	0	0	0	1	Θ	Θ	0	0	0	0	0	0	0	1	1	0
	0	0	0	1	1	0	0	0	1	1	Θ	Θ	0	0	0	0	0	0	0	1	1	0
c	0	0	1	1	0	0	0	0	1	0	Θ	Θ	0	0	0	0	0	0	1	1	0	0
	0	0	1	1	0	1	0	0	1	0	Θ	Θ	0	0	0	0	0	1	1	1	0	0
	0	0	1	1	1	0	0	0	1	0	Θ	Θ	0	0	0	0	1	0	1	1	0	0
	0	0	1	1	1	1	0	0	1	0	Θ	Θ	0	0	0	0	1	1	1	1	0	0
d	0	0	1	0	1	0	0	0	1	0	Θ	Θ	0	0	0	0	0	0	1	1	0	0
	0	0	1	0	0	1	0	1	0	1	Θ	Θ	0	0	0	0	0	0	Θ	1	0	0
	0	0	1	0	1	1	0	1	0	0	Θ	Θ	0	0	0	0	0	0	Θ	1	0	0
	0	0	1	0	0	0	0	1	0	1	Θ	Θ	0	0	0	0	0	0	Θ	1	0	0
e	0	1	0	0	1	1	0	1	0	0	Θ	Θ	0	0	0	0	0	0	Θ	1	0	0
	0	1	0	0	0	0	0	1	1	1	Θ	Θ	0	0	0	0	0	0	Θ	1	0	0
f	0	1	0	1	0	1	0	1	0	1	Θ	Θ	0	0	0	0	0	0	Θ	1	0	0
	0	1	0	1	0	0	0	1	1	1	Θ	Θ	0	0	0	0	0	0	Θ	1	0	0
g	0	1	1	1	0	0	0	1	1	0	Θ	Θ	0	1	0	0	0	0	Θ	1	0	0
h	0	1	1	0	0	0	1	1	0	0	Θ	Θ	1	0	0	1	0	0	Θ	1	0	0
i	1	1	0	0	0	0	1	1	0	0	Θ	Θ	0	0	1	0	0	0	Θ	1	0	0
	1	1	0	0	1	0	0	0	0	0	Θ	Θ	0	0	1	0	0	0	Θ	1	0	0

Figure 7.39 Indirect-addressed ROM contents for drag strip controller.

REFERENCES AND SUGGESTED READING

1. W.I. Fletcher, *An Engineering Approach to Digital Design*. Englewood Cliffs, N.J.: Prentice-Hall, 1980, chap. 7.
2. D. Winkel and F. Prosser, *The Art of Digital Design*. Englewood Cliffs, N.J.: Prentice-Hall, 1980, chap. 6.
3. N. Wirth, "On the composition of well-structured programs," *Computing Surveys*, vol. 6, no. 4, December 1974.

CHAP. 7 PROBLEMS

Work Probs. 7.1 to 7.4 using indirect-addressed MUXs for IFL.

7.1 Rework Prob. 6.1.

7.2 Rework Prob. 6.2.

7.3 Rework Prob. 6.3.

7.4 Rework Prob. 6.6.

Work Probs. 7.5 to 7.8 using a direct-addressed ROM.

7.5 Rework Prob. 6.1.

7.6 Rework Prob. 6.2.

7.7 Rework Prob. 6.3.

7.8 Rework Prob. 6.6.

Work Probs. 7.9 to 7.12 using an indirect-addressed ROM.
7.9 Rework Prob. 6.1.

7.10 Rework Prob. 6.2.

7.11 Rework Prob. 6.3.

7.12 Rework Prob. 6.6.

7.13 Develop a state diagram for a system that measures top speed for a drag strip. This system has sensors in each lane 131 ft before the finish line and at the finish line. The average speed of the vehicle over the last 131 ft of the strip is taken as the top speed. A 1 kHz clock drives the system, and a single ROM is used in a table look-up system.

* **7.14** Calculate the size of the ROM in Prob. 7.13 if the accuracy is to be 0.1 mph. Assume a minimum speed of 150 mph and a maximum speed of 300 mph.

APPENDIX

1

Asynchronous Systems

To this point in the text we have considered only those systems with clock-driven state changes. A major advantage of such a system is that all signals are stable when the state flip-flops change. If there is any conflict with unstable signals, certain variables can be delayed with respect to the clock signal to resolve this problem. There are two prices paid for the advantage of having this time reference. The first is in additional circuitry required to generate and control the clock signal. The second is that state changes must wait until a clock transition occurs even though the signals determining the state change may have stabilized long before the change takes place.

It is possible to design systems, called asynchronous state machines, to avoid these problems. Although the advantages of the asynchronous over the synchronous system are significant, there is one great disadvantage that leads to the continued popularity of the synchronous machine. It is considerably more difficult to design reliable asynchronous systems than to design reliable synchronous systems. For this reason, synchronous state machines abound while asynchronous systems are used much less often. This section will discuss some basic principles relative to the design of simple asynchronous state machines.

A1.1 THE FUNDAMENTAL-MODE MODEL

An asynchronous system uses feedback to produce memory elements as does the synchronous state machine. The synchronous system utilizes clocked flip-flops driven by the system clock. The asynchronous machine generally uses gates rather than flip-flops. Inputs to these gates are composed of system inputs and feedback signals from gate outputs which drive the system to change states. Figure A1.1 demonstrates a simple asynchronous circuit.

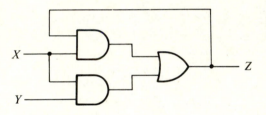

Figure A1.1 An asynchronous circuit.

X and Y are the system inputs while Z is the system output. The signal Z is fed back, however, to a gate input and in this way helps determine its own value. One difference between Z as an input signal and the system inputs X and Y is that there is always a delay from change in X or Y to change in Z. When an X or Y change dictates a change in Z, this change occurs only after the cumulative propagation delay time through the gates. We cannot be sure of the value of Z until the effect of this variable as an input has taken place.

It is characteristic of asynchronous circuits that the feedback variables along with system inputs determine the values of these same feedback variables. It is possible to force a change in feedback variable Z, by changing a system input, then Z may change a second time after a small delay time as feedback variables drive the input gates.

An idealized model has been proposed to reflect this behavior [2]. This fundamental-mode model is shown in Fig. A1.2 for the circuit under discussion. The gates are considered to have no delay in this model, while the delay element has an output that follows its input after a delay of Δt.

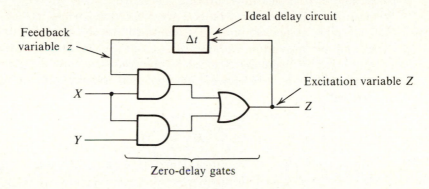

Figure A1.2 Fundamental-mode model for the circuit of Fig. A1.1.

The variable at the input of the delay element is called the excitation variable, while the feedback variable appears at the output of the delay element. This model produces the delay of the feedback variable but predicts no delay between input and output. While this is not completely accurate, the overall behavior of the actual circuit can be predicted rather accurately with this particular model.

In order to characterize the behavior of a circuit, we plot a map of excitation variable as a function of gate inputs. Of course, gate inputs consist of both system inputs and feedback variables. Figure A1.3 shows an excitation map for the circuit of Fig. A1.1. The system inputs are always plotted horizontally across the map, while the feedback variables are plotted vertically.

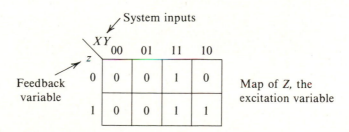

Figure A1.3 Excitation map for the circuit of Fig. A1.1.

It is important to realize that the value Z takes on will also be the value assumed by z after a delay of Δt. Thus, the information depicted by the map represents a dynamic situation. This can be demonstrated by supposing the system inputs are $X = Y = 1$ and $z = 1$ which leads to $Z = 1$. This is called a stable state since $z = Z$. If X is then changed to 0, the output Z

changes to 0 as indicated by the map location corresponding to $X = 0$, $Y = 1$, and $z = 1$. This condition will persist for only Δt since z will assume a value of 0 at this time, moving the system to the $X = 0$, $Y = 1$, and $z = 0$ location. The location 011 is a transient state, while 010 is a stable state. The stable states are normally identified on the map by drawing a circle around the excitation variable such as in Fig. A1.4. Whenever $z = Z$, the gate inputs will not be changed by the feedback variable; therefore the state is a stable state. To move from this state requires that a system input be changed.

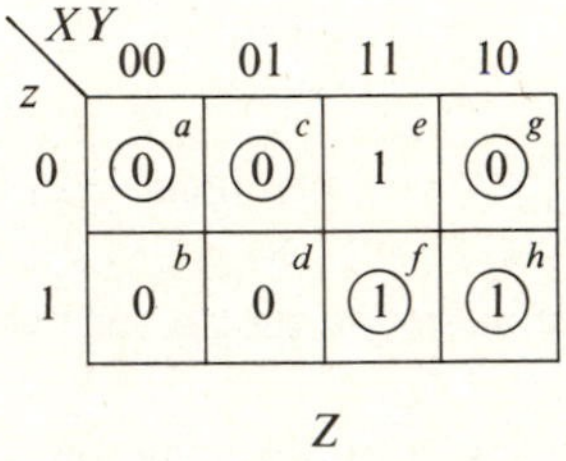

Figure A1.4 An excitation map showing stable states.

In examining the map of Fig. A1.4, we note that system input changes lead to a direct horizontal movement in the map. In an effort to build reliable systems, we impose the restriction that only one input can change at a given time. We may change from $XY = 10$ to 11 or 00, but never to 01. To reach this condition requires that we change to either 11 or 00, allow the system to stabilize, and then proceed to $XY = 01$.

Although system input changes force the direct horizontal movement in the map, vertical movement is indirectly influenced. For example, if the system of Fig. A1.4 exists in the stable state corresponding to $XYz = 010$, a change of X to 1 forces the system to move horizontally to location 110. This is a transient state causing Z to change from 0 to 1. After a delay of Δt, z will also change from 0 to 1, leading to a stable state located at $XYz = 111$.

We refer to every cell or location in the map as a state. In effect, the state is determined by the values of X, Y, and z. A primitive state diagram containing a number of states equal to the number of cells can be drawn from the map. Figure A1.5 contains the diagram for the map of Fig. A1.4. The format specifies all gate inputs, dividing the system inputs from the feedback variables by a slash. There are several points reflected by the primitive state diagram that deserve emphasis.

1. Only in a stable state can a system input be changed. All transient states have the same value of system inputs entering and exiting from the state. Transient states are shaded in the state diagram.

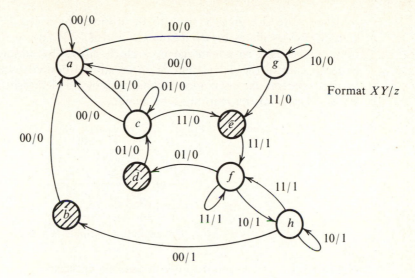

Figure A1.5 Primitive state diagram for map of Fig. A1.4.

2. In order to leave a stable state, a system input must be changed.

3. In order to cause a feedback variable change, the system must pass through a transient state.

4. Only one input can be changed in moving from one stable state to another.

It is not always possible to move from one stable state directly to all other stable states. To move from state a to state h, the system may move through states c, e, and f before reaching h. One input would change to exit from each stable state. The feedback variable only changes as a transient state is traversed.

A1.2 PROBLEMS OF ASYNCHRONOUS CIRCUITS

In Chap. 2 we discussed the problem of static and dynamic hazards. Any glitch occurring as a result of a hazard may cause an unwanted change of state in an asynchronous system. Since the elimination of hazards has been discussed previously, we will not discuss it further except to emphasize the necessity in asynchronous circuits of eliminating all hazards.

A second problem that can occur in a poorly designed circuit is that of oscillation. Consider the map of Fig. A1.6. If the system is in state a, a change of input from $B = 0$ to $B = 1$ sends the system to state c. State c is a transient state, and thus the excitation variable X changes to 1. A short time later x changes to 1, moving the system to state d. This state is also a transient state changing X back to 0, followed by a change in x to 0. The system now oscillates between states c and d. Of course, this type of situation can be used to advantage in a clock circuit by adding a delaying network to control the delay time Δt to create the desired oscillation frequency. In most systems the oscillation is unacceptable, and the situation depicted by states c and d of the map must be avoided.

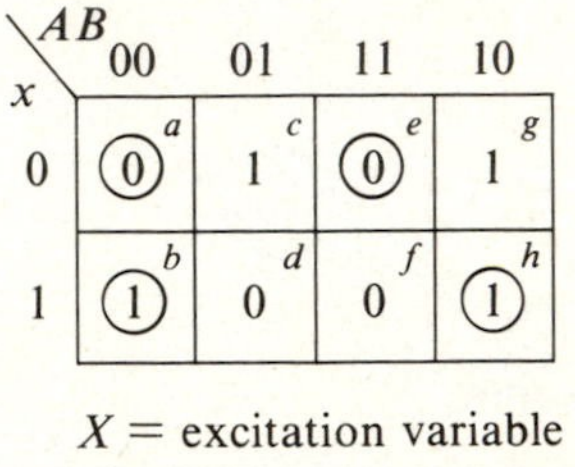

$X = $ excitation variable

Figure A1.6 An excitation map with a possible oscillation.

A third problem in asynchronous design is that of critical races. This situation can only occur when two or more feedback variables are present in the system. The excitation map of Fig. A1.7 demonstrates the critical race problem.

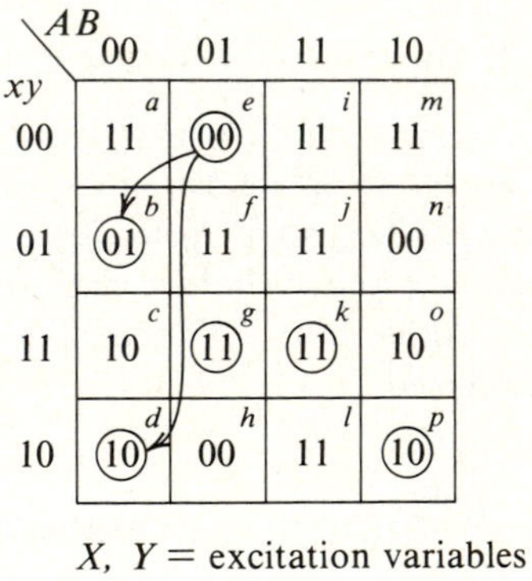

$X, Y = $ excitation variables

Figure A1.7 An excitation map with critical races.

One critical race occurs if the system starts in state e and input B changes from 1 to 0. The excitation variables begin to switch from $XY = 00$ toward $XY = 11$. Due to unequal propagation delays, one of the excitation variables will reach a value of 1 while the other has not changed from a value of 0. If the condition $XY = 10$ is reached, the system moves to stable state d. If the condition $XY = 01$ is reached rather than 10, the system moves to stable state b. The final stable state reached from this input condition depends on the relative speed of variables X and Y. This situation is referred to as a critical race. Although a particular circuit may always reach the same final state when AB switches from 01 to 00, an identical circuit may reach different states for these input values. In fact, if a gate requires replacing, this may result in a different circuit behavior. Critical races obviously must be avoided.

Noncritical races can be tolerated, as demonstrated by changing the input conditions from $AB = 01$ to 11 when the system starts in state e. The system again moves toward $XY = 11$, but will reach either 01 or 10, depending on gate speed. The system will move to either state j or state l, then to a final state of k. Regardless of which excitation variable wins the race, the same final state is reached.

Another critical race exists when the system starts in state b and the input changes to $AB = 10$. The system switches to transient state n with $XY = 00$. After a short period of time, xy changes to 00 moving the system to transient state m. A race situation again exists as the system moves toward $XY = 11$. If XY reaches 10, the system stops in stable state p. If XY reaches 01, however, the system moves back to state n and proceeds to oscillate between states m and n. Although a final state is never reached, we will also call this situation a critical race.

In designing asynchronous circuits, hazards, oscillations, and critical races must be avoided.

A1.3 BASIC DESIGN PRINCIPLES

When given an excitation map to analyze, all stable and transient states are determined. The system behavior for any combination of inputs can be precisely found. When we must design a system, however, we are given only the system behavior, and as usual, there may be several different circuits that can satisfy the specifications. In particular, we generally can only define the stable states in terms of the specs. We then must add transient or cycle states to cause the system to sequence through the proper states without introducing any critical races.

We will introduce a simple problem as a vehicle to discuss some design principles used for asynchronous circuits. Figure A1.8 shows the block diagram and timing chart of the circuit to be constructed.

The timing chart indicates that the input G gates the Osc input to the output. Additionally, if Osc is high when G is asserted, the output does not go high until the beginning of the next Osc cycle. If G is deasserted when Osc is high, the output Y remains high until Osc drops to a low level. We see that there must be a memory function included in the system to avoid cutting the length of output pulse when G changes level. We therefore can use a state machine to satisfy the specifications. A state diagram of this system appears in Fig. A1.9.

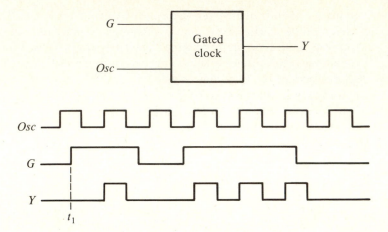

Figure A1.8 Gated clock system.

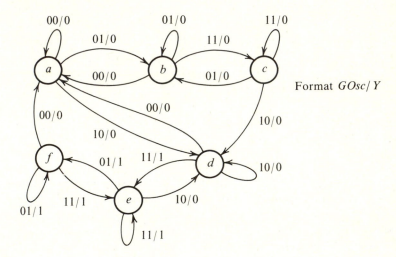

Figure A1.9 State diagram for gated clock.

In state a, GOsc = 00 and $Y = 0$. If Osc changes to 1, no output is generated, but the system moves to state b. When Osc drops to 0, the system moves back to state a. Osc is high when the system is in state b; thus if G goes high at this time, again no output should be generated. This corresponds to point t_1 on the timing chart and the system moves to state c. From this condition the system will move to state d when Osc drops to 0. As long as G remains at 1, all succeeding periods when Osc = 1 will cause the output Y to equal 1. The system will oscillate between states e and d in this case. If G is deasserted while in state e, the system moves to state f

keeping $Y = 1$ until the Osc signal drops to 0. At this time state a is again entered. We must recognize that this state diagram only shows the stable states and will require some transient states to determine the proper sequence of state change.

At this point, we do not know how many feedback variables are required to implement the circuit. In order to determine this number, we produce a primitive excitation table having only one stable state per row. With two inputs and six stable states, a 4-column, 6-row map is required as shown in Fig. A1.10.

$GOsc$ 00	01	11	10	Output Y
ⓐ	b	Θ	d	0
a	ⓑ	c	Θ	0
Θ	b	ⓒ	d	0
a	Θ	e	ⓓ	0
Θ	f	ⓔ	d	1
a	ⓕ	e	Θ	1

Figure A1.10 Primitive excitation map.

$GOsc$ 00	01	11	10	Output Y
ⓐ	ⓑ	ⓒ	d	0
a	Θ	e	ⓓ	0
a	ⓕ	ⓔ	d	1

Figure A1.11 Excitation map for diagram of Fig. A1.9.

This map is constructed by first inserting one stable state in each row. Then transient states are added to cause the proper movement between stable states as indicated by the state diagram. The requirement that inputs must not be changed simultaneously leads to the "don't care" conditions.

xz \ GOsc	00	01	11	10
00	(00)	(00)	(00)	01
01	00	(01)	11	(01)
11	ΘΘ	01	(11)	01
10	ΘΘ	ΘΘ	ΘΘ	ΘΘ

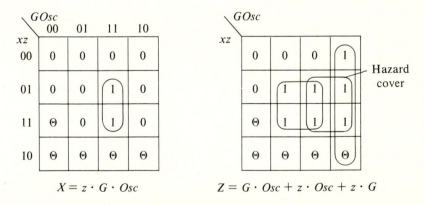

$X, Z =$ excitation variables

Figure A1.12 Excitation map with feedback variables.

It would be possible to synthesize a circuit from the primitive excitation map, but the circuit would not be minimal. This map has a great deal of redundancy that can be removed. Each row of an excitation map corresponds to a different combination of feedback variables. The primitive map has six rows and would require three feedback variables. By combining rows that contain the same states in corresponding locations, the number of feedback variables can be reduced. This process is called merging and we realize that a "don't care" condition can be taken as any state to allow a row to be merged. Before merging we must note which stable states produce an output and assign an output to each row. This is shown in the primitive state map of Fig. A1.10. As we merge rows, we will not merge similar rows if outputs are different.

$X = z \cdot G \cdot Osc$

$Z = G \cdot Osc + z \cdot Osc + z \cdot G$

Figure A1.13 *K*-maps for excitation variables.

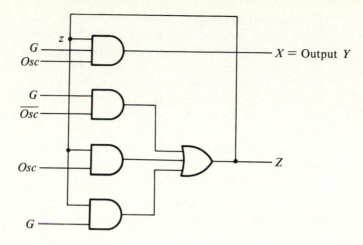

Figure A1.14 Gated clock circuit.

Following the rules mentioned, the excitation map of Fig. A1.11 is constructed. Comparing the state diagram with this map, we see that the same system behavior is predicted. With three rows, two feedback variables must be used resulting in the map of Fig. A1.12. The assignment of feedback variable value is arbitrary.

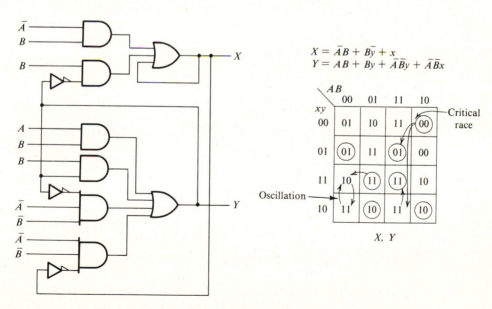

$$X = \bar{A}B + B\bar{y} + x$$
$$Y = AB + By + \bar{A}\bar{B}y + \bar{A}\bar{B}x$$

Figure A1.15 An asynchronous circuit with race problems.

Once the feedback variables are assigned, the excitation variables for the stable states are also determined. The transient states are not fixed at this point and can be selected to eliminate races or oscillations. To determine potential problems, we start in each stable state and then move to logically adjacent input combinations. If excitation variables remain the same or are logically adjacent to those of the stable state, critical races cannot occur. In this instance, there are no problems and the assignment in Fig. A1.12 is appropriate.

The next step is to implement the excitation variables X and Z. The separate K-maps of Fig. A1.13 are useful for this purpose. All hazards must be covered in this step. The final circuit is shown in Fig. A1.14.

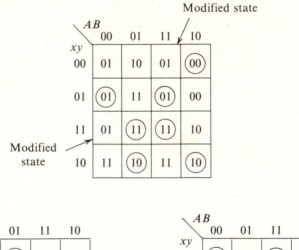

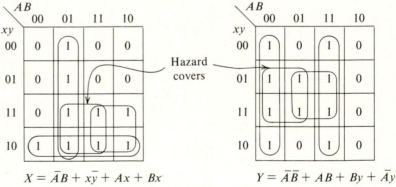

$$X = \bar{A}B + x\bar{y} + Ax + Bx$$

$$Y = \bar{A}\bar{B} + AB + By + \bar{A}y$$

Figure A1.16 Modified cycle states to eliminate races.

In order to demonstrate the use of transient state assignment, suppose we are given the circuit of Fig. A1.15 from which we derive the excitation map shown. The two problems indicated on the excitation map can be eliminated by modifying the cycle states as shown in Fig. A1.16. With the modifications shown, no new stable states are introduced. Figure A1.17 indicates the K-maps required to implement the circuit shown.

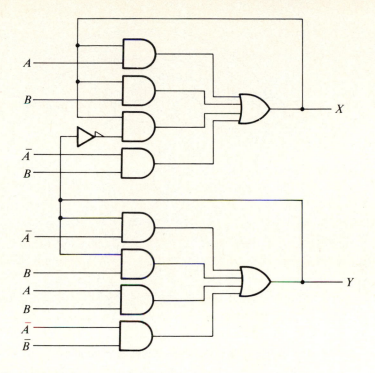

Figure A1.17 Asynchronous circuit with no race problems.

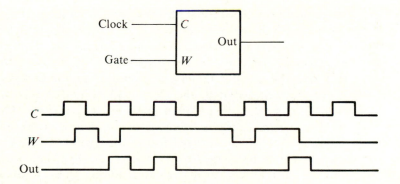

Figure A1.18 A pulse synchronizing system.

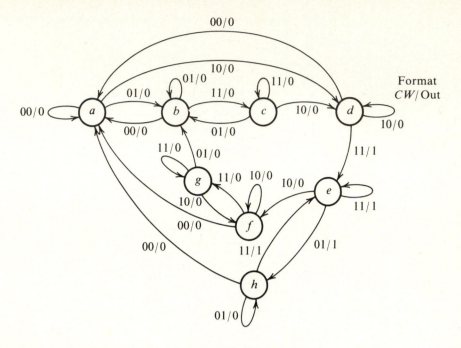

Figure A1.19 State diagram of the pulse synchronizer.

A1.4 AN ASYNCHRONOUS DESIGN EXAMPLE

In this section we will consider the design of a pulse synchronizer. This circuit is similar to the gated clock system of Fig. A1.8 except that a maximum of one clock pulse can appear at the output each time the gate goes high. If the gate remains high, no further output occurs after the first pulse. Figure A1.18 shows the block diagram and timing chart for this system. One output pulse synchronized with the clock will occur for each assertion of W, provided W is asserted longer than one-half clock period. The state diagram of Fig. A1.19 represents the behavior of the pulse synchronizer. The primitive excitation map is constructed from this diagram as shown in Fig. A1.20. The map can now be merged to produce a first approximation to the excitation map. This results in five rows requiring three feedback variables as shown in Fig. A1.21(a). These maps are then examined to find critical races and eliminate them as done in Fig. A1.21(b).

In modifying the transient states, we must be careful to send the system to the originally intended state. After the critical races have been removed, K-maps can be used to implement X, Y, and Z. This results in the equations

WC	00	01	11	10	Out
	(a)	b	Θ	d	0
	a	(b)	c	Θ	0
	Θ	b	(c)	d	0
	a	Θ	e	(d)	0
	Θ	h	(e)	f	1
	a	Θ	g	(f)	0
	Θ	b	(g)	f	1
	a	(h)	e	Θ	0

Figure A1.20 Primitive excitation map for pulse synchronizer.

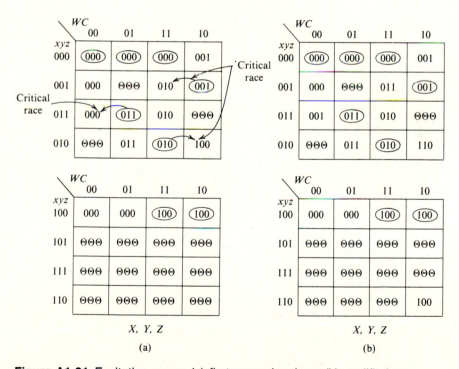

Figure A1.21 Excitation maps: (a) first approximations; (b) modified map.

$$X = xW + y\overline{C}W$$
$$Y = zC + yC + \overline{x}y\overline{z}$$
$$Z = y\overline{W} + \overline{y}zC + \overline{x}\overline{y}W\overline{C}$$

Implementing these equations results in the circuit of Fig. A1.22.

Before leaving the topic of asynchronous design, it should be mentioned that "don't care" states can also be used to minimize the final design.

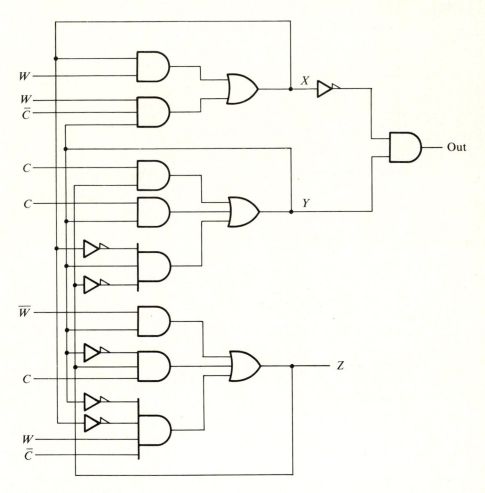

Figure A1.22 Pulse synchronizing circuit.

REFERENCES AND SUGGESTED READING

1. W.I. Fletcher, *An Engineering Approach to Digital Design*. Englewood Cliffs, N.J.: Prentice-Hall, 1980, chap. 10.
2. C.H. Roth, Jr., *Fundamentals of Logic Design*. St. Paul, Minn.: West, 1975, chap. 10.

APPENDIX 1 PROBLEMS

Secs. A1.1 and A1.2

A1.1 Given the excitation map shown in Fig. Prob. A1.1 for an asynchronous system:

a. Identify all stable states by circling them.

b. Identify all noncritical races that could occur. Indicate the starting stable state (AB/Y_1Y_2), the change of inputs initiating the race, and each state cycled through.

c. Identify any oscillations that could occur.

d. Identify all critical races that could occur.

Y_1Y_2 \ AB	00	01	11	10
00	11	01	11	00
01	01	10	11	01
11	01	11	10	11
10	11	10	10	00

Y_1Y_2

Figure PA1.1

A1.2 a. Draw the fundamental mode model for the asynchronous circuit of Fig. Prob. A1.2. Clearly label the feedback variables and the excitation variables.

b. Draw the excitation table for the circuit.

c. Identify any potential problems.

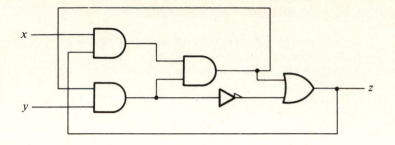

Figure PA1.2

Secs. A1.3 and A1.4

A1.3 Determine if the circuit shown in Fig. Prob. A1.3 has critical race problems. If so, do minimal redesign of the circuit to remove any problems.

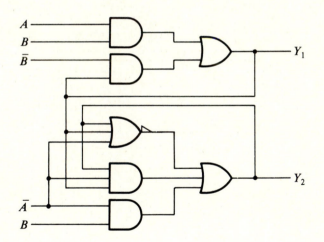

Figure PA1.3

A1.4 Design a system to generate the output shown in Fig. Prob. A1.4.

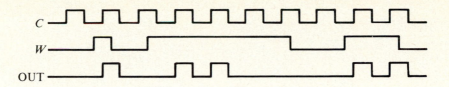

Figure PA1.4

APPENDIX

2

Logic Families[*]

A2.1 TRANSISTOR-TRANSISTOR LOGIC (TTL)

One of the most popular logic families at the present time is the TTL family. Since the late 1960s, this line has emerged as the most flexible and continues to be in great demand into the 1980s. This family possesses good fan-out figures and relatively high-speed switching. The Schottky-clamped TTL lowers switching times even further with propagation delay of gates in the area of 2 ns. The basic TTL gate is shown in Fig. A2.1.

* This material comes from the text by David J. Comer, *Electronic Design with Integrated Circuits*, © 1981, Addison-Wesley, Reading, Mass. Reprinted with permission.

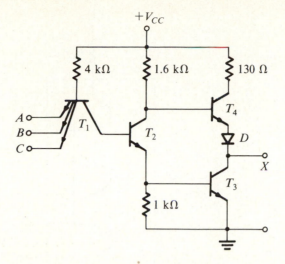

Figure A2.1 Basic TTL gate.

The TTL family is based on the multiemitter construction of transistors made possible by integrated circuit technology. The operation of the input transistor can be visualized with the help of the circuit of Fig. A2.2, which shows the bases of the three transistors connected in parallel, as are the collectors, whereas the emitters are separate.

If all emitters are at ground level, the transistors will be saturated by the large base drive. The collector voltage will be only a few tenths of a volt above ground. The base voltage will equal $V_{BE(on)}$, which may be 0.5 V. If one or two of the emitter voltages are raised, the corresponding transistors will shut off. The transistor with an emitter voltage of 0 V will still be saturated, however; and saturation will force the base voltage and collector voltage to remain low. If all three emitters are raised to a higher level, the base and collector voltages will tend to follow this signal.

Returning to the basic gate of Fig. A2.1, we see that when the low logic level appears at one or more of the inputs, T_1 will be saturated with a very small voltage appearing at the collector of this stage. Since at least $2V_{BE(on)}$ must appear at the base to T_2 in order to turn T_2 and T_3 on, we can conclude that these transistors are off. When T_2 is off, the current through the 1.6 kΩ resistance is diverted into the base of T_4, which then drives the load as an emitter follower.

When all inputs are at the high voltage level, the collector of T_1 attempts to rise to this level. This turns T_2 and T_3 on, which clamps the collector of T_1 to a voltage of approximately $2V_{BE(on)}$. The base-collector junction of T_1 appears as a forward-biased diode, whereas in this case the base-emitter junctions are reverse-biased diodes. As T_2 turns on, the base voltage of T_4

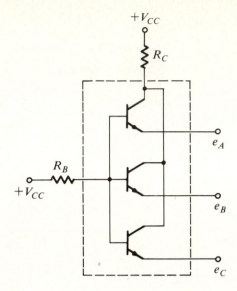

Figure A2.2 Discrete circuit equivalent to the multiemitter transistor.

drops, decreasing the current through the load. The load current tends to decrease even faster than it would if only T_4 were present, because T_3 is turning on to divert more current from the load. At the end of the transition, T_4 is off with T_2 and T_3 on. For positive logic the circuit behaves as a NAND gate.

This arrangement of the output transistors is called a totem pole. In the emitter follower the output impedance is asymmetrical with respect to emitter current. As the emitter follower turns on, the output impedance decreases. Turning the stage off increases the output impedance and can lead to distortion of the load voltage especially for capacitive loads. The totem-pole output stage overcomes this problem.

Transistor T_3 is called the pull-down transistor and T_4 is called the pull-up transistor. The circuit is designed such that these two transistors are never on at the same time. If this occurred, T_3 would be destroyed because it cannot sink as much current as T_4 can provide. Only one of these stages will be on at any given time. If T_3 is on, the output voltage is pulled down toward ground; if T_4 is on, the output voltage is pulled up toward $+5$ V.

There are two standard methods of improving the high-speed switching characteristics of TTL. The first is to add clamping diodes to the input emitters of the gate to reduce transmission line effects by providing more symmetrical impedances. This improvement is shown in Fig. A2.3 along with smaller resistors and a Darlington connection at the output. These gates have a typical propagation delay time of 6 ns.

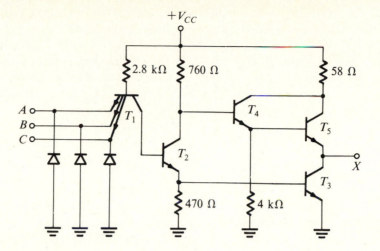

Figure A2.3 High-speed TTL gate.

A very significant improvement in TTL switching speed results from using Schottky barrier diodes to clamp the base-collector junctions of all transistors to avoid heavy saturation. Figure A2.4 shows this clamping diode arrangement.

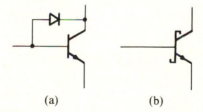

    (a)   (b)

Figure A2.4 (a) Schottky-clamped transistor. (b) Symbol for clamp.

A low forward voltage across the Schottky diode causes the diode to divert most of the excess base current around the base-collector junction. The transistor current can then decrease rapidly without the delay associated with excess base charge. The Schottky-clamped TTL gates have propagation delay times of 2 to 3 ns.

A very wide choice of logic circuits is available in the TTL family, making this line the most versatile of all families.

TTL circuits are generally called 54-series or 74-series circuits. For example, a 5400 or 7400 is a quad 2-input NAND gate. The 54 indicates the circuit is designed to meet military requirements by satisfying specified performance over a temperature range of –55 to 125°C. The 74 series is the commercial series with an operating temperature range of 0 to 70°C.

A typical TTL IC may be numbered 74S10. The letter or letters after the 74 indicate TTL type. If there is no letter, the element is conventional TTL, and its average propagation delay is 13 ns, while its average power dissipation is 10 to 15 mW per gate. The letter H designates high-speed TTL, such as the gate shown in Fig. A2.3, with propagation times that are approximately half that of conventional TTLs. The letter S indicates Schottky-clamped TTL, which is even faster than the H-type circuit with typical propagation delays of 2 to 3 ns and power dissipation slightly higher than conventional TTLs. Low-power TTL has a designator of L following the first two numbers. These circuits cut down system power dissipation by a factor of approximately 10 when replacing conventional TTL circuits, and their gate delays are increased by a factor of approximately 2.

The low-power, Schottky-clamped TTL is very popular at the present time. The power dissipation is about five times lower than conventional while delay times are comparable. An LS follows the first two numbers for low-power, Schottky TTL circuits.

The numbers after the type designator indicate what kind of circuit is contained in the IC. This number ranges from 00 up to a three-digit number in the hundreds. There may be a letter following the number of the IC that indicates package type. Another three- or four-digit number may appear on the package to identify the year and week the IC was manufactured.

A typical TTL IC, the 74S10, is a commercial Schottky TTL, with a triple 3-input NAND gate that has a propagation delay time of 3 ns and a 19 mW power dissipation. Its V_{IHmin} is specified as 2 V; V_{ILmax} is specified as 0.8 V; V_{OHmin} is specified as 2.7 V; and V_{OLmax} is specified as 0.5 V. The current I_{IHmax} is given as 50 μA when $V_{IH} = 2.7$ V, I_{ILmax} is specified as –2 mA when $V_{IL} = 0.5$ V. The short-circuit output current varies from a minimum of –40 mA to a maximum –11 mA.

A2.2 EMITTER-COUPLED LOGIC (ECL)

One of the older families is the ECL family. For many years this configuration was unrivaled in high-speed switching applications, but it now competes with Schottky-clamped TTL. The good switching characteristics of this family again result from the avoidance of saturation of any transistors within the gate. Figure A2.5 shows an ECL gate with two separate outputs. For positive logic X is the OR output while Y is the NOR output.

Often the positive supply voltage is taken as 0 V and V_{EE} as –5 V. The diodes and emitter follower T_5 establish a base reference voltage for T_4. When inputs A, B, and C are less than the voltage V_B, T_4 conducts while T_1, T_2, and T_3 are cut off. If any one of the inputs is switched to the 1 level, which exceeds V_B, the transistor turns on and pulls the emitter of T_4

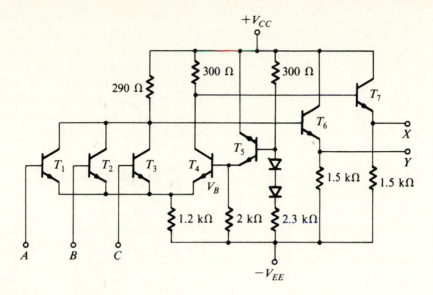

Figure A2.5 An ECL gate.

positive enough to cut this transistor off. Under this condition output Y goes negative while X goes positive. The relatively large resistor common to the emitters of T_1, T_2, T_3, and T_4 prevents these transistors from saturating. In fact, with nominal logic levels of -1.9 V and -1.1 V, the current through the emitter resistance is approximately equal before and after switching takes place. Thus, only the current path changes as the circuit switches. This type of operation is sometimes called current mode switching. Although the output stages are emitter followers, they conduct reasonable currents for both logic level outputs and, therefore, minimize the asymmetrical output impedance problem.

The ECL family has a disadvantage of requiring more input power than the TTL line. Furthermore, the great variety of logic circuits that can be realized with TTL cannot be duplicated with ECL. Propagation times of 0.5 ns can now be achieved with ECL circuits.

A2.3 MOSFET LOGIC

The p-channel (p-MOS) and n-channel (n-MOS) MOSFET devices offer several advantages over TTL and other bipolar logic families. Three big advantages of the MOS families are (1) the improved packing density because on-chip resistors are not required, (2) the lower power dissipation resulting from high-impedance active loads, and (3) the simplicity of a fabrication process that allows higher yields and the integration of more complex systems. Many microprocessors are based on MOS logic to make them cost-effective. The disadvantages of MOS logic compared with TTL are (1) a slower operating speed and (2) a lower drive-current capability.

CMOS or complementary-symmetry MOS is a well-established logic family that includes many SSI and MSI devices. Because both p- and n-channel devices must be fabricated on the same chip for CMOS, LSI systems are generally implemented by the simpler p-MOS or n-MOS methods. For less complex systems, CMOS offers the lowest power dissipation of any logic family and can also operate over a wide range of power supply voltages from a typical low of 3 V to a high of 18 V.

The inverter of Fig. A2.6 is the basic building block for a CMOS gate. Both p-channel and n-channel devices are enhancement-type MOS transistors. When the input voltage is near ground potential, the n-channel device T_1 is off. The voltage from gate to source of T_2 is approximately $-V_{DD}$ and, therefore, T_2 is on. With T_2 in the high conductance state and T_1 in the low conductance state, the power supply voltage drops across T_1 and appears at the output. When e_{in} increases and approaches V_{DD}, T_2 turns off and T_1 turns on, dropping the power supply voltage across T_2 and leading to zero output voltage. The high impedance of the off transistor results in negligible current drain on the power supply. The threshold voltage can be controlled during fabrication and is generally designed so that switching occurs at an input voltage of approximately $V_{DD}/2$.

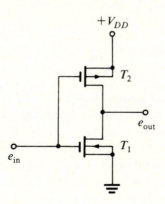

Figure A2.6 Basic CMOS inverter.

The output voltage levels of a gate are typically less than 0.01 V for the 0 state and 4.99 to 5.00 V for the 1 state ($V_{DD} = 5$ V). The required input voltages for this latter case might be 0.0 to 1.5 V for a 0 and 3.5 to 5.0 V for a 1. The noise margins in either state are then approximately 1.5 V.

Figure A2.7 shows a 2-input NOR gate using CMOS. If both A and B are held at the 0 logic level, both n-channel devices are off while both p-channel transistors are on. The output is near V_{DD}. If input A moves to the 1 state, the upper p-channel device turns off while the corresponding n-channel turns on. This leads to an output voltage near ground. If input B is at the upper logic level while A is at the 0 level, the lower p-channel device shuts off while the corresponding n-channel device turns on again, resulting in an output 0. When both A and B equal 1, the output is obviously 0.

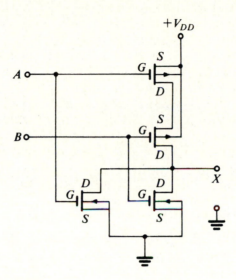

Figure A2.7 A CMOS NOR gate for positive logic.

A positive logic NAND gate is shown in Fig. A2.8. In this case at least one of the series n-channel devices will be off and one of the parallel p-channel devices will be on if A, B, or both A and B are at the low logic level. The output is then high. Only when A and B are raised to the high logic level will X equal the low level. For this input combination, both p-channels are off while both n-channels are on. These techniques can be used to create CMOS gates with more than two inputs.

There are several other logic circuits in the CMOS family, such as various types of flip-flops, memories, and shift registers. Although CMOS has several advantages compared with TTL, it has the disadvantage of longer switching times (by a factor of 5 to 10). The importance of this disadvantage is diminishing as better fabrication techniques allow smaller gate geometries and improve speeds. Typically a CMOS gate reduces the power dissipation per gate by a factor of 10 over low-power TTL. It also has a higher noise immunity than does TTL.

Because of the great popularity of TTL circuits, the 74C00 series CMOS family was designed to be equivalent to the 7400 series TTL family in terms of function and pin assignments. Although this CMOS family can drive low-power TTL, buffers are required to drive other TTL types.

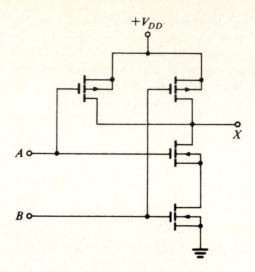

Figure A2.8 A CMOS NAND gate for positive logic.

A2.4 INTEGRATED-INJECTION LOGIC (I^2L)

I^2L is a form of bipolar circuit that can compete with n-MOS circuits in terms of the small amount of silicon required in fabrication. I^2L is especially suitable for large memories and can also be used to construct logic gates. A basic I^2L circuit is shown in Figure A2.9. The transistor T_1 acts as a current source in conjunction with the external resistance and power supply. When E_{in} goes low and sinks the collector current of T_1, T_2 has no base drive and is cut off. When E_{in} goes high, the collector current of T_1 drives the base of T_2 to saturation.

The collector of the lateral *pnp* transistor T_1 and the base of the *npn* transistor T_2 are fabricated as the same region. This, along with the fact that no resistors are fabricated on the chip, permits a greater density of transistors than chips of bipolar families such as TTL. In addition, the emitter region of T_1, which is called the injector, is shared by many gates.

The basic I^2L circuit of Fig. A2.9 functions as an inverter and also as a building block for any basic logic gate. Figure A2.10 shows an OR gate and a NAND gate. In Figure A2.10(a), if either T_1 or T_2 is turned on by a high-level voltage at inputs A or B, the collector current of T_3 is diverted from the base of T_4. Assuming that an external power supply and a resistive load are present at X, point X will go high. When both T_1 and T_2 are off, T_4 will be saturated, bringing X down to 0 V.

Figure A2.10(b) demonstrates a key feature of I^2L that allows high-density ICs for this family. The transistor T_6 and the resistor R_1 are used to

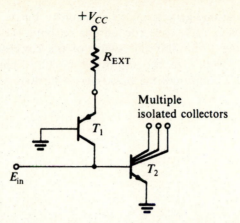

Figure A2.9 Basic I²L circuit.

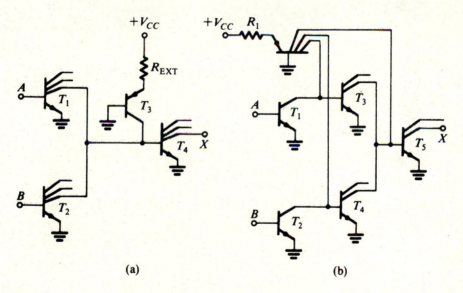

Figure A2.10 (a) An I²L OR gate. (b) An I²L NAND gate.

drive several other stages with the multicollector configuration of T_6. Few resistors are required, and many of these can be external to the chip. If A is driven high while B is low, T_3 will turn off and T_4 will remain on, preventing current from driving the base of T_5. Only if both A and B go high will T_3 and T_4 turn off to saturate T_5 and cause X to go low.

In second-generation I^2L, Schottky barrier diodes have been included in series with each collector output to limit the logic swing and speed up the switching. The 700 mV swing of the conventional I^2L circuit is reduced to about 250 mV in the improved circuit. The isoplanar integrated-injection logic (I^3L) improves the basic circuit further.

Answers to Drill Problems

Sec. 1.2

1. 1010101011_2, 1253_8, $2AB_{16}$.

2. 1010001011111_2, 5057_8, 2607_{10}.

3. 111001010_2, 0100 0101 1000.

4. 86_{10}, 1010110_2.

Sec. 1.4

1. The function table plots inputs and outputs in terms of high and low voltage levels. The truth table plots information in terms of 1s and 0s.

2.

A_{in}	B_{in}	Out
0	1	0
1	0	1

3.

A	B	NOR out	Out
0	0	1	0
0	1	0	1
1	0	0	1
1	1	0	1

Sec. 1.5

1. $V_{OHmin} > V_{IHmin}$, $V_{OLmax} < V_{ILmax}$.

2. $I_{OH} = -200 \ \mu A$, $I_{OL} = 9.6 \ mA$.

3. 0.2V

4. 24ns. Yes.

Sec. 1.6

1. $R_{max} = 2.78 \ k\Omega$, $R_{min} = 375 \ \Omega$.

2. The voltage at point A prior to connection of three-state output.

Secs. 2.1A—2.1C

3. $X = \bar{A} + \bar{B} + C$

4. $F = ABC$

5. $AC + \bar{A}C = (A + \bar{A})C = 1C = C$

6. $AB + A\bar{B} + \bar{A}C = A(B + \bar{B}) + \bar{A}C = A + \bar{A}C = A + C$

Sec. 2.1D

1. a. $F = \bar{A}\bar{C} + \bar{A}\bar{D} + \bar{B}\bar{C} + \bar{B}\bar{D}$ **c.** $F = ABCD$
    b. $F = AB\bar{C} + AB\bar{D}$ **d.** $F = ACD + BCD$

2. $X = 0$

Sec. 2.2A

1. $X = AB + \bar{B}\bar{C}$

2. $X = ABC + AB\bar{C} + A\bar{B}\bar{C} + \bar{A}\bar{B}\bar{C}$

3. $F = \bar{A}\bar{B}\bar{C} + \bar{A}BC + A\bar{B}\bar{C} + AB\bar{C}$

Sec. 2.2B

1.

C\AB	00	01	11	10
0	0	0	1	1
1	0	0	1	0

X

2.

CD\AB	00	01	11	10
00	1	0	1	0
01	0	0	1	1
11	0	0	1	0
10	0	0	1	0

Y

3.

C\AB	00	01	11	10
0	1	1	0	1
1	1	1	1	0

Z

Sec. 2.2C

1. $X = AB + \bar{B}\bar{C}$

2. $Y = ABC + \bar{B}\bar{C}D$

3. $Z = \bar{B}CD + BC\bar{D} + AB\bar{C}D$

Sec. 2.2D

1. $X = B\bar{C} + A\bar{C}D + AB\bar{D} + \bar{A}BD + \bar{A}C\bar{D}$

2. $F = \bar{C}D + BC$

3. $F = \bar{C}\bar{D} + \bar{B}C$

Sec. 2.2E

1. Hazard as A goes from 0 to 1

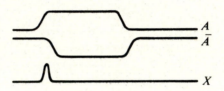

Sec. 2.3A—2.3B

1. $X = \bar{A}B + A\bar{C}$

2. $Y = A\bar{B} + \bar{B}\bar{D}$

3. $F = \bar{C}D + BC$

Sec. 2.3C

1.

C \ AB	00	01	11	10
0	$\bar{D}\bar{E} + \bar{D}E$ $+ DE$	0	DE	$DE + \bar{D}\bar{E}$
1	$\bar{D}\bar{E}$	DE	$D\bar{E}$	$\bar{D}E$

F

2.

B \ A	0	1
0	$\bar{C}D$	$C\bar{D}+CD$
1	CD	$\bar{C}\bar{D}+\bar{C}D+$ $C\bar{D}+CD$ $(=1)$

$$X = \overline{AB}\bar{C}D + AC\bar{D} + ACD + AB + BCD$$

Sec. 2.4B

1. $X = \overline{AB} + AB$. 2 AND gates plus an OR or 1 NEXOR gate.

2. $X = \overline{AB} + AB + C$. 2 AND gates plus an OR or 1 NEXOR plus an OR.

Sec. 3.1A

1. Connect: A to S_2, B to S_1, C to S_0; logic 0 to lines 3, 5, 6; logic 1 to lines 0, 1, 2, 4, 7.

2. Connect: A to S_1, B to S_0, logic 1 to line 0, $\bar{C}$ to lines 1 and 2, C to line 3.

3.

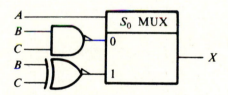

Sec. 3.1B

1. Inputs: Connect A to line 2, B to line 1, C to line 0.
Outputs: Connect to inputs of NAND gate lines 0, 1, 2, 4, 7.

2. Inputs: Connect A to line 2, B to line 1, C to line 0.
Outputs: Connect to inputs of NAND gate 1 lines 0, 1, 2, 3, 5, 6. Connect to inputs of NAND gate 2 lines 1, 2, 4. Connect to inputs of NAND gate 3 lines 0, 7.

3.

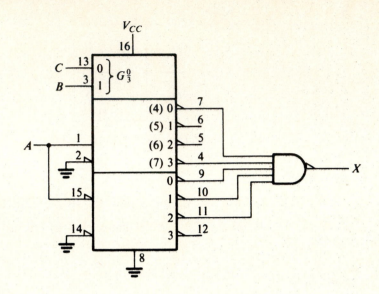

Sec. 3.1C

1.

	Inputs					Outputs							
Gnd	A	B	C	D						F4	F3	F2	F1
I_4	I_3	I_2	I_1	I_0	O_8	O_7	O_6	O_5	O_4	O_3	O_2	O_1
0	0	0	0	0					0	1	1	0
	0	0	0	1		Ө			0	1	1	0
	0	0	1	0					0	0	0	0
	0	0	1	1					0	0	0	1
	0	1	0	0					0	1	0	0
	0	1	0	1		Ө			1	1	0	1
	0	1	1	0					0	1	0	1
	0	1	1	1					1	0	0	1
	1	0	0	0					0	1	1	0
	1	0	0	1		Ө			1	0	0	1
	1	0	1	0					1	1	0	0
	1	0	1	1					0	1	1	0
	1	1	0	0					1	0	0	1
	1	1	0	1		Ө			0	0	0	0
	1	1	1	0					0	1	1	0
0	1	1	1	1					0	0	1	1

2.

Inputs			*Outputs*			
A	*B*	*C*		*F*3	*F*2	*F*1
I_2	I_1	I_0	O_4	O_3	O_2	O_1
0	0	0	Θ	1	0	0
0	0	1		0	1	0
0	1	0		0	1	0
0	1	1		0	0	1
1	0	0		0	1	0
1	0	1		0	0	1
1	1	0	↓	0	0	1
1	1	1	Θ	1	0	0

Sec. 3.2B

1.

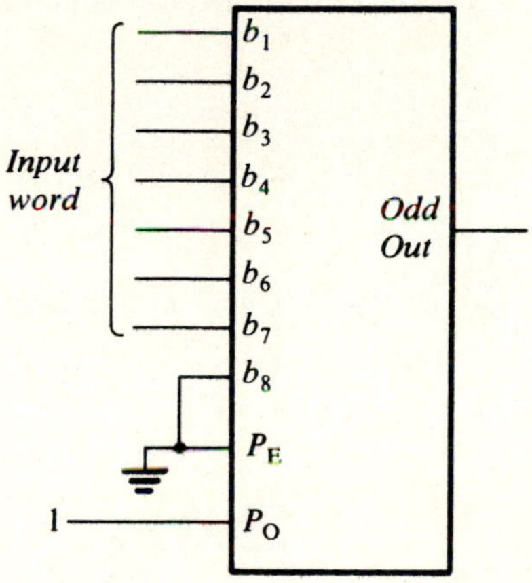

Sec. 3.3B

1. 1's comp. = 00111110; 2's comp. = 00111111.

2. 1's comp. = 00101110; 2's comp. = 00101111.

3. 7360

4. 7361

Sec. 3.3C

1. $S_7 - S_0$ = 11011100. $0_7 - 0_0$ = 00100011. $G = 0$.

2. $S_7 - S_0$ = 00100010. $0_7 - 0_0$ = 00100011. $G = 1$.

Sec. 3.3D

1. $S_7 - S_0$ = 10101100, OF = 0.

2. $S_7 - S_0$ = 10101100, OF = 0.

3. $S_7 - S_0$ = 10001001, OF = 0.

Sec. 4.1A

2. a. $Q = 0$ **b.** $Q = 1$ **c.** $Q = 0$

3. 0, 0, 1, 1, 1, 1.

Sec. 4.1B

2. Let $C = 1\ \mu F$, $R_A = 5.82\ k\Omega$. Many values will work.

Sec. 4.1C

1. Solving eqn. previous to Eq. 4.4 gives result.

2. Let $C = 1\ \mu F$, $R_B = 1.44\ k\Omega$, $R_A = 2.02\ k\Omega$.

3. Let $C = 1\ \mu F$, $R_B = 3.17\ k\Omega$, $R_A = 288\ k\Omega$.

Sec. 4.2A

1.

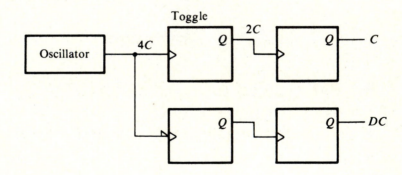

Sec. 4.2C

1.

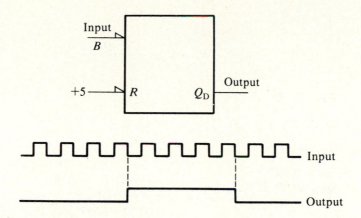

Sec. 4.2D

1. First bit exists from time that a 1 is set into this *FF* until first possible clock pulse.

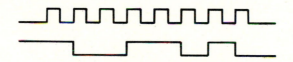

2. Extra *FF* makes first bit last only one cycle.

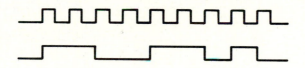

Sec. 4.2E

1. Make one-shot 1 period = 11 1/2 ms, one-shot 2 period = 0.25 ms, and one-shot 3 period = 0.75 ms.

Sec. 5.3A

1. $D(t_0) = 1, D(t_1) = 1, D(t_2) = 0, D(t_3) = 1$.

2. $J(t_0) = \Theta$, $K(t_0) = 0$, $J(t_1) = \Theta$, $K(t_1) = 0$, $J(t_2) = \Theta$, $K(t_2) = 1$, $J(t_3) = 1$, $K(t_3) = \Theta$.

3.

Q_n	Q_{n+1}	T
0	0	0
0	1	1
1	0	1
1	1	0

Sec. 5.3C

1.

			Next state	
A	B	X	A	B
0	0	0	0	0
0	0	1	0	1
0	1	0	1	0
0	1	1	1	0
1	0	0	0	0
1	0	1	0	1
1	1	0	0	0
1	1	1	0	0

2.

X \ AB	00	01	11	10
0	0	1	0	0
1	0	1	0	0

D_A

X \ AB	00	01	11	10
0	0	0	0	0
1	1	0	0	1

D_B

Sec. 5.3D

1.

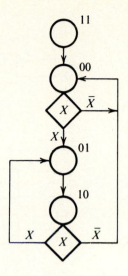

Sec. 5.4A

1.

A	B	Next A	B
0	0	0	1
0	1	1	1
1	0	0	1
1	1	1	0

D_A: B\A	0	1
0	0	0
1	1	1

D_A

$D_A = B$

D_B: B\A	0	1
0	1	1
1	1	0

D_B

$D_B = \bar{A} + \bar{B}$

Sec. 5.4D

1. 3 flip-flops.

2. Assign states b, c adj.; d, e adj. from Principle 1. Assign f, h adj. and g, h adj., and b, c, d, e adj. from Principle 2.

C\AB	00	01	11	10
0	d	e	f	h
1	b	c	a	g

one possible assignment

Sec. 5.4F

1.

2.

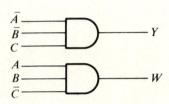

3.

4.

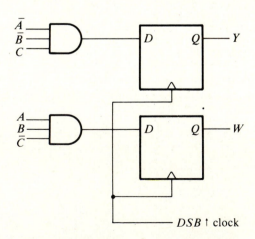

5.

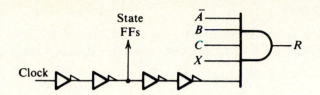

Sec. 5.5A

1. 111,001.

2. Once state 001 is reached, the *IFL* is driven to cause a state change to either 101 or 011.

Sec. 7.2A

1. See Prob. 2

2. Three 4:1 MUXs are needed for *IFL*. On all three MUXs, assign input line 0 to $\overline{X}$, input line 1 to X, and lines 2 and 3 to logic 0. This is an arbitrary choice. Then eqns. are $S_{0A} = AB + C$, $S_{1A} = B$, $S_{0B} = AB + C$, $S_{1B} = \overline{A}$, $S_{0C} = AB + \overline{A}\overline{B}$, and $S_{1C} = A$.

Sec. 7.2B

1. See Prob. 2.

2. A 16×5 ROM is required, so a 16×8 might be used.

Inputs				A	B	C	W	Y	Indirect-addressed ROM (Prob. 7.2C-1)		
A	B	C	X								
I_3	I_2	I_1	I_0	O_5	O_4	O_3	O_2	O_1	I_2	I_1	I_0
0	0	0	0	1	0	0	0	0	0	0	0
0	0	0	1	0	1	0	0	0	0	0	1
0	0	1	0	0	0	0	0	1	0	1	0
0	0	1	1	0	0	0	0	1	0	1	0
0	1	0	0	0	0	1	0	0	0	1	1
0	1	0	1	1	1	0	0	0	1	0	0
0	1	1	0	θ	θ	θ	θ	θ	θ	θ	θ
0	1	1	1	θ	θ	θ	θ	θ	θ	θ	θ
1	0	0	0	1	1	0	0	0	1	0	0
1	0	0	1	0	0	1	0	0	0	1	1
1	0	1	0	θ	θ	θ	θ	θ	θ	θ	θ
1	0	1	1	θ	θ	θ	θ	θ	θ	θ	θ
1	1	0	0	0	0	0	1	0	1	0	1
1	1	0	1	0	0	0	1	0	1	0	1
1	1	1	0	θ	θ	θ	θ	θ	θ	θ	θ
1	1	1	1	θ	θ	θ	θ	θ	θ	θ	θ

Sec. 7.2C

1. An 8 × 5 ROM can be used. Using the assignment shown in previous problem for $I_2 I_1 I_0$, the code converter equation becomes

$$I_2 = BX + A\overline{X}, \quad I_1 = C + \overline{A}B\overline{X} + A\overline{B}X, \quad I_0 = B\overline{X} + AX + \overline{B}\overline{C}X$$

The ROM contents are

			A	B	C	W	Y
I_2	I_1	I_0	O_5	O_4	O_3	O_2	O_1
0	0	0	1	0	0	0	0
0	0	1	0	1	0	0	0
0	1	0	0	0	0	0	1
0	1	1	0	0	1	0	0
1	0	0	1	1	0	0	0
1	0	1	0	0	0	1	0
1	1	0	θ	θ	θ	θ	θ
1	1	1	θ	θ	θ	θ	θ

Answers to Selected Problems

Chapter 1.

      1.2   250

      1.4 $841_{10} = 1000\ 0100\ 0001$

      1.7   **a.** 255   **c.** 1.375

             b. 150   **d.** 3.8125

1.11

A	B	X
0	0	1
0	1	1
1	0	1
1	1	0

1.15

A	B	X
0	0	1
0	1	0
1	0	1
1	1	1

1.18

1.21 30 buffer inputs

Chapter 2.

2.1 True

2.5 False

2.9 True

2.14

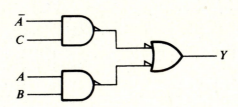

2.19 $F = \overline{A}\,\overline{B}CD$

2.21 $F = C(\overline{A} + B)$

2.24 $F = A + BC$

2.28 $F = B(\overline{A} + C)$

2.33

C \ AB	00	01	11	10
0	0	$D + \bar{D}$	D	0
1	0	$\bar{D}$	0	D

$F = \bar{A}B\bar{D} + B\bar{C}D + A\bar{B}CD$

2.39 $X = (A \oplus B) \oplus C$
$= A\bar{B}\bar{C} + \bar{A}B\bar{C} + \bar{A}\bar{B}C + ABC$

Chapter 3.

3.1 Logic 1 to input lines 0, 5, 6, 7, 8, 9, 12, 14. Logic 0 to all others. A to S_3, B to S_2, C to S_1, and D to S_0.

3.5

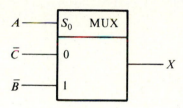

3.8 $ABCD$ to input with $A = MSB$. Output lines 0, 5, 6, 7, 8, 9, 12, and 14 to NAND inputs. Y is NAND output.

3.13

A	B	C	D	$F4$	$F3$	$F2$	$F1$		A	B	C	D	$F4$	$F3$	$F2$	$F1$
0	0	0	0	0	1	1	0		1	0	0	0	0	1	1	0
0	0	0	1	0	1	1	0		1	0	0	1	1	0	0	1
0	0	1	0	0	0	0	0		1	0	1	0	1	1	0	0
0	0	1	1	0	0	1	0		1	0	1	1	0	1	1	0
0	1	0	0	0	1	0	0		1	1	0	0	1	0	0	1
0	1	0	1	1	1	0	1		1	1	0	1	0	0	0	0
0	1	1	0	0	1	0	1		1	1	1	0	0	1	1	0
0	1	1	1	1	0	0	1		1	1	1	1	0	0	1	1

3.16

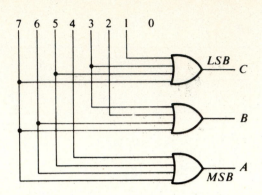

3.20 8033

3.22 11011111

3.25 Overflow

Chapter 4.

4.1

A	B	X
0	0	Q1
0	1	1
1	0	0
1	1	Avoid

4.3 For TTL the circuit would function if resistors were deleted, but would be more susceptible to noise.

4.6

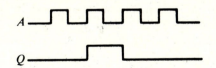

4.9

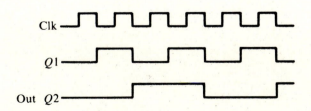

4.12

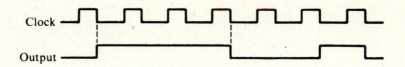

4.16 Choose $C = 1\ \mu F$, $R = 910\ \Omega$.

4.18 Choose $C_D = 0.01\ \mu F$
$R_D = 5\ k\Omega$

4.22 Choose $C = 1\ \mu F$, $R_B = 58\ k\Omega$, $R_A = 14.5\ k\Omega$.

4.25

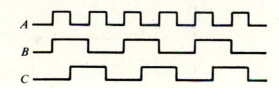

4.29

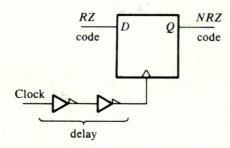

Chapter 5.

5.2 A system with clock-driven state changes and at least one asynchronous input.

5.3

Q_n	Q_{n+1}	A	B
0	0	θ	1
0	1	1	0
1	0	0	1
1	1	1	θ

5.7

	t_1	t_2	t_3	t_4
$T1$	0	1	0	1
$T2$	1	1	1	0

5.10 $T1 = I2$
$T2 = \bar{I}1 + \bar{I}2$

5.16

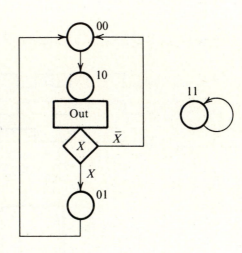

5.19 $D_A = \bar{A}B, D_B = \bar{A}C + A\bar{B}\bar{C}, D_C = \bar{B}\bar{C}.$

5.22

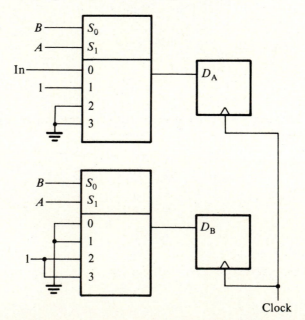

5.23

$\overset{AB}{\underset{C}{}}$	00	01	11	10
0	a		b	
1	e	c		d

5.35 1111, 0101

Chapter 6.

6.3

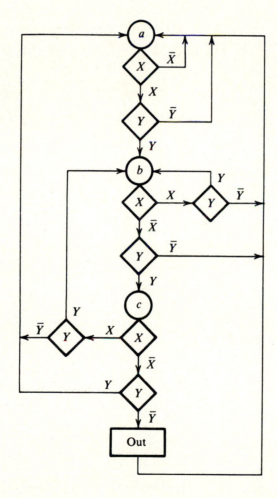

Chapter 7.

7.14 This requires at least 1500 locations (2K) of 14 bits each.

Index